191
216
224

HARRY O. YATES III
SIGMA CHI
ORONO, MAINE

McGRAW-HILL PUBLICATIONS IN THE
BOTANICAL SCIENCES
EDMUND W. SINNOTT, Consulting Editor

AN INTRODUCTION TO
PLANT ANATOMY

Selected Titles From

McGRAW-HILL PUBLICATIONS IN THE BOTANICAL SCIENCES

Edmund W. Sinnott, *Consulting Editor*

Arnold—An Introduction to Paleobotany
Avery et al.—Hormones and Horticulture
Babcock and Clausen—Genetics
Braun-Blanquet and Fuller and Conard—Plant Sociology
Curtis and Clark—An Introduction to Plant Physiology
Eames—Morphology of Vascular Plants
Eames and MacDaniels—An Introduction to Plant Anatomy
Fitzpatrick—The Lower Fungi
Gates—Field Manual of Plant Ecology
Gäumann and Dodge—Comparative Morphology of Fungi
Haupt—An Introduction to Botany
Haupt—Laboratory Manual of Elementary Botany
Hill—Economic Botany
Hill, Overholts, and Popp—Botany
Johansen—Plant Microtechnique
Kramer—Plant and Soil Water Relationships
Lilly and Barnett—Physiology of the Fungi
Maheshwari—An Introduction to the Embryology of Angiosperms
Maximov—Plant Physiology
Miller—Plant Physiology
Pool—Flowers and Flowering Plants
Sharp—Fundamentals of Cytology
Sharp—Introduction to Cytology
Sinnott—Botany: Principles and Problems
Sinnott—Laboratory Manual for Elementary Botany
Sinnott, Dunn, and Dobzhansky—Principles of Genetics
Smith—Cryptogamic Botany
  Vol. I, Algae and Fungi
  Vol. II, Bryophytes and Pteridophytes
Smith—The Fresh-water Algae of the United States
Swingle—Textbook of Systematic Botany
Weaver—Root Development of Field Crops
Weaver and Clements—Plant Ecology
Wodehouse—Pollen Grains

There are also the related series of McGraw-Hill Publications in the Zoological Sciences, of which E. J. Boell is Consulting Editor, and in the Agricultural Sciences, of which R. A. Brink is Consulting Editor.

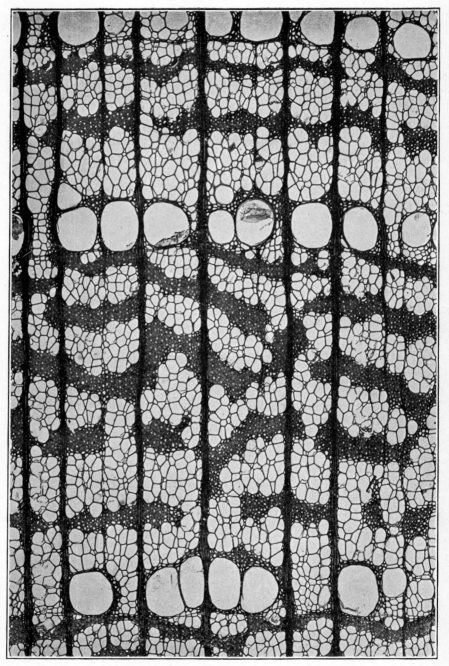

Frontispiece.—Transverse section of secondary wood of *Ulmus americana*. × 50. (*Courtesy, U. S. Forest Products Laboratory.*)

"The Staple of the Stuff is so exquisitely fine, that no Silk-worm is able to draw anything near so fine a Thread. So that one who walks about with the meanest Stick, holds a piece of Natures Handicraft, which far surpasses the most elaborate Woof or Needle Work in the World."—Nehemiah Grew.

AN INTRODUCTION TO
PLANT ANATOMY

BY

ARTHUR J. EAMES
Professor of Botany, Cornell University

AND

LAURENCE H. MacDANIELS
Professor of Horticulture, Cornell University

SECOND EDITION
FOURTH IMPRESSION

McGRAW-HILL BOOK COMPANY, INC.
NEW YORK AND LONDON
1947

AN INTRODUCTION TO PLANT ANATOMY

Copyright, 1925, 1947, by the
McGraw-Hill Book Company, Inc.
PRINTED IN THE UNITED STATES OF AMERICA

*All rights reserved. This book, or
parts thereof, may not be reproduced
in any form without permission of
the publishers.*

PREFACE TO THE SECOND EDITION

In the second edition of this textbook the authors have maintained the viewpoint and the aims of the first edition: a textbook, introductory in level, that will serve primarily for classroom use, and only secondarily as a reference text. They have been mindful of the request that the new edition present greater amounts of detailed description and that it cover aspects of the field not treated in the first edition—that it include, for example, "full details of meristem development" and a discussion of the structure of chimeras and their significance. The inclusion of such additions is obviously impossible in a book of this type and size.

Since the preparation of the first edition, there has been marked progress in our knowledge of certain fields of plant anatomy, especially cellwall origin, development, and structure; meristems; phloem; leaf development; abscission; flower and fruit structure. The authors have incorporated into the new edition as much from this material as in their opinion properly belongs in this book. These additions have increased the length of most of the chapters, and in an effort to keep the book at approximately its original size, Chap. XV—A Sketch of the History of Plant Anatomy—has been omitted because, although of general interest, it is not essential to the student in gaining a working knowledge of plant anatomy.

A few changes have been made in the terminology used in the first edition. Some changes have been made because the structure of certain cells and tissues, such as the phloem, is more fully understood today than in 1925; other changes have been made so that the terminology of the textbook conforms with a glossary of terms used in describing wood, adopted by the International Association of Wood Anatomists.

The authors wish to acknowledge the help of Dr. Antoinette M. Wilkinson and Dr. H. W. Blaser, who were especially helpful in the preparation and reading of the manuscript. They are also greatly indebted to Mrs. Rita B. Eames for her assistance with the preparation and improvement of the illustrations and her supervision of the manuscript. The authors also wish to thank W. R. Fisher for his interest and cooperation in making many of the photographs in this book.

ARTHUR J. EAMES
LAURENCE H. MACDANIELS

ITHACA, N.Y.,
July, 1947.

PREFACE TO THE FIRST EDITION

In presenting this book the authors hope to fill a need for a textbook in plant anatomy of a type at present not available—a need which they, as teachers in this field, have keenly felt. Not only, however, in their opinion, is there need for a book for class study and guidance, but also for one which shall serve as a reference text for workers in fields of applied botany, and for teachers and students in other fields of pure botany. A double purpose, therefore, has been kept in mind in the preparation of the book. In the treatment of the subject matter, however, emphasis has been placed on adaptability to classroom use from the standpoint of the student beginning anatomical study. Thus, the book is, first of all, a textbook in the elements of plant anatomy—an introduction to the field. It presupposes an acquaintance only with the fundamental structure and activities of plants—an acquaintance such as is ordinarily obtained from a first course in botany.

Though the book is thus introductory in nature, it is believed to embody a fairly comprehensive treatment of the fundamental facts and aspects of anatomy—to be, in fact, so inclusive as to provide a working basis for independent study. Yet it does not lay claim to the exposition, in detail, of the known facts and the theories concerning any structural features. So great is the number of recorded facts, and so confused is the terminology of anatomy, that a treatise approaching completeness in the presentation thereof would not be usable as a textbook. Further, the anatomy of vascular plants, especially that of the angiosperms is, in detail and in some broader features, still largely unknown. It is thus obviously impossible to present data covering the facts and structural features which will be met by the student in later work. It is, further, the firm opinion of the authors that the student of anatomy should not learn facts primarily, but should be taught self-reliance in the study of plant structure through training in power of observation and interpretation. Therefore, the book is not a compendium of facts.

Training which results in independence in the study of anatomy is, of course, secured only by laboratory practice. On such practice the authors believe emphasis must be placed, and not on lectures, text study, nor, in the beginning, on reading. For laboratory teaching the present book should provide a background of facts, terms, history, etc.; it may, indeed, be used, in part, as a laboratory guide. The sequence of subjects adopted is that which in the experience of the authors has been found most satisfactory in laboratory work. Such material as is specifically mentioned, or is used for illustration, suggests only in a general way the range and the amount of material that may be used in a first course in plant anatomy. It is not necessary, nor is it even desirable, that the same plants be used in the classroom. Any available material may be used, and, by comparison with the descriptions and illustrations of the text, the teaching can be made more effective. The authors in their own classes use considerable material, sufficient to cover so far as possible the range and type of variation in each structure. An acquaintance with variation is thus acquired by the student, and a power of interpretation is given by practice with many examples, so that he is enabled thereafter to interpret wholly new material.

As a reference book, the synoptical treatment of the more important facts, usage of terms, present status of opinion, etc. should render the book generally useful. More detailed information may be obtained, of course, from the larger reference works, though often it is only to be found in papers and articles of limited scope. A certain small amount of material embodied in the book represents the result of unpublished research and observation on the part of the authors, or represents their personal opinions.

Except for an occasional mention of lower forms, the structure of vascular plants only is considered, since the histological structure of the thallophytes is usually not complex. Where it is, the method of study and the terms applied to cells and tissues in higher plants may generally be used. In the selection of forms for illustration there have been chosen, so far as available material has permitted, well-known or economically important plants.

The viewpoint of the treatment is fundamentally that of descriptive morphology, that is, of existent form and structure. Physiological anatomy regards form but little; yet an understanding of form and of structural relationship must precede all valuable anatomical study. The physiological aspects and the practical bearing of the subject matter are discussed briefly, and incidentally to the general treatment. Comparative morphology is made use of whenever an understanding of phylogenetic development helps to make structural complexity clear. A textbook written on the basis of descriptive morphology the authors believe to be most generally useful. For students going beyond the introductory steps in anatomy, however—either into the various fields of applied subjects, such as pathology and horticulture, or into any field of the pure science—a complete understanding of morphological modification and variation can be obtained only through the consideration of the phylogenetic history of the structure in question.

The book does not pretend to present the historical development of our knowledge of the field, or of any phase or part thereof; nor are the contributions of prominent students brought out as the work of individuals. The present status of knowledge and opinion is made the first aspect of treatment; secondarily, other viewpoints are considered. Chapter XV, however, outlines the history of the subject, and deals briefly with the contributions of some of the earlier prominent students of anatomy. This historical sketch is placed at the end of the book, since the beginning student may best make use of it only when he has acquired an understanding of the subject matter.

Owing to the state of confusion which exists in the terminology of anatomy, it has in many cases been necessary to evaluate the different uses of terms. For use in the book, those terms have been accepted which seem best on a basis first of morphological usefulness and secondly, of priority; in a few instances the history and the use of a term are briefly discussed.

Since the book is primarily an elementary text and not a text for research reference nor for the use of advanced students, the bibliography has been kept at a minimum. After each chapter there are listed a few of the more recent and more important books and articles—and sometimes those valuable for their own bibliographies—dealing with the subjects discussed in that chapter. By reference to these, students may obtain longer lists. Such a method of citation naturally often excludes the older, "classic" treatments. To the first chapter is appended a list of texts which are generally useful for some or for many phases of the subject. Reference to these texts is not repeated after the various chapters, except where the book in question deals particularly with the subject matter of a chapter.

The common names of plants have for the most part been omitted from the text. They may be sought in the index, as such, and also under the generic names with which they are associated.

In present-day botany, the terms "anatomy" and "histology" are often loosely used. To many botanists the study of the internal structure of plants is unfortunately known as histology. This is doubtless due in part to the fact that histology deals with the structure of cells and tissues—internal structure meaning this and little more—and in part to the fact that the grosser internal structures, such as steles and traces, have been consistently neglected in courses in "histology." The study of these features of grosser internal structure, and sometimes of those of external make-up also, have been looked upon as "anatomy." Anatomy, however, deals with the structure of organisms, structure both gross and minute, external and internal. Histology, which deals with the minute structure of organisms, is, therefore, a part of the broader field of anatomy. An understanding of the structure of a plant obviously cannot be obtained from the study of minute features alone. Thus, a treatment such as is here presented is anatomical rather than histological, and the book is, therefore, a text in Plant Anatomy.

In so far as the treatment in its histological aspects deals with the structure of cells—especially wherever the protoplast is concerned—it enters the field of cytology; and cytology in recent years has become an independent division of biological science. Cytological aspects of anatomy, therefore, need not be considered, and have been omitted from the discussion except in so far as they are essential to histological study. An arbitrary limit has necessarily been set to the description of the cell. The protoplast is very briefly discussed, except for the paragraphs on plasmodesma, plastids, and cell inclusions; nuclear division is omitted since it is ordinarily taught in first courses in botany, and again in greater detail in cytology. The wall, however, which, aside from aspects of origin and early development, is usually considered by cytologists as a histological feature, is more fully treated.

The illustrations have in large part been made directly from the material itself. The explanation of the figures is placed chiefly in the legends. With a few exceptions, the drawings are the work of the authors themselves and of Mrs. Rita Ballard Eames to whom the authors are greatly indebted for invaluable assistance and suggestions. The helpful criticism of their colleagues the authors also desire to acknowledge.

ARTHUR J. EAMES
LAURENCE H. MACDANIELS

ITHACA, N.Y.,
July, 1925.

CONTENTS

PREFACE TO THE SECOND EDITION vii

PREFACE TO THE FIRST EDITION ix

INTRODUCTORY

Chapter I

GENERAL STRUCTURE OF THE PLANT BODY—AN OUTLINE 1
Fundamental parts—The axis—The stele—Primary and secondary growth—Constitution of the plant body—Methods of study—General references for plant anatomy.

GENERAL HISTOLOGY

Chapter II

THE CELL 6
Uses of the term "cell"—Cellular complexity of plants: arrangment, shape, size, development, adjustment during growth—The protoplast: organization, plasma membrane, plastids, vacuoles and cell sap, inclusions—The cell wall: nature, origin, gross structure, minute structure of the wall and middle lamella, sculpture and modification, chemical nature—Cuticle—References for Chap. II: general, the protoplast, plamodesmata, ergastic substances, the cell wall.

Chapter III

MERISTEMS 60
Meristematic and permanent cells and tissues—Classification based on: stage or method of development, history of initiating cells, position in plant body, function—Theories of structural development and differentiation: apical cell theory, histogen theory, tunica-corpus theory—Types of stem apices—Discussion of tunica-corpus theory—The floral apex—The root apex—Types of root-tip development—References for Chap. III.

Chapter IV

TISSUES AND TISSUE SYSTEMS 81
    Tissues based on stage of development: meristematic, permanent—Tissues based on kind of constituent cells: simple (parenchyma, collenchyma, sclerenchyma); the important complex tissues (xylem, phloem, "transfusion tissue")—Tissue systems—Secretory tissue—References for Chap. IV.

THE PRIMARY BODY

Chapter V

THE PRIMARY BODY 123
    Primary tissues and tissue systems—Ontogeny of the axis—Primary vascular tissue: procambium, protophloem and protoxylem, primary phloem, primary xylem, arrangement of cells in primary vascular tissues, primary xylem types, types of vascular bundles—The primary vascular skeleton: the stele, stelar types, leaf traces, branch traces, leaf and branch gaps, the dissection of the vascular cylinder by reduction, general structure of the primary vascular cylinder—The pith: structure, the medullary sheath, of roots, duration—The pericycle—The endodermis: occurrence and position, function as related to structure—The cortex—The epidermis: ontogeny and duration, function, root hairs, the stoma, hairs—References for Chap. V.

THE SECONDARY BODY

Chapter VI

THE ORIGIN AND DEVELOPMENT OF THE SECONDARY BODY AND ITS RELATION TO THE PRIMARY BODY. THE CAMBIUM 175
    Origin from procambium—Fascicular and interfascicular cambium—Time of development: in stems, in roots—Extent—Duration—Effect of cambial activity upon the primary body—Relation of secondary growth to leaf traces—Relation of secondary growth to leaf and branch gaps—Function of cambium—Structure of cambium: size of cells, structure of cells, cell division—Gliding and intrusive growth of cambial cells and cambial derivatives—Ontogeny of secondary vascular tissues—

Time of cambial activity—Burial of branch bases—Cambium growth about wounds—Cambium in budding and grafting—Cambium in monocotyledons—References for Chap. VI.

Chapter VII

SECONDARY XYLEM. 204
Gross structure: xylem rays, annual rings, growth rings, early and late wood, false annual rings, ring-porous and diffuse-porous wood—Histological structure: cell types and cell arrangement, wood-parenchyma distribution, structure of gymnosperm wood, structure of angiosperm wood, xylem rays, ray tracheids, marginal ray cells of angiosperms, tyloses—Sapwood and heartwood—Relation of microscopic structure to properties and uses of wood: weight, strength, durability, other properties—Penetrability by preservatives—Grain of wood—Compression wood—Pith-ray flecks—Gummosis—References for Chap. VII.

Chapter VIII

SECONDARY PHLOEM 231
Extent and amount—Structure: sieve cells and sieve tubes, companion cells, phloem parenchyma, phloem fibers and sclereids, phloem rays, seasonal rings—Function—Cessation of function—Economic importance—References for Chap. VIII.

Chapter IX

PERIDERM AND ABSCISSION 248
PERIDERM. Structure: phellogen, phellem, phelloderm—Origin—Extent—Duration—Commercial cork—Morphology of the periderm—Protective layers in the monocotyledons—Storied cork—Function of periderm—Wound cork—Lenticels: distribution, origin, structure, duration.
ABSCISSION. Of leaves—Of floral parts—Of stems: immature and herbaceous, woody—References for Chap. IX.

THE ORGANS

Chapter X

THE ROOT. 277
Function—Gross morphology—Primary and secondary roots—Ontogeny—Root cap—Root hairs—Cortex—Pericycle—Pri-

mary vascular tissues—Secondary growth—Formation of lateral roots—Adventitious roots—Periderm—References for Chap. X.

Chapter XI

The Stem 293
Origin—Root-stem transition—Types: woody, herbaceous, monocotyledonous, vine—"Medullary rays" of vines and herbs—Internal (or intraxylary) phloem—The vascular bundle: size, shape, histological structure—Anomalous structure in stems—Interxylary phloem—Accessory cambium: formation and activity—References for Chap. XI.

Chapter XII

The Leaf 317
Morphology—Arrangement—Ontogeny—Duration of leaf development—Vascular tissue: development, orientation, xylem and phloem elements, bundle ends—Bundle sheath—Mechanical support—Mesophyll—Epidermis—Distribution of stomata—The monocotyledonous leaf—The grass leaf: sclerenchyma, bundle sheaths, mesophyll, epidermis, arrangement of stomata—Persistence of leaves—References for Chap. XII.

Chapter XIII

The Flower—The Fruit—The Seed 341
THE FLOWER. Ontogeny—Vascular skeleton: the floral stele, traces of the floral appendages—Anatomy of simple flowers—Anatomy of more complex flowers: fusion in the floral skeleton, fusion under cohesion and under adnation, the inferior ovary, adnation of flowers to other flowers and other organs—Placental vascular supply—Vestigial vascular tissue—Reduction of vascular supply within an organ—Sepal and petal—Stamen and carpel.
THE FRUIT. Morphology—Ontogeny—Structure—Vascular skeleton—Epidermis—Periderm—Fleshy fruits—Dry fruits—The placenta—Accessory fruit parts—Dehiscence.
THE SEED. Seed coats: morphology, histological structure, vascular bundles—Embryo and endosperm—References for Chap. XIII: flower, fruit, seed.

ECOLOGY

Chapter XIV

ECOLOGICAL ANATOMY 380
  Xerophytes: xerophytic environments, structural modifications (lignification and cutinization, sclerenchyma, hairs, rolling of leaves, stomatal structure, reduced leaf surface, needle leaves of gymnosperms, fleshy xerophytes)—Epiphytes—Hydrophytes: epidermis, dissected leaves, air chambers, absence of sclerenchyma, reduction of vascular and absorbing tissues—Shade leaves—Parasites—"Saprophytes"—References for Chap. XIV.

INDEX 397

"I know it will be difficult to make observations of this kind upon the *Organical Parts* of *Plants*, severally. . . . For what we obtain of *Nature*, we must not do it by commanding, but by courting of Her . . . I mean, that where ever Men will go beyond Phansie and Imagination, . . . they must Labour, Hope and Persevere . . . And as the means propounded, are all necessary, so they may, in some measure, prove effectual. How far, I promise not; the Way is long and dark . . . If but little should be effected, yet to design more, can do us no harm: For although a Man shall never be able to hit *Stars* by shooting at them; yet he shall come much nearer to them, than another that throws at *Apples*."—Nehemiah Grew, *The Anatomy of Plants*, 1682.

CHAPTER I

GENERAL STRUCTURE OF THE PLANT BODY—AN OUTLINE

Among vascular plants there is very great diversity in size, form, and structure; yet, underlying the variations in form and the complexities in structure in the plant body, there is a simple, uniform, structural plan. The body consists fundamentally of a cylindrical *axis* which bears lateral appendages. The more or less free branching of the axis and the variety and complexity of the appendages, however, often conceal this simplicity of plan.

Fundamental Parts of the Plant Body.—The axis, though a continuous structure, consists of two parts, different structurally and physiologically, and clearly morphologically distinct: that portion which is normally aerial is known as the *stem*, and that portion which is subterranean is called the *root* (Fig. 1). The appendages are of three ranks. Those into which pass strands of vascular tissue may be said to be of the first rank, and are known as *leaves*. Appendages of this type are characteristic of the stem and do not occur on the root. They are arranged in a definite manner, and bear an intimate structural relation to the skeleton of the axis. The leaf may be looked upon from the standpoint of the present treatment as a lateral expansion of the stem, continuous with it, in the formation of which all fundamental parts of the stem are concerned. In the appendages of the second rank only the outermost layers of the stem, the cortex and the epidermis, are usually present. These are known as *emergences;* the prickles of the rose are a familiar example. In appendages of the third rank, projections of the outermost layer of cells only are present; these form *hairs*. Emergences and hairs occur on both axis and leaves, usually without definite arrangement.

The Axis.—The axis itself consists of a central core with a surrounding, ensheathing layer. This core serves chiefly the important functions of conduction and support; it contains the *vascular tissue* and the larger part of all the supporting and conducting cells of the mature axis. Because of its shape and its position in the axis, this central unit mass is known as the *central cylinder*, or *stele* (Fig. 2). The surrounding layer, which serves for protection, support, storage, and for other purposes, is the *cortex;* the outermost layer of cells is the *epidermis*.

The Stele.—Primarily, the stele is composed of vascular tissue of two types: that which conducts water and other substances absorbed from the soil, the *xylem;* and that which carries the food (and possibly mineral nutrients), the *phloem.* Xylem and phloem nearly always occur together, usually side by side radially, the phloem outermost (Fig. 2). These tissues together may form a solid rod, a hollow cylinder, a sheath

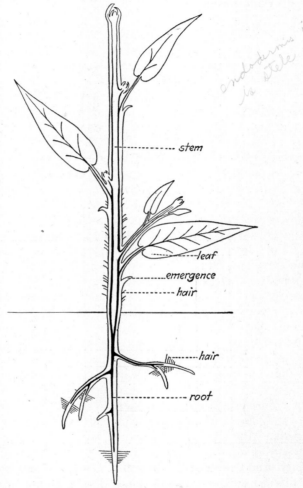

Fig. 1.—Diagram of plant body, showing fundamental parts.

of more or less symmetrically placed strands (Fig. 66), or a group of scattered strands (each consisting of xylem and phloem). Where the arrangement of the vascular tissues is such that they enclose tissue of a different type, usually soft and loose, a central portion, the *pith*, is set off. Outside the external conducting cells and forming the outermost part of the stele are a few layers of nonconducting cells, the *pericycle*. The pericycle is usually limited externally by a definite uniseriate sheet of cells of peculiar structure, the *endodermis*. The vascular core is thus ensheathed by the pericycle in a way similar to that in which the stele is enveloped by the cortex.

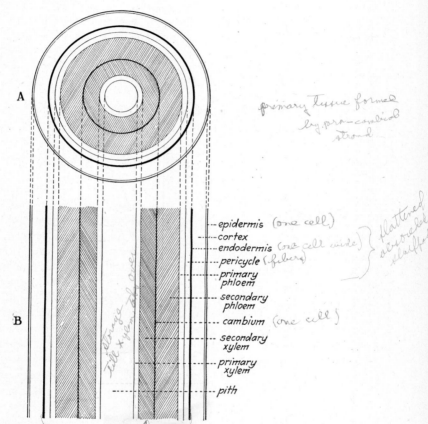

Fig. 2.—Diagram showing structure of axis. *A*, transverse section; *B*, longitudinal section.

Primary and Secondary Growth.—An axis complete in all the structural features above mentioned and with complete appendages is built up by growth at the growing points, situated at the tips of the axis. This first-formed body is known as the *primary body*, since it is built up by first, or *primary, growth*. Its tissues are, for the same reason, known as *primary tissues;* for example, the first-formed xylem is called *primary xylem*. In many vascular plants the primary body is reinforced by a different sort of growth, which because it begins later and adds to the original primary tissues is called *secondary growth*. The tissues thus formed are termed *secondary tissues*. Secondary growth does not usually form new types of cells, but merely increases the bulk of the plant, especially of the vascular tissues, providing new conducting cells and additional support and protection. It does not fundamentally change the structure of the primary body. Primary growth increases the length of the axis, laying down its branching system and adding its appendages; that is, it builds up the new, or young, parts of the plant body. After the parts thus formed have attained full size, additional increase in diameter is secured only by secondary growth.

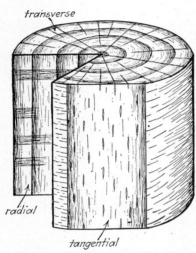

Fig. 3.—Diagrammatic sketch of a cylinder of wood, to show transverse, radial, and tangential planes of section.

The secondary vascular tissues are formed by a specialized growing layer, the *cambium*, which arises between the primary xylem and the primary phloem, and lays down new xylem and phloem adjacent to these. The secondary masses of xylem and phloem lie, therefore, entirely within the central cylinder and between the primary phloem and primary xylem. The newly formed xylem cloaks and ultimately completely surrounds the primary xylem and the pith, not changing the primary structure within, but burying it intact. The primary phloem and all other tissues outside the cambium are forced outward by secondary growth and may be ultimately more or less distorted or destroyed. The primary growth of a given region is completed in a relatively brief period, whereas secondary growth continues for a longer time, and in perennial axes may persist indefinitely.

Constitution of the Plant Body.—The root, stem, and leaves of a plant constitute its *organs*. These perform distinct, general functions for which they are adapted by the kinds, proportion, and arrangement of the *tissues* of which they are composed. The tissues have more restricted functions, which are determined by the kinds of *cells* that constitute them. The plant body thus consists of cells that, aggregated, form tissues; these grouped together, form organs.

Methods of Studying the Anatomy of the Plant.—The principal methods of studying the minute structure of the plant body are by means of thin sections of plant material and macerations in which the individual cells are freed from one another. Moreover, for adequate comprehension of the complex structure of most parts of the plant, study of sections cut in more than one plane is necessary. For the axis—a cylindrical structure—three planes, each at right angles to the other two, are most useful. One plane is transverse to the long axis; the others are parallel with the long axis, that is, longitudinal. Of the two longitudinal planes, that dividing the cylinder radially is the radial plane, and that at right angles to the radial plane is the tangential plane (Fig. 3). Sections cut in these planes are known as *transverse* (or *cross*), *radial*, and *tangential sections* respectively.

General References for Use with All Chapters

BONNIER, G., and LECLERC DU SABLON: "Cours de botanique," Paris, 1905.
DE BARY, A.: "Comparative Anatomy of the Phanerogams and Ferns," Engl. transl., Oxford, 1884.
BOWER, F. O.: "Size and Form in Plants (with Special References to the Primary Conducting Tracts)," London, 1930.
COMMITTEE ON NOMENCLATURE, INTERNATIONAL ASSOCIATION OF WOOD ANATOMISTS: Glossary of terms used in describing woods, *Trop. Woods*, **36**, Dec., 1933.
FOSTER, A. S.: "Practical Plant Anatomy," New York, 1942.
GUILLIERMOND, A.: "The Cytoplasm of the Plant Cell," Engl. transl. by L. R. Atkinson, Waltham, Mass., 1941.
HABERLANDT, G.: "Physiological Plant Anatomy," Engl. transl. of 4th Germ. ed., London, 1914.
———: "Physiologische Pflanzenanatomie," 5th Germ. ed., Leipzig, 1918.
HAYWARD, H. E.: "The Structure of Economic Plants," New York, 1938.
JEFFREY, E. C.: "The Anatomy of Woody Plants," Chicago, 1917.
SHARP, L. W.: "Fundamentals of Cytology," New York, 1943.
SIFTON, H. B.: Developmental morphology of vascular plants, *New Phyt.*, **43**, 87–129, 1944.
SOLEREDER, H.: "Systematic Anatomy of the Dicotyledons," Engl. transl., Oxford, 1908.
———, and F. J. MEYER: "Systematische Anatomie der Monocotyledonen," Berlin. I, 1928; III, 1929; IV, 1930; VI, 1933.
STRASBURGER, E.: "Histologische Beiträge," III, Jena, 1891.
TROLL, W.: "Vergleichende Morphologie der höheren Pflanzen. I. Vegetatsionsorgane." Lief. 1. Berlin, 1935.
TSCHIRCH, A.: "Angewandte Pflanzenanatomie," Vienna, 1889.
VAN TIEGHEM, P.: "Traité de botanique," Paris, 1891.

CHAPTER II

THE CELL

Plants and animals are made up of living substance, *protoplasm*, and its secretions. The body of an organism, fundamentally a protoplasmic structure, is of complex organization in that, except in the simpler forms, it consists of many more or less independent parts or units. These parts, which are clearly units both of structure and of function, are termed *cells*. The structural distinctness of the cell—in plants especially—is due in large part to the presence of an outer layer, or coat, the *cell wall*, which in plants is usually firm, and often hard and thick; in animals, delicate or lacking. The functional distinctness of the cell depends in part upon the presence and properties of an outer, limiting layer of specialized protoplasm lying just within the wall, the *plasma membrane* (page 13). The cell wall closely invests the protoplasmic unit, thus separating the protoplasm of one cell from that of adjacent cells. This separation is incomplete, however, since very minute perforations occur in the wall. Through these perforations extend thread-like processes of protoplasm, the *plasmodesmata* (page 35), which secure the continuity of the living matter of one cell with that of contiguous cells. The protoplasmic body of an organism is thus a continuous living system, though its individual parts, the cells, are definitely set apart by cell walls.

Uses of the Term "Cell."—The term "cell" has, in its different uses, sometimes included and sometimes excluded the cell wall. Early students of plant anatomy, seeing cells only when the walls were thick and visible at low magnification, called these structures "cells," "pores," "bladders," and described their contents as nutritive juices. The walls were the prominent part of these "receptacles." When the cell content was recognized as the essential part of the cell, attention turned to it as "the cell." Following this recognition of the protoplast as the fundamental unit, the term "cell" was often used to designate the protoplast alone. In the more common sense, however, the term implies protoplast plus wall—these together constituting the obvious unit of structure (Fig. 4), a unit for which there exists no term if "cell" is restricted in use to the protoplast alone. The term is further applied even to units of wall alone, as, for example, to the tracheid where the protoplast has disappeared, leaving within the walls a cavity, the *lumen*. The use of the term "cell" to indicate both protoplast and wall is desirable in view of the difficulties of expression otherwise involved and also in view of the intimate relationship between the two. The closeness of this relationship is not yet fully understood. Possibly the wall of a living cell is not merely a nonliving, external secretion of the protoplast, but is itself partly protoplasmic in nature. The use of the term to include both protoplast and wall is, of course, less strict than that underlying the frequently used definition of the cell as "an organized and more or less independent mass of protoplasm constituting the structural unit of an organism." Nevertheless, from the standpoint of anatomy, the term is better used in the looser and more comprehensive sense; it is so used in this text.

Cellular Complexity of Plants.—The primitive organism was doubtless unicellular or, in a sense, noncellular. The larger body of more advanced types is multicellular and the cells are limited by walls. The walls perhaps separate regions of activity of individual nuclei but certainly serve also for protection and support to the individual protoplasts and to the entire organism. Apparently, this walled condition is of great importance in large plant bodies both from the standpoint of mechanical stability and of physiological delimitation. Clearly, adaptation to a terrestrial and aerial habitat and the maintenance of a large body under conditions unfavorable to protoplasm have resulted in the specialization of the cell along various lines, especially those involving elaboration of the wall. Hence, among higher plants there are many kinds of cells with much variety in function, structure, arrangement, and with a great deal of complexity of wall structure. Variety, both structural and functional, in kind of cell; variety in relation of cells to one another; variety in arrangement of cells and of systems of cells with relation to one another and to the body as a whole—these give great cellular complexity to the more highly organized plants. Indeed, it may be said that in a general way the higher the plant, phylogenetically, the more complex its cellular structure.

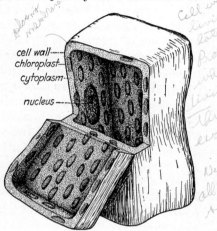

Fig. 4.—Diagrammatic representation of mature mesophyll cell opened to show central vacuole (filled with cell sap), nucleus and plastids embedded in the peripheral cytoplasm; wall thin. (Based on Zea.)

Cell Arrangement.—Regularity of arrangement of cells is brought about by successive divisions in the same plane (Figs. 41, 92). Such arrangement is characteristic of secondary tissues (those formed by a cambium or similar meristem) (Chap. VI) and is occasional in primary tissues (page 62), especially those of pith and cortex. But most primary tissues lack symmetrical arrangement and secondary tissues may lose this during development. Regardless of regularity or irregularity of arrangement and of the method of origin, a group of cells may be compactly arranged forming a close tissue continuous in one or more planes or they may be more or less free from one another, with the resulting formation of *intercellular spaces* (page 29) between them (Figs. 43, 173). Such spaces vary in shape with the shape and arrangement of the surrounding cells, in continuity, in abundance, and in size from microscopically minute openings to very large spaces which render the tissue in which they lie loose, spongy, and light (Figs. 149, 183). Such large intercellular spaces are known as *air chambers, canals, lacunae,* etc. These terms are applied rather loosely—the larger spaces being called chambers and the much elongated ones canals. Spaces formed by the separation of walls followed by the retraction of the separated parts or by spatial movements of the cells are said to be *schizogenous;* for example, the resin canal of *Pinus* (Fig. 55B) and the spaces of the aerenchyma of *Decodon* (Fig. 184). Others, developed by the destruction of cells formerly occupying the position of the cavity are called *lysigenous;* for example, the oil cavities of citrus fruits (Fig. 55F). Some cavities are formed by a combination of these two methods and have been called *schizolysigenous.* Some protoxylem lacunae (Fig. 64c) are of the last type. The intercellular spaces of a region may be without uniformity in size and shape or in arrangement of the cells surrounding them; they may, on the other hand, form definite structural features of a tissue or organ because of the development about them of specialized limiting or supporting "walls," or layers of cells, such as the diaphragms of many aquatic plants. Intercellular spaces commonly form definite systems and constitute, perhaps, functionally conducting or aerating systems. Special types of intercellular-space systems form ducts and canals (Chap. IV) with cells regularly arranged about them forming a sort of epithelial lining.

Cell Shape.—Because protoplasm is of semifluid nature, cells when free and independent tend to be spherical. But where young cells of the same type and essentially the same age lie together and grow in size, mutual contact and compression render them polyhedral with diameters nearly alike. Further growth toward individual or group specialization in relation to specific function leads to great variety in form—ovoid, ellipsoid, cylindrical, tabular, prismatic, lamelliform, fiber-like, and stellate or otherwise lobed or branched. There are, however, two chief types: the *subglobose,* or *polyhedral,* with diameters equal or only slightly different; and the *elongate*, with one diameter many times that of the others. Transitional types are, of course, numerous.

Undifferentiated or unspecialized cells, in continuous mass, when surrounded and closely compressed by other cells of similar type and size, are on the average 14-sided. They approach in shape the orthic tetrakaidecahedron, a mathematically determined form for bodies of uniform volume and minimal surface that fill space without interstices. This form has 14 surfaces of contact—8 hexagonal and 6 quadrilateral. In plants, uniform cells of this shape are probably never formed. The closest approach is perhaps in parenchyma cells of pith that are arranged in columns. The surface cells of a mass of such cells have an average of 11 faces. Epidermal cells, for example, have about that number of contact faces. When intercellular spaces are present, the number of sides is less—the larger the spaces, the fewer the sides in contact with other cells. Where similar cells of different sizes lie together under these conditions, the larger cells have more than 14, the smaller less than 14 faces. The shape of newly formed cells is dependent upon that of the mother cell and upon the plane and frequency of division.

Cell Size.—Cells vary greatly in size; size, like shape, is in part related to function. Extremely small cells do not occur among the higher plants. Parenchyma cells serving the usual functions, with normal protoplasm, have a transverse diameter of 0.01 to 0.1 mm. In pith and fleshy fruits, the diameter of parenchyma cells may reach 1 mm or more, and the cells then become readily visible to the naked eye. Fibers of wood and phloem range in length chiefly from 1 to 3 mm in the angiosperms, and from 2 to 8 mm in the gymnosperms. Cortical, pericyclic, and primary-phloem fibers are often much longer, and—partly because of their considerable length—are of particular economic importance, as, for example, those of flax and hemp. Fibers of excessive length—20 to 550 mm—occur in the Urticaceae and in certain monocotyledons. The largest cells known are latex cells of the type which form branching systems throughout a plant body (Chap. IV). However, such cells are perhaps not individuals morphologically, since they are coenocytic and continue to grow almost indefinitely. Latex vessels consist of cells united ontogenetically in series.

Cell Development.—All cells are formed from preexisting cells—or from nucleated protoplasmic masses—by division. Cell division is a complex process during which the nucleus and cytoplasm are each divided into two parts, usually equal and alike. In this division, the wall is not directly involved. In the newly formed cells, plasma membranes and intercellular substance are not readily distinguishable, and hence a visible plane of separation is at first lacking or scarcely detectable.

But very early, new walls appear along this plane as delicate membranes. Cells, therefore, while still very young, are provided with walls. The small daughter cells soon enlarge to about the size of the mother cell.

Young cells are of more nearly uniform size and shape than mature cells and are simple in structure. Elaborate shape, complex structure, and very large size are not found among young cells. Growth of cells involves increase in size and the development of special shape and structure. Since the wall is present from a very early stage, both protoplast and wall are involved in these changes. As the active part of the cell, the protoplast initiates the changes, the wall being modified with the changing protoplast. While changes in size and shape are taking place, the physical and chemical nature of the wall is usually such that these accommodations are readily made; only after cell maturity is attained is a fully developed and unchanging wall present. In the maturing of the protoplast many changes occur. These are, briefly, the reduction of the proportionate size of the nucleus; the development, usually, of less dense or less richly granular cytoplasm; the appearance of vacuoles that increase in size, and, in most cases, ultimately fuse to form one large central vacuole which restricts the cytoplasm to a peripheral position adjacent to the wall; changes in size and type of vacuole, where vacuoles are already present in the mother cell, for example, in cambium initials and their daughter cells; the increase of plastids in size and number, and the development of special types of plastids. In the maturing of the wall there occurs increase in extent and thickness accompanied by complexity of form and changes in chemical and physical structure (see Cell Wall).

Cellular Adjustment during Growth.—Growth frequently brings about changes in the relationships of contact and position among neighboring cells. The changes may involve all surfaces of a cell or be localized. In many cells the rate of growth of the wall is unequal in different areas, and in areas of greater or more rapid growth there are commonly greater changes in intercellular contacts. Development of elaborate or extreme shape and size—especially in such cells as large vessel elements, fibers, and some types of sclereids—brings about great changes. These may involve the deep penetration of cell tips or lobes between adjacent or nearby cells and into intercellular chambers. The steps in these changes are not fully understood. As new contacts are made by a penetrating tip and cell walls are separated, the intercellular cementing substance must be modified or removed. The plasmodesmata of the region are ruptured and the primary pit-fields and even partly developed pit-pairs are divided.

There is lack of agreement as to the manner in which these new wall contacts are made. *Gliding* (or *sliding*) *growth* has commonly been described as the basis of these changes. *Intrusive growth* and *symplastic growth* have recently been proposed as supplanting gliding growth in part or wholly. None of these three methods alone seems to cover the conditions met in all tissue and cell types.

By *gliding growth* is understood the slipping, during growth, of the wall of a cell over the wall of a contiguous cell along the surface of contact so that new areas of contact are made with the same cell and with nearby cells with which no contact previously existed. It has been generally assumed that real movement takes place in this type of growth, that the wall is increasing in area over more than the points of new contact, sometimes throughout, and that a sliding therefore occurs. Examples of apparent gliding growth are to be found (a) in the formation of new cambium initials with increase in the girth of the cambium—size, form, and position of the new cells are evidence of growth throughout the cell walls and of slipping and the making of new contacts (Fig. 89); (b) in the development from cambium derivatives of mature phloem and xylem elements—nonmatching pit-fields in the phloem and pits without complementary pits in the xylem are evidence here. The development of the lobes or arms of some types of sclereids is an example of gliding growth of parts of the cell wall.

Gliding growth has also been assumed to cover extreme changes in cell contacts brought about by the forcing, under pressure, of cells into new positions and distorted forms (Fig. 46D). As an example of this, the distortion in developing secondary xylem of cell arrangement and form about enlarging vessel elements has been cited. In ring-porous woods such as that of *Quercus*, the rapid and very great increase in diameter of a vessel element results in compression, or stretching, even tearing apart, of the smaller cells nearby (Fig. 92A). These cells find new positions, and their new contacts must often involve large areas of their wall surfaces. A sliding of cells on one another is the obvious method of change. Continuing growth cannot explain so great a change of position. It has been claimed that these changes are the result of cell multiplication among the smaller cells, but no cell division is found at this stage and the addition of new cells in the area would not explain extreme distortion of form.

By *intrusive growth* is understood *differential* expansion of the wall—the increase of cell size locally, with the projection of the newly formed parts of the cell between adjacent cells or their extension into intercellular spaces. The parts of other cells with which there is new contact may or may not also be enlarging. In this type of change it is understood that no real slipping occurs; the new contacts of a cell are those made by new areas of the wall as they form. The elongation of the apices of fibers, tracheids, and vessel elements has been called intrusive growth, but it is perhaps gliding rather than intrusive growth, or, more likely, it involves changes of both these types. An example of extreme continuation of intrusive growth is found in the latex cells of the milkweeds—Apocynaceae and Asclepiadaceae—whose rapidly growing apices maintain their position in apical meristems throughout the plant.

Most gliding and intrusive growth undoubtedly occurs while the cells are very young, but some apparently takes place after the walls are partly mature. In secondary xylem, walls of contiguous cells, already so mature that secondary walls and the beginnings of pits are present, may be split apart by the developing tips of adjacent partly mature cells. Such pits become functionless and are present in mature cells as *blind pits* (pits without complementary pits in the adjacent cell wall). Blind pits may even lead toward an intercellular space; the matching pit may be found across the space. (In sections pit-pairs often appear to extend only part way through the wall; these are usually parts of a pit-pair cut obliquely.) Intrusive growth may bring about a wrinkling of the wall of the intruding cell tip.

By *symplastic growth* is understood growth in size throughout a group of young cells with mutual adjustment in shape by all members of the group. The entire walls are growing. New shapes are formed and new spatial positions taken, but there are no new contacts and no movement either gliding or intrusive. The group of cells grows thoughout with their walls forming a framework that is continuously adjusted under pressure and tension. Differences in rate of division and rate of enlargement play a prominent part in symplastic growth. Growth of the derivatives of the initials in an apical meristem is an example of symplastic growth.

All three types of growth doubtless occur. Symplastic growth is characteristic of groups or masses of cells in early stages of growth. Evidence in support of the occurrence of typical gliding growth in many tissues is strong. Distinction between gliding and intrusive growth is doubtless one of degree rather than kind—the difference depends upon the definition of "tips" of the cell. If only a small part of the cell at the apex is growing and making new contacts with the walls of other cells, growth is intrusive; if there is growth in regions other than the tips, there must be movement of the gliding type.

THE PROTOPLAST

The cell may be considered separable, both structurally and functionally, into the central protoplasmic unit, the *protoplast*, and the surrounding *cell wall*. The protoplast is the cellular unit of the body of protoplasm that fundamentally constitutes the organism. The protoplast may be further described as an organized protoplasmic unit which contains specialized protoplasmic structures of various kinds, and also nonliving *inclusions*, organic or inorganic. Under the heading of inclusions are starch grains, oil globules, protein granules, crystals of many kinds, and the cell sap. These nonprotoplasmic materials are commonly known as *ergastic substances*. They are often considered not a part of the protoplast but, aside from the intimate relation of these substances to the physiological activities of the protoplast, they constitute a part of the protoplast as a structural unit, if only as inclusions. The cell wall is usually considered definitely distinct from the protoplast; that it may not be so, however, is possible. In the discussion of the nature of the wall (page 23), this aspect of cell structure is further treated.

Organization of the Protoplast.—The protoplast nearly always possesses a highly organized part, the *nucleus*. This is proportionately small, usually more or less nearly spherical or disk-shaped, and functionally of the greatest importance in the activities of the cell. Nuclei of extreme shape—linear, spindle-shape, vermiform—are frequent in elongate, slender cells. Bi- and multinucleate cells are occasionally found, but most reports of such cells (in apical meristems and in cambium initials and their derivatives) are based on misinterpretation of material. To the remaining protoplasmic material the term *cytoplasm* is given. The cytoplasm itself has portions of different degrees of specialization which are associated with segregated functions.

Plasma Membrane.—A film of extreme delicacy—of ultramicroscopic thinness—the *plasma membrane*, forms a limiting layer on the surface of the cytoplasm. This has been likened to the film that limits a free drop of water and forms the surface of a quiet body of water, preventing the immediate merging of the drop with the larger body upon which it falls. The plasma membrane completely covers the protoplast, including its plasmodesmata extensions, where it probably merges with that of adjoining cells. It has a close, perhaps intimate, association with the wall, possibly even penetrating its inner layers. (It is not shown in Figs. 4 and 8.) The semipermeable nature of the plasma membrane is of the greatest physiological significance in the functions of the cell.

Plastids.—The *plastids* are differentiated bits of protoplasm—"organs," or areas, of metabolic activity associated with particular functions. They have a limiting, apparently semipermeable membrane and complex internal structure. They are often colored and usually conspicuous. In size they are small, and generally many occur in a cell (Figs. 4, 5, 6, 7). They are variable in shape, but rounded types are most common. Spherical, ovoid, diskoid, granular, rod-like plastids all occur frequently. Large plastids of peculiar shape are present in many of the algae and rarely elsewhere. They may be found in all living cells of a plant and probably are present in every cell in its early stages of development. Later they become restricted to certain cells and are abundant only in those which have specialized functions, such as photosynthesis, storage, and color manifestation.

Origin.—Plastids are present in large numbers in young meristematic cells (Fig. 5) where they are minute, the smallest being at the limit of

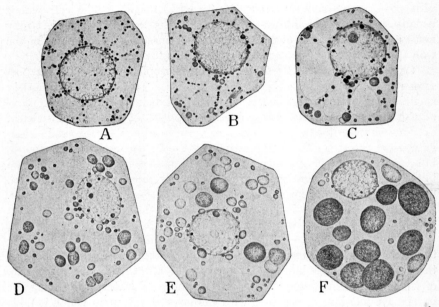

FIG. 5.—The development of a mesophyll cell in *Zea*. *A–F*, successive stages: the nucleus decreases in proportionate size; vacuoles appear, enlarge, and fuse; chloroplasts develop from proplastids. (*After Randolph.*)

microscopic visibility. At this stage, when they are known as *proplastids*, they are rounded bodies that do not resemble plastids. As the cell grows, proplastids multiply freely and mature plastids gradually develop. Increase in number by division continues at all stages but less frequently in mature plastids. Probably plastids arise only from preexisting plastids.

Types of Plastids.—Several fairly distinct types of plastids occur, but all are of similar fundamental nature. They fall into two chief classes: colored plastids, *chromoplasts*, and colorless plastids, *leucoplasts*. Green, chlorophyll-bearing chromoplasts are known as *chloroplasts* and are commonly set apart as a third major type because of their important function in food manufacture. This leaves the term "chromoplast" to cover all colored plastids except chloroplasts, and this is its general use.

Chloroplasts.—The chloroplasts of the higher plants are mostly uniform in size and chiefly flattened-ellipsoid or disk-like in shape. They are numerous—from a few to very many in a cell—and small, averaging 5 micra in diameter. They multiply by constriction, and may at other times change shape, appearing as though semiliquid. The chlorophyll is wholly or largely restricted to numerous small granules known as *grana* (Fig. 6). Apparently these deeply colored bodies are commonly so arranged and so closely packed that the plastid appears structurally homogeneous. Sometimes an obscure layering can be seen (Fig. 6C).

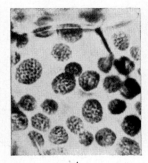 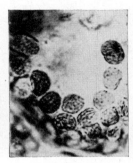 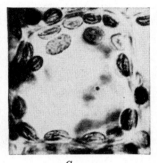

           A   B   C

Fig. 6.—Minute structure of chloroplasts. *A, Aponogeton; B,C, Todea superba. A,B,* grana large and free; *C,* grana small and aggregated in plate-like layers. (*After Heitz.*)

Chloroplasts of shade leaves are somewhat larger than those of leaves in full light on the same plant and their chlorophyll content is possibly greater per unit volume.

Chromoplasts.—Nongreen chromoplasts range in color through yellow, orange, and yellow-red. The color is given chiefly by xanthophyll, carotin, and carotinoids. Chromoplasts show great variety in shape but are chiefly irregular; granular, angular, acicular, and forked types occur (Fig. 7). The irregular and sharp-pointed shapes are believed to be caused in part by the presence of the colored substances, especially carotin and carotinoids, in crystalline form, as in the root of *Daucus* (Fig. 7A). The functions of these plastids are obscure. They are associated with color in flowers and fruits but occur also in other regions such as roots. Chromoplasts commonly represent transformed chloroplasts, but may form directly from small leucoplasts.

Leucoplasts.—The term "leucoplasts" covers various types of colorless plastids. The early stages of all plastids are called leucoplasts but such young plastids are best called *proplastids*, and only mature colorless plastids should be called leucoplasts. Like chromoplasts, leucoplasts vary in shape; extreme forms are rod-like. They change shape readily and are probably always highly plastic. They are concerned with food storage and doubtless have other unknown functions. The type of leucoplast associated with starch-grain formation in storage regions is known as *amyloplast*. Leucoplasts related to the formation and storage of oils and fatty substances are known as *elaioplasts*. These are at least

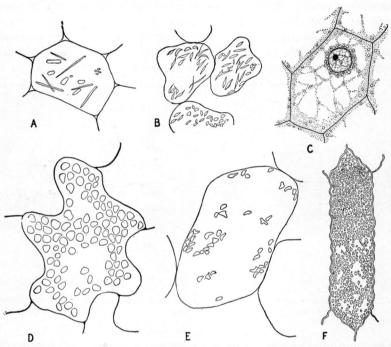

FIG. 7.—Plastids. Chromoplasts: *A*, in cortex cell of root of *Daucus*; *B*, in pulp cells of fruit of *Arisaema*; *D*, in cell of petal of *Forsythia*; *E*, in pulp cells of fruit of *Lycopersicon*; *F*, in cell of corolla of *Taraxacum*. Leucoplasts: *C*, in young endosperm cells of *Zea*.

in part amyloplasts that vary in function at certain times. The leucoplasts of hairs and of other epidermal cells are probably degenerate or dormant plastids of other types.

That all plastids are alike in nature is clear from the readiness with which one type is transformed into another. For example, the chloroplasts of young fruits and of developing petals may become the chromoplasts of the ripe fruit and of the mature flower respectively; the leucoplasts of a potato tuber become chloroplasts on exposure to light.

Occurrence of Plastid Types.—Chloroplasts may occur in any part of a plant that is exposed to the light; they also occur in some tissues that are apparently without light, such as the wood of many Rosaceae and

Ericaceae, and in embryos and endosperm, as in the seeds of some citrus fruits. Red and yellow chromoplasts likewise may occur in any organ of the plant, and their presence is not related to the presence of light. They are chiefly to be found, of course, in flowers and fruits. Leucoplasts occur mostly in parts not exposed to light. Further statements concerning the distribution and the function of plastids are to be found in the discussion of various tissues, tissue systems, and organs, such as collenchyma, the cortex, the leaf, and the petal.

Many protoplasmic bodies smaller than plastids occur in cytoplasm; these are known as *chondriosomes* and *mitochondria*. Some of those known as chondriosomes are doubtless proplastids. The nature and distribution of these smaller cytoplasmic structures lie outside the field of the present treatment.

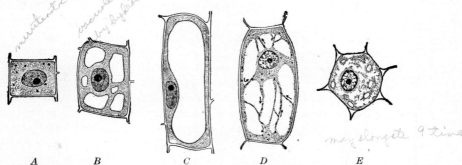

A   B   C   D   E

FIG. 8.—*A–C*, diagram of a plant cell in three successive stages of development: the vacuoles increase in volume and fuse and the cytoplasm becomes limited to the parietal region. *D*, cell of stamen hair of *Tradescantia*, indicating direction of streaming movements in the cytoplasmic strands. *E*, parenchyma cell from cortex of *Polygonella*, showing nucleus, plastids, and scanty cytoplasm. (*After Sharp.*)

Vacuoles and Cell Sap.—The cytoplasm of mature and of many young cells contains one or more cavities known as *vacuoles*. Conspicuous cavities usually develop as the cell matures, beginning as one or several small cavities and enlarging and fusing until in the mature cell there is usually one large central vacuole which restricts the cytoplasm to a thin layer lining the wall (Figs. 4, 8*C*). The nucleus, plastids, and most inclusions are contained in this peripheral layer. Frequently strands of the cytoplasm extend from the outer layer as irregularly anastomosing branches through the central vacuole; under these conditions the nucleus may occupy a median position in the cell. The vacuoles and their liquid content, the *cell sap*, constitute the cell *vacuome*.

The cell sap is a watery, nonprotoplasmic liquid consisting of water with various substances in solution—inorganic salts, carbohydrates, proteins, amides, alkaloids, pigments, etc. These substances may be mineral nutrients, elaborated food, waste products, or substances of unknown relation to metabolism. In the physiology of the cell the significance of these substances dissolved in the cell sap is not well understood.

Color in Cells.—Substances that give color to cells are found chiefly in the plastids and the cell sap. The cell wall, cytoplasm, and nucleus are commonly nearly colorless. The green coloring substance of chloroplasts is *chlorophyll*, in two forms known as a and b. With the chlorophyll, but masked by it, are yellow pigments, *carotinoids*. Included in the carotinoids are carotins, carotenes, and xanthophylls. These substances give the yellow color of "golden" varieties of plants (where the chlorophyll content is low) and of leaves variegated with yellow. Leaves variegated with white have little or no color in the plastids of the pale areas. Abundant carotinoids (sometimes in crystalline form) in chromoplasts give yellow, orange, and yellow-red colors to flowers and fruits. Another group of pigments, the *flavones*, are water-soluble and color the cell sap. In some genera, for example, *Verbascum*, they give yellow color to corollas. The *anthocyanins*, oxidation products of flavones, also water-soluble, color cell sap and give red, purple, and blue color to many flowers and fruits. A reddish, anthocyanin-colored cell sap, together with the green color of the chloroplasts, gives the bronze or purple color to the leaves of purple-leaved forms of many plants. Anthocyanins are also present in young twigs and leaves, especially if growth is taking place at low temperatures. Other color-bearing substances may be present, and lower groups of plants—algae and bacteria—possess some distinctly different substances. In white floral parts there is no color, and light is reflected from the many semitransparent cells separated by intercellular spaces.

Autumn Color.—In slowly dying leaves chlorophyll breaks down into colorless substances. Green leaves become yellow as the xanthophyll is thus unmasked. Xanthophyll in the disorganizing chloroplasts in this way gives much of the yellow color seen in autumn foliage. Reds and purples are cell-sap colors, resulting from oxidation of flavones, and are especially brilliant when formed in the presence of sugars and abundant sunshine. These cell-sap colors, in combination with some persisting chlorophyll, xanthophyll, browning cell walls, tannins, and less common color-giving substances, give great variety in autumn color. The changes from green foliage are not caused by frost, although frost may hasten the process. Many plants whose leaves color brilliantly in cool temperate climates show equally bright color when grown in frostless climates. Dying branches of trees often show full "autumn" color in midsummer.

Inclusions of the Protoplast (Ergastic Substances).—Many kinds of solid particles, organic and inorganic, as well as such substances as oils, gums, resins, etc., are frequently present within the protoplast, either in the cytoplasm or in the vacuole. These represent, like the dissolved substances, food products, such as starch and aleurone grains; waste products, such as crystals; and other substances of doubtful or unknown function, such as rubber, mucilages, tannins, latex, alkaloids. Gums, resins, tannins, etc., are often present in the lumina of nonliving cells; for example, in the heartwood of *Sequoia* (Fig. 99) and mahogany (*Swietenia*), and in cork cells of many trees. Some of these substances, for example, starch, occur in most plants; others are characteristic of certain large or small groups of plants, and are lacking in others.

Crystals.—Crystals of various chemical nature occur freely in plant cells—of these, salts of calcium, chiefly calcium oxalate, constitute the majority. Crystals of other salts of calcium and of various other inorganic substances, such as silica and gypsum, occur less frequently; crystals of many organic substances, such as carotin, berberin, and saponin, are frequent. All parts of the plant may contain crystals, though these structures are more abundant in certain regions, as in the pith, cortex, and phloem, than in others. There are many forms of crystals (Fig. 9). Of these, solitary, rhombohedral crystals, sheaf-like bundles of long acicular crystals known as *raphides* (Fig. 9*E*), and clustered crystals in globose masses called *druses* (Fig. 9*A,B,D,F*) are most common. The individual crystals of druses may be raphides. Solitary raphides, small prismatic crystals, and minute crystals called *crystal sand* are other frequent types. *Cystoliths*, often considered crystalline bodies, consist in large part of cell-wall substance.

One form of crystal may occur in a given cell—the usual condition where crystals are clustered—or two or more types may be found in the same protoplast. Many inorganic crystals appear to involve organic matter in the course of their formation. The larger crystals of xylem often show this condition (Fig. 9*G,H,K,L*), and druses frequently contain a prominent organic center (Fig. 9*F*).

The large crystals of somewhat various shapes—prismatic, rhomboid, etc.—occur chiefly in fibers of xylem and of phloem and in parenchyma cells associated with fibers. Raphides occur in thin-walled, mucilage-containing parenchyma-cells of soft tissues, such as storage parenchyma of underground parts, fruit pulp, and the tissues of aquatic plants generally. They are present in monocotyledons chiefly. Druses are characteristic of parenchyma cells of cortex and pith, especially of stems and petioles; they also are abundant in phloem.

Crystals lie in the protoplast—usually in the vacuoles—or in the lumen

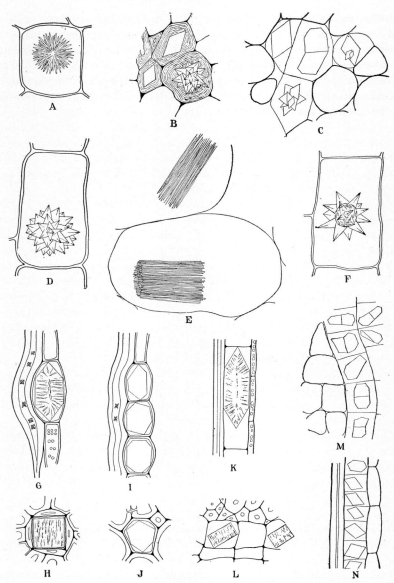

Fig. 9.—Crystals. *A*, druse in cortical cell of stem of *Viburnum Lentago*; *B*, druse and rhombohedral crystals in stone cells of nutshell of *Carya glabra*; *C*, solitary and grouped crystals in pith cells of *Populus grandidentata*; *D*, druse in cortical cells of *Carica Papaya*; *E*, "bundles" of raphides in pulp cells of fruit of *Smilacina racemosa*; *F*, druse with organic center, in phloem parenchyma cell of *Juglans nigra*; *G, H*, longitudinal and transverse sections of crystal in wood parenchyma of *Carya Pecan*; *I, J*, longitudinal and transverse sections of crystals in wood parenchyma of *Juglans nigra*; *K, L*, longitudinal and transverse sections of crystals in phloem parenchyma of *Tilia americana*; *M*, various forms of crystals in phloem parenchyma of *Malus pumila*; *N*, rhombohedral crystals in phloem parenchyma of *Salix nigra*.

when the protoplast has disappeared, as in fibers. Occasionally, they are partly or wholly embedded in the cell wall; more frequently they are suspended in the lumen by projections of the walls. These projections are rod-like; or they are sac-like, covering the crystal and holding it in a central position. Large crystals may fill the lumen and then determine the interior contour of the wall (Fig. 9*B*); in elongate cells the lumen may be filled at a given level only (Fig. 9*K*).

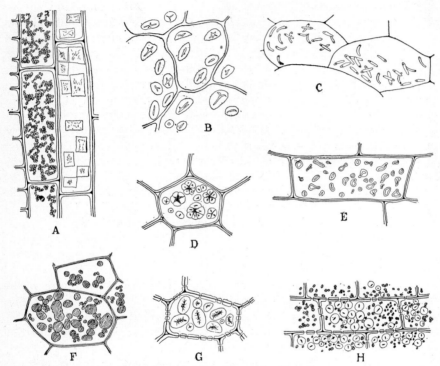

Fig. 10.—Starch grains and tannin. Tannin: *A*, in phloem parenchyma of *Pinus* (also crystals); *F*, in pith cells of *Fragaria*; *H*, in ray cells of wood of *Malus pumila* (also starch grains). Starch grains: *B*, in pith cells of *Alsophila*; *C*, in outer pericarp of *Musa*; *D*, in cotyledon of *Pisum*; *E*, in ray cell of phloem of *Ailanthus*; *G*, in cotyledon of *Phaseolus*.

Cells may be given over entirely to crystal storage, and the protoplasts become much reduced or disappear, but typical active cells may also contain abundant crystals. The early stages of crystals are found in very young cells; apical meristems often contain young druses.

In large part, inorganic crystals are probably waste products, the result of metabolic processes. Their development in tissues which soon cease to be functional, such as pith, cortex, and secondary phloem, is suggestive of this.

Starch.—Food materials are present either as transitory material or in the more or less permanent form of storage particles. Newly formed food may be in solution or exist as solid particles. Starch grains are the most common kind of solid food material found. They are of numerous types and vary in size and in form over a considerable range (Fig. 10). In shape they are mostly rounded or oval. Crowding results in the formation of angular shapes, and symmetrical, polyhedral grains are characteristic of some plants. Other plants possess compound grains which closely resemble simple grains, but show their compound nature when broken up into their constituent parts which are minute simple grains. All grains in a cell may be simple or compound, or both types may occur in the same cell (Fig. 10*D*). Structurally the grain is made up of radially arranged tapering needles of amylose. The needles lie with their tips at the center, their bases fused to form a border. Both the central area, the *hilum*, and the border are highly refractive. Concentric layering about the hilum is commonly seen, although this may be obscure without special treatment. The layering may become strongly excentric in the larger grains of some plants, for example, the potato. The layered appearance is probably the result of variation in the amount of water in the amylose; and this variation is due to alternations in growing conditions—in light intensity, temperature, and humidity. Grains formed under constant growth conditions are not lamellated. The hilum is rounded or angular, sometimes lobed, forked, or stellate, and is highly refractive to light. By the presence of hila and by staining properties, starch grains may be distinguished from plastids and other protoplasmic bodies, and from other solid particles.

Starch grains are first formed within chloroplasts and are often found in that position; leucoplasts in storage cells build up starch grains within themselves from translocated food brought from green cells. Starch grains are thus formed both primarily and secondarily within plastids. It is not known with certainty whether, when a grain is mature, it is freed from the plastid or is still surrounded by a very delicate layer of plastid substance.

Nitrogenous Inclusions.—Solid nitrogenous particles such as *crystalloids*, or *protein crystals*, are frequent in seeds and in other storage organs, such as bulbs and tubers, as in the potato tuber; and *aleurone* grains are often found in seeds, where definite layers of cells are filled with them, as in the corn fruit (Fig. 11). Crystalloids and aleurone grains are rare elsewhere. Tanniferous bodies are abundant in many plants, especially in the cortex and phloem (Fig. 10*A*). They occur chiefly in parenchyma cells but may be found in other cells, such as those of collenchyma and cork. Their abundance in the phloem of oak, hemlock, and other trees makes the bark of these trees valuable for tanning. Tanniferous bodies are small, granular, or rounded particles, often more or less fused in masses. Tannins may be mixed with or dissolved in gummy or mucilaginous masses in the protoplast. Other solid or semisolid substances, such as resins, gums, mucilages, fats, oil globules, occur frequently.

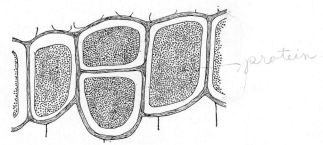

Fig. 11.—Aleurone grains in endosperm of *Zea*.

THE CELL WALL

In vascular plants only certain cells related to reproductive processes and early embryology are naked; all others have cell walls. This wall, at first very thin and delicate, is modified in various ways as the cell matures. The more important changes are increase in extent and in thickness; in chemical nature; and in modification of grosser physical structure, such as the absorption of the end walls of vessels. The presence of a prominent wall, especially after the earliest stages of development, makes cellular structure in plants distinct—a feature in strong contrast with the condition in animals where the limits of the protoplast are less readily discernible.

In recent years the origin and structure of the cell wall have received critical attention and are now much better understood. A new interpretation is placed on the obscure earliest stages of wall formation, the structure of the wall is known in detail, and necessary changes have been made in the terminology applied to the wall.

Nature of the Wall.—The cell wall is commonly looked upon as a secretion of the protoplast, laid down upon its surface. Strictly as such, the wall would be a nonliving layer. Although this is the view now generally accepted, the opinion is held by some that the wall in early stages, and the layers contiguous with the protoplast in later stages, may contain protoplasmic material and therefore form a part of the living unit. The middle lamella may also contain living substance in early stages and possibly as long as the protoplast remains active. If the wall is a living adjunct of the protoplast, certain facts of structure and behavior of cells are more readily understood whereas others are difficult of interpretation.

Origin of the Wall.—Wall formation is first evident at the close of nuclear division. As the daughter nuclei reach the telophase, the spindle fibers become stronger along the equatorial line and weaker near the nuclei (Figs. 14A,15A). With the surrounding, somewhat denser cytoplasm

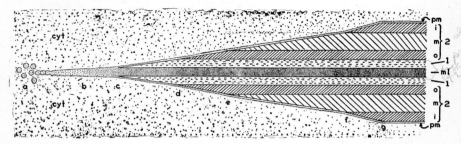

Fig. 12.—Diagrammatic representation of successive stages (left to right) in formation of cell wall with secondary layers. (Walls of both daughter cells are shown.) a, origin of cell plate; b, beginning of transformation of cell plate into middle lamella (ml); c, beginning of deposition of primary wall (1, 1); d, e, f, beginning of deposition of outer, middle, and inner layers (o, m, i) of secondary wall (2, 2); g, completion of wall thickening; pm, plasma membrane; cyt, cytoplasm. (*Based on researches of W. A. Becker, I. W. Bailey, T. Kerr, and others; from Sharp.*)

(the kinoplasm) they form a rather poorly limited, bulging-barrel-shaped structure, the *phragmoplast*. Along the median line of the figure, droplet-like particles appear, increase in size, and merge to form a fluid *plate* that divides the phragmoplast (Figs. 12, 14B). The plate is first evident at the center of the phragmoplast and then extends marginally in all directions toward the mother-cell wall. As the plate forms, the central spindle fibers disappear, and with the extension of the plate, the peripheral spindle fibers also progressively disappear—the outer ones persisting longest. When the plate is complete, reaching the wall on all sides, its substance becomes less fluid, and delicate membranes appear on both its surfaces. These membranes are probably deposited by the daughter protoplasts. They constitute the first stages of the new cell walls of the daughter cells. The cell-plate material is gradually transformed into intercellular substance, which becomes an *intercellular layer* (page 28). The nature of the substance forming the cell plate is unknown; it is possibly cytoplasmic, and the spindle fibers may be "lines of flow" in the cytoplasm.

Where the mother cell is small and isodiametric and the nucleus large and central, the division figure is large, and no complex structural conditions are involved in the division of the protoplast. But cell division and the completion of the new wall are frequently much more complex.

Division in cells with a large central vacuole involves predivision changes in form and position of nucleus and in structure of the vacuome (Fig. 13) —the cytoplasm increases in amount, extending in strands through the vacuole; the nucleus becomes more rounded and moves away from the wall along one of the cytoplasmic strands; the strands gradually become

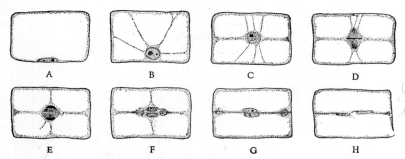

Fig. 13.—Division in cell with large central vacuole. Cytoplasm increases in amount forming layer in position of new wall; nucleus becomes globose and migrates to central position; normal nuclear division with extending phragmoplast. (*After Sinnott and Bloch, from Sharp.*)

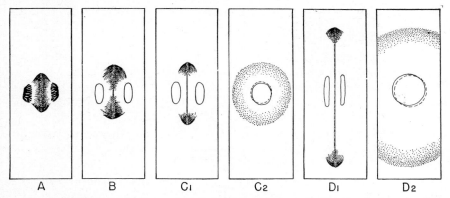

Fig. 14.—Diagrams to show cell-wall formation during cell division of elongate cells, as seen in longitudinal sections. *A*, at late telophase stage: the phragmoplast enlarging equatorially, with droplets of cell-plate substance forming along the median line; the spindle fibers disappearing in the polar and central regions. *B*, the beginning of the cell plate (lying at right angles to the page), formed by fusion of the droplets in the center of the phragmoplast; new fibers forming on the equatorial periphery of the phragmoplast, the older ones progressively disappearing. *C1*, the phragmoplast, now an annular band, continuously expanding peripherally, has extended the cell plate in all directions. *C2*, section at right angles to *C1*, the phragmoplast (lying in the plane of the page), has extended nearly to the mother-cell wall in the central part of the cell; the cell plate, a delicate membrane in the plane of page, not shown. *D1*, the cell plate and phragmoplast ring extended much farther than in *C1*. *D2*, section at right angles to *D1*, the phragmoplast ring has reached the lateral walls in the central part of the cell and disappeared there, leaving two arcs (kinoplasmasomes), which will continue to progress to the ends of the cell; the cell plate has also reached the lateral cell walls in the center, dividing the cell in the central part, while both ends remain undivided.

concentrated in a layer (the "phragmosome") in the plane in which the new wall is to be formed. Nuclear division and wall formation then take place much as in cells with abundant cytoplasm.

In cells in which one diameter is markedly greater than the others, especially in those that are very long, and when division is longitudinal,

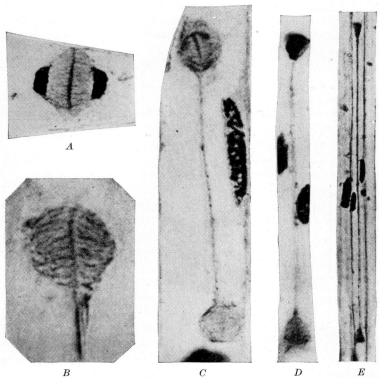

Fig. 15.—Cell-wall formation during cell division in cambium initials of *Pinus Strobus*, seen in radial section. *A*, Late telophase stage in nuclear division and beginning of cell-plate formation. *B*, One section of phragmoplast ring, seen in *C*. *C*, Stage later than *A* in cell-plate formation, showing the cell plate extended, two circular sections of the phragmoplast ring. Only one daughter nucleus seen in this plane of section. *D,E*, The cell plate farther extended than in *C*. (The largest nucleus in *E* lies in an adjacent cell.) (*After Bailey*.)

as in cambium cells, the phragmoplast persists long after nuclear division is complete and builds the cell plate far beyond the daughter nuclei and to the ends of the mother cell (Figs. 13, 14, 15). Its polar parts and the central fibers disappear, as in isodiametric cells, with the formation of the first part of the cell plate (that between the daughter nuclei) and the plate increases in diameter, expanding as an equatorial belt in all directions. New fibers are added progressively at the periphery as the older ones disappear. The new fibers are short, forming only close to the plate, and recurve toward the nuclei (Fig. 14C,D). The barrel-shaped phragmoplast thus becomes a continuously expanding, annular band of short fibers and dense cytoplasm. As the band expands, it is broken up into arcs (which have been called "kinoplasmasomes") as the parts that reach the nearer walls disappear. The persisting arcs continue to progress to the farthest parts of the cell, adding to the cell plate on its margins until the cytoplasmic body is completely divided. The constant presence of the fibers as the plate is formed and their renewal on its progressing margins far from the nuclei suggest close relation of these structures to the building of the plate.

As the phragmoplast ring moves away from the daughter nuclei, it appears, when seen in face view, as a halo-like band surrounding the two nuclei. This circular band, which has been called the "phragmosphere," is merely the ring of kinoplasm which at this stage constitutes the phragmoplast. With the two nuclei, it has been mistakenly interpreted as a binucleate cell and made the basis for the statement that meristematic cells are commonly binucleate. The nuclei are not in the same cell but are separated by the very young cell plate which lies in the plane of section and is not readily detected. It is merely a polar view of a late stage in cell division. Longitudinal cross sections show the phragmoplast as two rounded or wedge-shaped masses at the ends of the weakly developed cell plate (Figs. 14C,D, 15).

In such cells as cambium initials, procambium cells, young sieve-tube elements, young fibers, and other elongate cells, cell-plate extension is of extraordinary degree and the wall often of unusual form. Curved and even cup-shaped walls may be built by the changing plane of the ring as the plate is formed. The breaking up of the phragmoplast ring into arcs and the slow completion of the division of the protoplast are common but neglected features of cell division. This represents not a special type of cell division but an adaptation to cell form.

Gross Structure of the Wall

In mature, thick-walled cells, a concentric layering is usually evident in the wall. The layers represent successive deposits of wall substance by the protoplast upon the cell plate. The layers differ from one another in physical and chemical nature, their capacity for adaptation to changes in cell size and shape, and in extent over the earlier formed wall. They fall into two major groups, constituting the *primary wall* and the *secondary wall*. The primary wall usually consists of a single layer. The secondary wall is made up of one to many layers, most frequently of three.

The Primary Wall.—The membrane formed on the surface of the cell plate is the first stage of the primary wall. As cell enlargement begins, this membrane becomes somewhat unevenly thickened but remains delicate and mesh-like in finer structure. During rapid increase in area, the wall may maintain its thickness and unevenness, or be alternately thickened and thinned. The uneven thickening soon sets off more or less distinct thinner areas, *primary pit-fields*, sometimes known as *primordial pits*. Primary pit-fields are conspicuous in cambium initials (Fig. 88*B*). At this stage abundant plasmodesmata (page 35), protoplasmic connections between the protoplasts separated by the new walls, are evident, clustered chiefly in the primary pit-fields. The presence of large numbers of plasmodesmata perhaps determines the location of the pit-fields. The mesh-like earlier wall is probably also freely penetrated by cytoplasmic projections, the protoplasts remaining in continuity from the division stage, but such connections have not been demonstrated. At this early stage the wall substance is possibly highly protoplasmic, as may also be the middle lamella.

The young wall remains plastic and adaptable to changing volume and shape as the young protoplast continues growth. Inequalities in thickness persist, and alternate thinning and thickening may continue. With attainment of final size and form in the cell, the primary wall takes on mature structure and may thicken further. Small, well-defined thin areas, *pits* (page 37) are formed, one or more in each of the primary pit-fields. The plasmodesmata are largely restricted to the pits, but scattered, even solitary, strands may occur outside these areas. The mature primary wall varies greatly in thickness and in chemical and physical structure in cells of different types and in cells of the same type in different tissues and different plants.

The Middle Lamella.—As the first stages of the primary wall are developing, the substance of the cell plate becomes, by physical and chemical changes, a firm *intercellular layer*, a *true middle lamella*, which serves to bind the cells together. The term *middle lamella* has been in common use for a middle layer in the walls separating two protoplasts, or two cell lumina, that differs from the rest of the walls in staining and light-refracting qualities. Such a layer is, however, structurally and morphologically complex; according to this loose usage the "middle lamella" includes the true middle lamella (intercellular substance) plus the contiguous primary walls of the two cells and often also the thin, first-formed layers of the secondary wall. The intercellular layer is the middle layer of this group of three or five layers and the term middle lamella should be restricted to this. In stained preparations the true middle lamella is commonly difficult to differentiate because of its close resemblance in chemical nature to the contiguous walls. For convenience, the term "middle lamella" will doubtless continue to be used for the compound layer, but it should be remembered that this is a loose and inaccurate use.

Along the line of contact of the new wall with the mother-cell wall the new and the old middle lamellas are separated by the primary wall of the mother cell (Fig. 16). Union of the layers of middle lamella is secured by the development, along the line of contact of the old and new walls, of a cavity—triangular in cross section—in the mother-cell primary wall, which enlarges and extends to the middle lamella, and the extension into this cavity of the intercellular substance of the new wall. When the cavity continues to enlarge, and the intercellular substance does not fill it, an intercellular space lined by intercellular substance is formed.

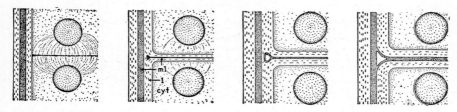

Fig. 16.—Stages in the establishment of the connection between the middle lamella of a newly formed wall with that of the lateral wall of the divided cell. *ml*, middle lamella; *l*, primary wall; *cyt*, cytoplasm. (*After Martens, from Sharp.*)

Commonly the formation of intercellular spaces is more complex than this. More than two walls are concerned and the development of the interwall space is preceded, accompanied, or followed by a splitting of the middle lamella of the older wall nearby. The interwall space and the space formed by the splitting merge, and the cavity so formed enlarges by further splitting and by pulling away of surrounding cells under growth stresses. Increase in size of spaces sometimes results from shrinkage of cells. When two new walls are formed opposite one another, the two triangular spaces merge—sometimes with a split which appears between them—and a diamond-shaped space is formed. In meristems where cell divisions succeed one another rapidly and all the walls are thin, intercellular spaces seem to be formed by a simple splitting apart of cells through the middle lamella.

The Secondary Wall.—In many cell types further wall-thickening occurs after cell size and shape are fully attained. The wall then formed is the *secondary wall*. This differs from the primary wall especially in that it is incapable of increase in area and that such modifications as occur are irreversible. The primary wall covers the protoplast except where the plasmodesmata occur; the secondary wall is formed over the primary wall except over the pit membranes. As this wall thickens, the pit cavities become deeper, and in some kinds of cells, partly enclosed and complex in structure (bordered pits, page 39). The secondary wall tends to be more massive than the primary, and in most thick-walled cells it constitutes the major part of the wall. In the tracheids and vessels of the protoxylem (page 132) the secondary wall covers much less of the primary wall; it forms only as rings, spiral bands, and bars over the delicate primary wall (Fig. 62).

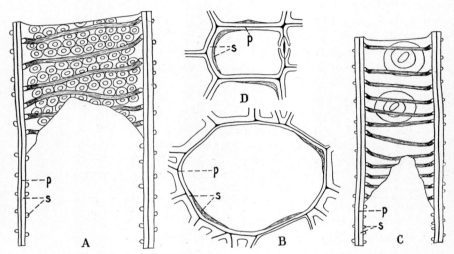

Fig. 17.—Primary wall and secondary wall with tertiary spirals. *A,B*, longitudinal and transverse sections of vessel of *Tilia americana*; *C,D*, longitudinal and transverse sections of tracheid of *Taxus brevifolia*.

The last-deposited layer of the secondary wall of some wood elements—laid down over a wall with fully developed pits—has the form of delicate spiral bands (Fig. 17). Because these bands appear to represent thickening after the secondary wall is complete they have been called a *tertiary wall*, but they represent only the innermost layer of the secondary wall and are now called *tertiary spirals* or *spiral thickenings*. The term "tertiary spirals" better distinguishes them from the spiral (secondary) bands of protoxylem which they somewhat resemble. Tertiary spirals lie upon a thick secondary wall; protoxylem spirals lie upon a delicate primary wall.

Minute Structure of the Wall and Middle Lamella

Both primary and secondary walls are composed fundamentally of a complex mesh-like matrix of cellulose. Other substances fill the interstices of this framework and may be replaced later by different ones. In

Fig. 18.—Minute structure of secondary wall of fiber tracheid of *Siparuna*. Above, after delignification, showing radiating pattern in the remaining cellulose; below, after removal of cellulose, showing radiating pattern of the lignin. (*After Bailey and Kerr, from Sharp.*)

the primary wall pectic substances accompany the cellulose in early stages; the secondary wall is at first composed mostly of cellulose or of cellulose and hemicelluloses. Substances that at later stages replace, or are added to these are lignin, cutin, suberin, various inorganic materials, tannin, oils, and many others.

In minute structure the cellulose matrix of both walls is made up of aggregates of delicate fibrils that grade down in size to the limits of microscopic visibility. The fibrils are believed to be made up of *micellae*, aggregates of molecules, probably in crystalline form. The fibrils form complex, three-dimensional anastomosing systems (Fig. 18) which cohere firmly and are continuous throughout a wall. Variations in size, number, and arrangement of the fibrils give different structural patterns. The pattern may be dominantly radial or dominantly concentric. It is probably usually a combination of these—radial arrangement intensified at intervals during formation by increase in number or in size of fibrils so that a concentric layering is apparent.

Within the meshes of this matrix, channels and pores form another system in which are deposited various substances. The substances filling the interstices form a 'secondary' structural system which in pattern necessarily follows that of the primary system. A wall may thus be built up, for example, chiefly of two substances such as cellulose and lignin, closely intermeshed. Either of these substances can be dissolved out leaving intact the system formed by the other (Fig. 18). The orientation and size of fibrils and pores varies in similar cells, in different parts of a cell and in different layers of the same wall.

The fibrils are commonly oriented with their long axes parallel with one another and with the surface of the layer they constitute. When the aggregates of fibrils are sufficiently coarse they may be seen on the face of the cell wall as delicate *striations*. These lines are most frequently spiral but may run in any direction. The fibrillar arrangement varies in different layers of the secondary wall (Fig. 19), and that of the primary wall differs from that of the adjacent layer of the secondary wall and may aid in distinguishing these walls. In successive layers the fibrils commonly run in different directions, and the major lines of structure in a thick cell wall run at an angle with those of the abutting wall of the neighbor cell suggesting that such arrangement gives mechanical support (Fig. 19A, lower part of figure). In fiber-tracheids and wood fibers, the long axis of elongate bordered-pit apertures is parallel with the fibril aggregates of the thick median layer of the secondary wall. The pit cavities lie between radial plates of fibrils. The form of pit apertures thus may give clues to the orientation of the fibrils in thick walls.

The mature primary wall of many cells remains thin. When young, uneven thickening may give it reticulate structure, and when mature, abundant small pits may make it appear sieve-like. Where this wall is a place for food storage, as in endosperm and some sieve tubes, it becomes greatly thickened. Probably the thick wall of collenchyma cells, with its high percentage of water, is wholly primary. A distinction between

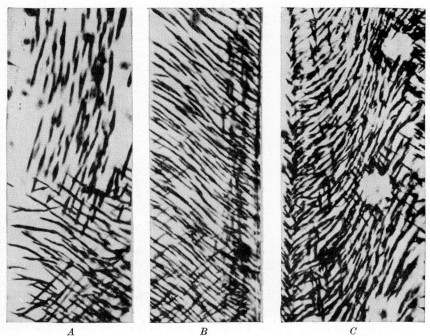

A   *B*   *C*

Fig. 19.—Minute structure of the secondary wall. (Crystalline aggregates of iodine have been induced to form in the elongate interstices of the cellulosic matrix. The crystals are oriented parallel to the fibrils and so show variations in orientation of cellulose in the different wall layers.) Latewood tracheids of *Larix occidentalis*. *A*, lower part of figure, spiral arrangement in outer layers of two abutting cells, the fibrils crossing at right angles; upper part of figure the fibrils running nearly longitudinally. *B*, as in lower part of *A*, the crystals in the wall (of the adjacent cell) below showing faintly. *C*, deviations in the spiral arrangement due to the presence of bordered pits. (*After Bailey and Vestal.*)

a strongly thickened primary wall and a primary wall plus a secondary wall may be difficult to make without special study of optical characteristics.

Layering is not usually apparent in primary walls except with special treatment. In wood cells, especially in heartwood, the primary walls of adjacent cells and the middle lamella between them may be chemically closely alike and difficult to distinguish. This triple layer, together sometimes with parts of the secondary walls, has for this reason passed, in loose use, as "the middle lamella."

The secondary wall, which constitutes the larger portion of the wall in most thick-walled cells, commonly shows marked lamellate structure, the layers ranging in number up to many. A division into three major layers is frequently seen, especially in wood cells: an outermost thin layer lying against the primary wall; a thick median layer; and an innermost thin layer. Constituting the "wall" between two protoplasts or lumina among such cells, there are nine major layers—six secondary layers, two primary walls, and the middle lamella (Fig. 20). The heavy median secondary layer itself frequently shows many minor layers varying

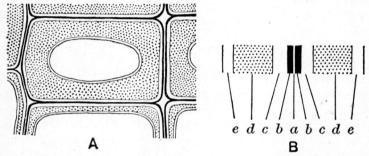

Fig. 20.—Diagrams showing wall structure and middle lamella in thick-walled cells (wood fibers). *A*, cross section of one fiber and parts of seven adjacent fibers; *B*, cross section of adjacent walls and middle lamella more highly magnified. *a*, middle lamella; *b*, primary wall; *c,d,e*, outer, median, and inner layers, respectively, of secondary wall. (*After Bailey.*)

in distinctness. The function of the thick secondary wall is primarily mechanical.

The chemical and physical nature of all wall layers and of the middle lamella undergoes various changes as the walls mature, and may undergo even further changes at a later period, for example, during the transformation of sapwood into heartwood. The middle lamella, at first containing much pectic material, becomes heavily lignified in thick-walled tissues and cutinized in epidermal regions. It shows no evidence of layering. The primary wall likewise becomes strongly lignified in wood and most sclerenchyma. Much of the mineral content of the cell wall is found here. In epidermal cells of seeds and fruits this wall is often gelatinous or mucilaginous. Lignification, cutinization, and suberization are the major changes in secondary walls. Gelatinous layers are frequent in the secondary wall of wood.

Method of Wall Building.—The method of addition of building material as the wall increases in area and thickness is not wholly understood. Two processes have been described: *apposition*, the placing of new particles in layers on the surface of the earlier formed wall; and *intussusception*, the addition of new particles among those already in position. When the wall is considered a matrix to which new material is added throughout, intussusception must play a prominent part in increasing the extent and thickness of the wall. Particularly in the increase in area of primary walls, intussusception must be the chief method. The layering of the secondary wall and the varying direction of the fibrils of its successive layers suggest that the addition of particles is periodically restricted to the layers next to the protoplast. Hence, considering the major layers alone, increase in thickness is by apposition; but each layer itself may be built by both apposition and intussusception. When the original matrix is increased by the addition internally of new micellae or fibrils, and when it is chemically transformed by replacement of particles or the deposition of new substances in the interstices of the meshwork, the method must be intussusception.

Plasmodesmata.—That the relation between cell wall and protoplast is very close is evident by the presence in the wall of many delicate threads of cytoplasm which connect the protoplasts of adjacent cells (Fig. 21). These threads, known as *plasmodesmata*, fill minute passages which constitute the only breaks in the primary walls of the two cells. They are characteristic of living cells and are undoubtedly present in all living parts of the higher plants, maintaining the continuity of the protoplasm of the plant although its units, the cells, are sharply set apart by firm walls. Their extreme delicacy and protoplasmic nature make special technique necessary for their demonstration in most tissues. Where the fibrils of contiguous cells meet at the middle lamella there are often seen slight knob-like thickenings. These may be enlargements of the threads owing their form to the nature of the intercellular substance, which in early stages may be partly protoplasmic or they may be merely artifacts. Plasmodesmata are most readily seen in the endosperm of some seeds (*Diospyros, Aesculus, Phoenix*) where food storage has greatly increased wall thickness, and in some cotyledons. They are extremely fine in gymnosperms. They have not been demonstrated in the earliest stages of wall formation, and they disappear with the death of the cell. In such cells as tracheids and vessel elements even the channels in the wall in which they lie are apparently filled when they disappear with the protoplast.

They occur in groups or are scattered. Usually many of them are concentrated in restricted thin areas of the young wall, the primary pit-fields. In mature walls with secondary layers, the larger groups occur only in the pit membranes (the thin wall areas between two pits, page 38). It has been suggested that the presence of clustered plasmodesmata in the primary wall determines the position of a pit.

36 AN INTRODUCTION TO PLANT ANATOMY

Origin.—Plasmodesmata probably represent persisting connections of the protoplasts of daughter cells. Although their presence in the very delicate first walls has not been demonstrated, meristematic cells may

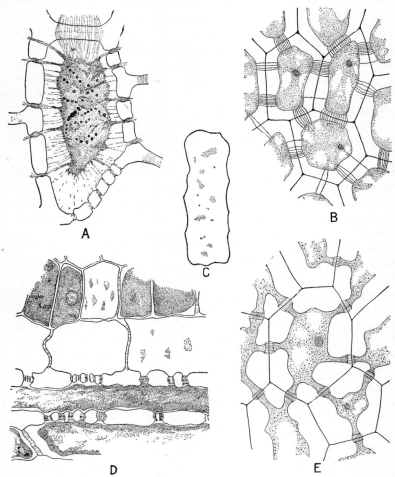

Fig. 21.—Plasmodesmata. *A*, in a cell from the petiole of *Marattia* (the cytoplasm plasmolized); *C*, in a cortical cell of *Ophioglossum*, the strands seen "end on"; *D*, in cells of root cap and adjacent cortex of *Vicia Faba*; *B*, in endosperm of *Diospyros* (persimmon); *E*, in endosperm of *Phoenix* (date). (*A* and *C* after Poirault; *D* after Gardiner and Hill.)

show them well. In early stages of wall formation the wall is mesh-like, with pores and interstices which doubtless contain cytoplasm. The nature of the cell-plate fluid at this time is unknown but it is doubtless in part protoplasmic. As cellulosic and pectic materials are laid down in the three layers between the two protoplasts the cytoplasmic connections are restricted to slender threads and at this stage become detectable.

As the wall increases in area with growth of the cell, the number of plasmodesmata seems to increase. The increase possibly results from the splitting of the original strands. New connections between protoplasts are apparently not freely made but when cells during growth establish new contacts, secondary plasmodesmata may form in these areas. During symplastic growth the strands are distributed over larger areas, accompanied perhaps by multiplication. In intrusive growth they are ruptured and when the new-contact areas of both walls are young enough are replaced by new strands. It has been claimed that no secondary strands are formed; that evidence for this is seen, for example, in their absence on the new-contact walls of wood fibers and developing vessel elements. But where such new contacts of enlarging vessels are with parenchyma cells new plasmodesmata are formed. The presence of plasmodesmata in graft hybrids and periclinal chimaeras between cells of different genetic origin seems to be sufficient evidence that plasmodesmata may develop secondarily.

Function.—It has been suggested that plasmodesmata are "channels of translocation of solids," that they "conduct stimuli," but no demonstration of such functions has been made. Their existence is doubtless important in many ways because they unite the otherwise isolated protoplasts of the plant body into a single protoplasmic structure.

SCULPTURE AND MODIFICATION OF THE WALL

Pits.—During early stages in the thickening of the primary cell wall, small, sharply defined areas in the primary pit-fields are left thin and uncovered. As thickening continues, recesses or cavities are left in the wall at these places. These cavities, together with the walls that surround them—the exposed area of primary wall that forms the "bottom" or external wall, and the side walls and roof—are known as *pits*. (The term pit is often applied to the cavity alone but best covers not only this but the modified wall area in which the pit lies.) Pits are conspicuous features of mature cell-wall structure. They vary in type, in form, and in abundance. They are characteristic of all kinds of cells and are probably lacking only in cells with very thin walls. Functionally, pits seem to be diffusion areas, regions where better interchange between cells may occur. That pits may be areas of interchange of another type also is apparent from the fact that in living cells groups of plasmodesmata are frequently or always present in the pits in larger numbers than elsewhere (Fig. 21).

Pits are structural features either of the primary wall alone or of the primary plus the secondary wall. Those that involve both wall layers are more readily seen; those of the primary wall alone are conspicuous only where the wall is fairly thick.

Pit-pairs.—A pit has a complementary or opposing pit in the wall of the contiguous cell. The two pits form a structural and functional unit, a *pit-pair*. (The use of the term "pit" for both the pit and the pit-pair has long caused difficulty and confusion in descriptions. The

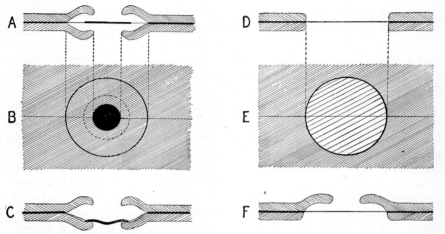

Fig. 22.—Diagrams of three kinds of pit-pairs. *A,B*, section and face view of bordered pit-pair, showing the overarching borders of secondary wall which enclose the pit cavities, between which passes the closing membrane whose central portion is thickened to form the torus; *C*, section of same showing the closing membrane in a lateral position, the torus closely appressed to the pit aperture; *D,E*, section and face view of a simple pit-pair in which the secondary wall does not overhang the pit cavities and the closing membrane has no torus; *F*, section of half-bordered pit-pair, the secondary wall overarching only one pit cavity, no torus. (*A–C, from Sharp, after Bailey.*)

adoption of "pit-pair" for the double unit corrects the morphological error.)

The recess of the pit is the *pit cavity;* the membrane that separates the two cavities of a pit-pair is the *pit membrane,* or *closing membrane* of the pit-pair; the opening or mouth of the pit at the inner wall surface is the *pit aperture.* A pit-pair has two cavities, two apertures, and one closing membrane. The closing membrane consists of three layers—the primary wall of each of the two cells (different in thickness and chemical nature from adjacent parts of the primary walls) and, between these, the middle lamella which also is different from its adjacent parts in this area.

Two types of pits are recognized: *simple* and *bordered* (Fig. 22). These form *simple pit-pairs* when both members of the pair are simple; *bordered pit-pairs* when both are bordered; and *half-bordered pit-pairs* when one pit is simple and the other bordered.

Simple pits are those in which, as the wall is built up in thickness, the cavity remains of the same diameter, or is enlarged, or is only slightly constricted, and the closing membrane is simple in form and structure. In section, such pits show the side walls of the pit at right angles to the primary wall, sloping away from the margin of the exposed closing membrane, or sloping only slightly and gradually toward the center (Fig. 23). Simple pits with chambers slightly narrowing to the aperture somewhat resemble some forms of bordered pits—showing evidence of a narrow border in face view—but the closing membrane is without a torus.

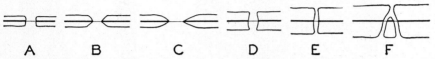

FIG. 23.—Diagrams showing varieties of simple pit-pair structure as seen in section. *A*, the walls of the cavities at right angles to the closing membrane; *B*, *C*, the walls sloping away from the closing membrane; *D*, *E*, the cavities narrowing slightly to the apertures; *F*, two pit-pairs, with two pits fused to form one cavity at one aperture. In no form is the pit cavity differentiated into cavity and canal as in bordered pits.

In *bordered pits*, the cavity, as the wall thickens, is so constricted that a rim or roof is built, overarching the recess. The closing membrane is elaborate in form and structure (Figs. 24, 25). The pit is called "bordered" because in face view the rim forms a border around the aperture. In section such pits show the cavity more or less nearly enclosed by the secondary wall (Fig. 24). Where the secondary wall is thick (Fig. 30), the pit cavity is divided into parts, an external *pit chamber*, and a passageway leading from the lumen to the chamber, the *pit canal;* where the secondary wall is thin (Fig. 22), no distinction of chamber and canal can be made. The pit canal in bordered pits has an *outer aperture*, the opening into the pit chamber, and an *inner aperture*, the opening into the cell lumen. The cavity of a simple pit is not divided into chamber and canal, but deep simple pits in thick walls have "*canal-like cavities.*" Such a pit has only one aperture.

In *half-bordered pit-pairs* both pits vary in type as do the pits of simple and bordered pairs. The closing membrane may have a weak torus or no torus.

The bordered pit-pair is complex not only in the form of its cavity but in the structure and behavior of its membrane. The membrane of a bordered pit-pair (consisting of parts of two primary walls and the middle lamella) has a thickened central part, the *torus*, surrounded by a delicate marginal area. The triple nature of this membrane is not readily evident because all the layers are much alike chemically (mainly cellulosic) and structurally. In some of the gymnosperms the thin area of the closing membrane is perforated by minute openings (Fig. 25), or even by larger

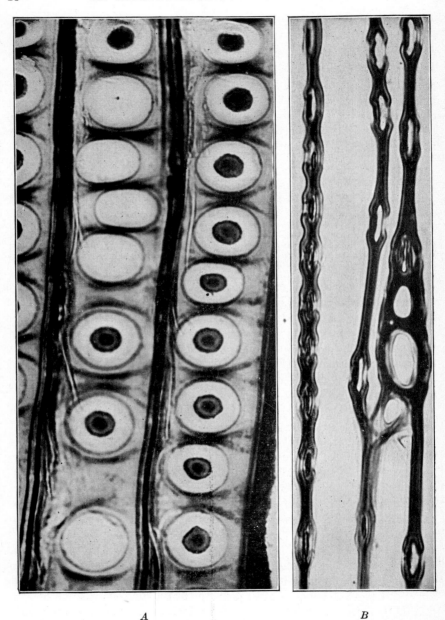

FIG. 24.—Bordered pit-pairs in coniferous tracheids. *A*, in face view, showing crassulae and limits of apertures, tori, and cavities; *B*, in section, showing apertures, cavities, and tori in lateral position. (*Courtesy of the U.S. Forest Products Laboratory.*)

pores, so that the torus is suspended by a meshwork or, rarely, by a few strap-like pieces. The existence of these perforations has been clearly demonstrated by the passage of minute solid particles (finely ground India ink) from tracheid to tracheid.

The closing membrane of many bordered pit-pairs may change position within the cavities, apparently because of changes of pressure within the cells. The torus may occupy the median position (Fig. 22A); or lie close against the aperture on either side (Figs. 22C, 25B). When in the lateral position it may be distorted by the forcing of its central part into, or even partly through, the aperture (Fig. 25C). The diameter of the

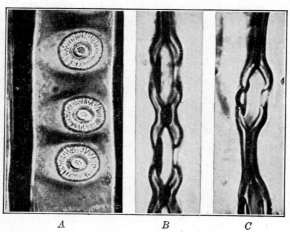

FIG. 25.—Bordered pit-pairs. A, in face view, showing perforations of the peripheral part of the pit membrane; *Larix laricina*. ×520. B, C, in vertical section, showing torus in lateral position—in C, pressed into the pit aperture; *Pinus* sp. ×1000. (*After Bailey.*)

torus is always greater than that of the aperture so that, when the torus is in the lateral position, passage through the pit-pair can be only by diffusion through the torus. With the torus in the median position, the thinner margin provides passage of a perhaps different type. In the conifers with actual perforations, free passage is possible when the torus is in median position, but this has not been demonstrated in other groups. Membranes capable of these changes of position are characteristic of living spring wood. Those of summer wood are inflexible (in some woods 20 to 60 per cent are said to retain median position) and all membranes of heartwood are "frozen" in position by the changes incident to heartwood formation. These pit-membrane characteristics are important from the standpoint of impregnation of wood under pressure with preservatives and other substances. Under pressure the membranes of heartwood and those of the summer layers of sapwood, held rigidly in position (perhaps in heartwood often cracked by shrinkage), can be ruptured to permit the passage of fluids; those of living spring wood are forced into the lateral position with the torus jammed into the aperture and cannot be ruptured.

The simple pit-pair has no torus; the half-bordered pit-pair has either no torus or a weakly developed one. The closing membrane of simple pit-pairs retains its median position; that of half-bordered pit-pairs may bulge well into the lumen on either side when the apertures are large.

Departures from normal pit-pair structure are few. A large pit in one cell may have two or more smaller pits opposite it in the adjacent cell. This is *unilateral compound pitting*. A pit without a complementary pit is a *blind pit*. Such pits are possibly halves of pit-pairs that have been split apart in the readjustment of cell contacts during the cell growth. (The pits without complementary pits commonly seen in sections of wood cells and other thick-walled cells are halves of pit-pairs that have been cut obliquely in sectioning.)

Simple pit-pairs, with the canals slightly and gradually narrowing toward the aperture, are frequent in thick-walled parenchyma, in some sclereids, and elsewhere. Some of these have been incorrectly called "bordered." These do not have the pit membranes of bordered pits or the distinction of cavity into chambers and canals. The bordered pit, as a morphological structure, is characteristic of and restricted to water-conducting cells and their phylogenetic derivatives. If pits with slightly overhanging borders are called "bordered," important morphological structure is ignored.

Simple pits occur in walls that are wholly primary and those that are both primary and secondary; bordered pits occur only in primary plus secondary walls. Simple pits occur in many kinds of cells; bordered pits are characteristic of water-conducting cells and of kinds of cells morphologically derived from these cells. In a given cell or kind of cells, the type is constant.

Variations in Size and Structure of the Bordered Pit.—The bordered pit-pair varies greatly in size and structure (Figs. 26, 27, 28). The prominent modifications of its typical form and size are those associated with the specialization and reduction of the pit as it occurs in fiber tracheids and fibers. In the typical bordered pit-pair, the cavities, which are circular in outline, are large, and the apertures are also circular (Figs. 24, 29*A*). Where such pit-pairs occur in thin walls, the borders may arch and bulge into the lumen, as in the spring-wood tracheids of *Larix* and species of *Pinus* (Figs. 28*E*, 29*B* 1). These pit-pairs are large in comparison with the cells to which they belong (Fig. 24) and, in section, may resemble small cells (Fig. 28*E*). Where the wall is thicker the presence of a pit causes no mounding or bulging of the wall, and the aperture opens into the pit canal, not directly into the chamber.

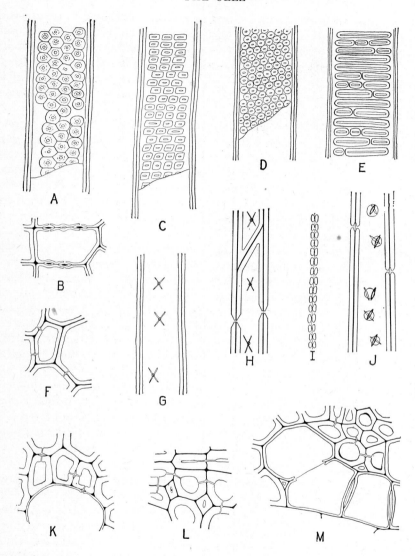

Fig. 26.—Bordered pit-pairs. *A, B*, crowded angular pit-pairs in face view and in section in tracheid of *Agathis australis; C*, rectangular pit-pairs, face view, in horizontal rows in vessel of *Liriodendron; D*, crowded pit-pairs, face view, in spiral rows in vessel of *Acer rubrum; E*, elongate pit-pairs in face view in vessel of *Magnolia acuminata*—the same pit-pairs in vertical section in *I*, and in transverse section in *M; F, G*, pit-pairs with narrow, inner apertures, in section and face view in thick walls of wood fiber of *Populus tremuloides; H*, similar pit-pairs in wood fibers of *Magnolia*, face view and longitudinal section—the same pit-pairs in transverse section in *L; J, K*, pit-pairs with narrow, unsymmetrically placed inner apertures in wood fiber of *Malus pumila* in face view and in longitudinal and transverse section; *M*, large, elongate pit-pairs (*E*) in vessels and small pit-pairs in fibers, both in transverse section in *Magnolia*.

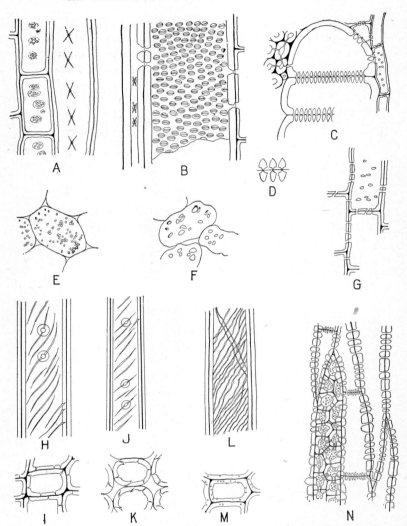

FIG. 27.—Pit-pairs and other wall sculpturing. *A*, clustered simple pit-pairs in wood parenchyma, and very small bordered pit-pairs in fibers, face view, in *Fraxinus americana*; *B*, bordered pit-pairs of two forms, face view, in vessel and in fiber, and half-bordered pit-pairs, in section, between vessel and parenchyma cell, in *Diospyros* (ebony); *C*, the same pit-pairs as in *B*, in transverse section, also simple pit-pairs, face view and section in ray and parenchyma cells; *D*, detail section of bordered pit-pairs between vessels, as shown in *C*; *E* and *F*, simple pit-pairs in thin-walled parenchyma cells—*E*, in pith of *Chenopodium*, *F*, in pericarp of *Citrullus*; *G*, simple pit-pairs in thick-walled parenchyma of pith of *Clematis*; *H* and *I*, *J* and *K*, microscopic checking, face view and section, of secondary wall in tracheids of *Pinus Strobus* and *Sequoia sempervirens*—the cracking extending through pit apertures simulates elongate apertures; *L*, *M*, erosion of secondary wall by fungus hyphae, simulating checking, face view and section, in tracheid of *Podocarpus*; *N*, simple pit-pairs in thick-walled ray cells and wood parenchyma, face view and section, in late summer wood of *Magnolia*.

THE CELL 45

In general, the thicker the wall in bordered pit-pairs, the smaller the chamber and the smaller the aperture. In thick walls, although the chambers and outer apertures remain always circular in outline, the inner apertures become narrowed and elongated, being elliptical to slit-like

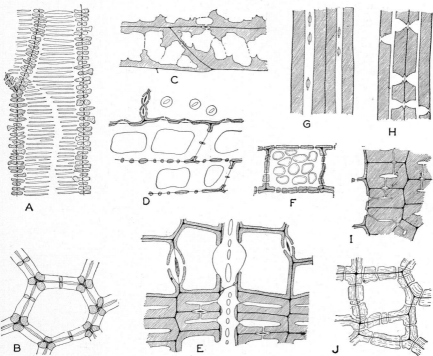

FIG. 28.—Pit-pairs. *A,B*, elongate bordered pit-pairs of tracheid of *Osmunda Claytoniana*, face view and section; *C*, bordered pit-pairs and "dentations" of wall in section in marginal ray cell of *Pinus Banksiana; D*, full-bordered, half-bordered, and simple pit-pairs, face view and section, in wood ray cells, radial section of *Pinus Strobus; E*, similar, in part the same, pit-pairs, in face view and section, in tracheids and ray cells in transverse section of wood of *Pinus Strobus; F*, simple pit-pairs in section and half-bordered pit-pairs in face view in ray cell of wood of *Salix alba; G,H*, and *I*, small bordered pit-pairs with narrow, vertical inner apertures coinciding in position, and long, narrow pit-canals between pit chamber and lumen, face view, longitudinal section, and transverse section, respectively, in wood fibers of *Eucalyptus globulus; J*, simple pit-pairs in stone cells of phloem of *Platanus*.

in outline (Fig. 29). When only slightly narrowed, the inner aperture, seen in face view, may extend only part way to the margin of the cavity (Fig. 29*A* 2), or may reach the very limit of the cavity (Fig. 29*A* 3), so that there are no borders at those points. Where the inner aperture is much narrowed, the slit usually extends far beyond the limit of the chamber as seen in face view (Fig. 29*A* 4–6). In reduction of pit size in thick walls, the greater the reduction in diameter of the chamber, the greater the increase in length of the inner aperture. These much elongated apertures occur only in thick walls, and hence the pit canal through the thick wall to the small inner chamber is in shape like a much flattened, or laterally compressed, funnel (Fig. 30). The outer aperture of this funnel-like canal remains circular. Where the inner pit apertures

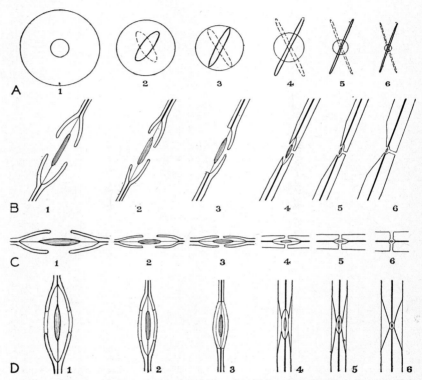

Fig. 29.—Diagrams of various types of bordered pit-pairs, showing changes of form accompanying reduction in size and function, and occurrence in walls of various thickness. *A*, face views; *B*, sections along plane of an elongated aperture; *C*, transverse sections; *D*, view, "edge on," of entire pit-pair lying in wall—the view of small pit-pairs commonly seen, because of their small size—in longitudinal sections of wood.

are circular, the two apertures of a pit-pair lie directly opposite one another (Fig. 29*A* 1), coinciding in position; where the inner apertures are elongate, the apertures are commonly crossed, usually symmetrically (Fig. 29*A* 2–6). (The angles at which the inner apertures lie are apparently controlled by the orientation of the strands of fibrils that make up the major layer of the secondary wall.) These crossed pit apertures are conspicuous features of pit marking in the walls of wood cells (**Figs. 26***G***,***H***,***J***, 27***A***). The pit cavity in such pits is complicated in form

(Figs. 29, 30). With extreme reduction the borders of the pits nearly disappear, and slit-like pit-pairs, essentially simple in structure, are formed (Fig. 29A 6). Morphologically, such pit-pairs are, however, clearly bordered pit-pairs. The pits of libriform wood fibers are often

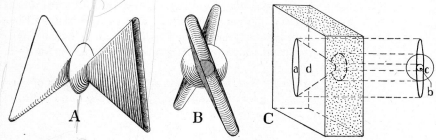

Fig. 30.—Diagrams showing bordered-pit structure of the type shown in Fig. 29A,4. A,B, showing the form of the cavities of a pit-pair—the inner apertures are slit-like; the small chambers, flat-dome shape; the canals like flattened funnels. C, a pit shown in position in a thick cell wall. a, the inner aperture; b, the outer aperture; c, the chamber; d, the canal. (C, after Record, modified.)

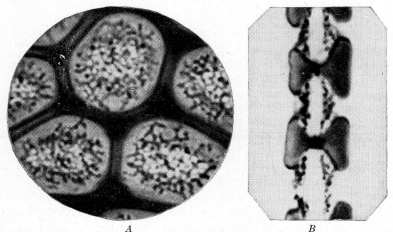

Fig. 31.—Vestured pit-pairs in secondary wood of *Duabanga moluccana*. A, face view; B, in section. (After Bailey.)

wholly vestigial, being simple and hardly more than markings on the wall. In much reduced bordered pit-pairs the closing membrane is of simpler structure, with no torus and no capacity to change position. Such pits are probably physiologically functionless or nearly so.

Vestured Pits.—Pits that have minute outgrowths from the surface of the side wall of the pit chamber or along the rims of the apertures are known as *vestured pits* (Fig. 31). The projections that form the vesture are of many types—filamentous, papillose, coralloid; branched and anastomosing, or simple. They may be scattered or closely clothe the surface of the wall. In some pits they cover also the surface of the wall of the vessel in which the pits occur. When long, they may fill the pit cavity and project into the cell lumen. Only bordered pits are vestured; in half-bordered pit-pairs the simple pit is without vesture.

The refractive and deep-staining qualities of these minute structures give a punctate or reticulate appearance to pits seen in face view. For this reason such pits were formerly called "cribriform" and "sieve-like" in the mistaken interpretation that the closing membrane was perforated like a sieve plate.

Vestured pits are probably restricted to the tracheary elements of the secondary wood of certain angiosperm families, for example, the Legumi-

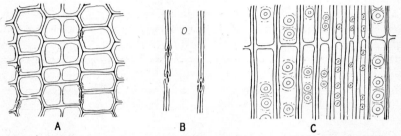

FIG. 32.—Trabeculae in the tracheids of *Abies balsamea*. *A*, transverse section; *B*, tangential section; *C*, radial section.

nosae, Cruciferae, Myrtaceae, Caprifoliaceae. When present, they are constant throughout the tissue in the species or genus and probably also in subfamilies and families, and so may be important in systematic and phylogenetic studies. No function has been suggested for them. Their presence is associated with highly specialized wood types and they appear to indicate a highly specialized pit type.

Other Wall Sculpturing.—The modeling of the inner surface of the cell wall is in large measure due to the presence of pits. Pits and perforations of vessel walls (Chap. IV) form the conspicuous structural modifications. Also prominent are the secondary-wall thickenings of protoxylem—rings, spirals, and reticulations (Fig. 62). Other internal thickenings or projections occur—the "dentations," really ridges, of the ray tracheids of hard pines (Fig. 28*C*); the Casparian strips of some endodermal cells (page 158); the trabeculae of conifer tracheids (Fig. 32); the supporting bars or meshes of cells of the velamen and anther wall.

Trabeculae.—Rod-like or spool-shaped wall projections crossing the lumen radially are known as *trabeculae*. They are frequent in the secondary wood of conifers, rare elsewhere. They usually occur in rows radially, and the rows may extend across the cambium and into the phloem.

THE CELL

Crassulae.—Structural variations within the wall itself may appear like surface features. Bar-like or crescent-shaped thickenings of the primary wall and middle lamella that separate or partly encircle individual bordered pits (Fig. 24A) or small groups of pits may be visible through the secondary wall when properly stained. These *crassulae* are the persisting marginal rims of the primary pit-fields of the young cell wall. The secondary wall covers but does not form a projection over these primary ridges so that they are not features of the sculpture of the wall surface. The names "bars of Sanio" and "rims of Sanio" have long been applied to crassulae but these terms have been confused in usage and in part applied also to trabeculae. The new term "crassulae" should replace both older ones.

Centrifugal Thickening.—In cells that have free surfaces the wall is apparently sometimes thickened centrifugally. Some hairs and other epidermal cells, some sclereids, and certain cells abutting on large intercellular spaces have minute nodules, granules, or ridges on the outside of the wall. Some of these are parts of the wall; others are deposits on the surface, as is the cuticle. The external wall layers and surface projections of spores and pollen grains are formed in part by tapetal fluid or mother-cell cytoplasm.

Chemical Nature of the Cell Wall.—The chemical nature of the early stages of the cell wall have already been briefly discussed (page 30). Later modifications and additions are of many kinds. These are brought about by infiltration of the interstices of the original wall matrix by new substances, or perhaps in part by replacement or transformation of part of that first-deposited wall. *Lignification,* infiltration by lignin, is prominent in these changes. It takes place in both primary and secondary walls of many types of cells, especially those of wood. Primary walls of wood cells, together with the intercellular layer between, are characteristically heavily lignified; the secondary walls of the same cells are more lightly lignified. In angiosperm wood the so-called "middle lamella"—intercellular layer, primary walls, and thin outermost layers of the secondary walls—is usually the most heavily lignified part of the cell wall. The secondary wall, or its inner layers, is *gelatinous* in the fibers of many angiosperm genera. The function of such gelatinous wall layers is not understood. Gelatinous or mucilaginous cell walls are found also in other tissues, such as phloem and especially the outer layers of seeds and fruits. A general breakdown of wall structure, known as *gummosis,* may occur even in woody tissues under certain pathological conditions.

Cutinization and suberization are processes in which the wall and often the middle lamella are infiltrated by chemically allied substances, *cutin* and *suberin,* respectively. Cutin and suberin are mixtures of substances, chiefly fatty acids, that are impermeable to water. These waterproof the cells in which they are deposited. Cutinization is found chiefly in the epidermis, and suberization in the periderm, but both occur less prominently elsewhere. Cutinization is further discussed under "cuticle" (page 53) and suberization under "phellem" (Chap. IX).

"*Mineralization*," the deposition of large amounts of inorganic salts in the wall is only rarely a prominent feature of wall modification. All mature walls contain small amounts of mineral matter, sometimes in the form of visible minute crystals. Silica and the oxalate and carbonate of calcium are commonly present; the first heavily infiltrates the walls of the outer tissues of grasses, sedges, horsetails, and some other families. Large crystals may become attached to, or even embedded in, the secondary wall by the formation about them of the later layers of wall substance (Fig. 9B).

Other changes in the wall involve resins, gums, oils, tannins and tannin-decomposition products, various aromatic and coloring substances, and many others of minor importance. Tannin, together with other substances, is responsible to a large extent for the durable and useful qualities of certain woods, and of heartwood (Chap. VII) as contrasted with sapwood. These substances are present not only in the walls, but often fill the lumina and pit cavities as well. They are formed in part from the contents of the living cells of the immediate region, but doubtless also in part from additional material brought to the region.

The modifications discussed above may involve the middle lamella and both chief wall layers or any of these alone. The primary wall is perhaps less changed than the secondary.

Cystoliths.—In a few dicotyledonous families peculiar structures known as *cystoliths* project into the protoplasts in large specialized cells. These structures are lime-encrusted or -infiltrated, stalked projections of the wall. The foundation of the cystolith is a stalked, stratified, cellulosic body which arises early in cell development as a local wall thickening. With the addition of large amounts of calcium carbonate this becomes an irregular body which may nearly fill the cell. In shape, cystoliths vary greatly in different genera and families. They are characteristic of the leaves of the Urticaceae, Moraceae, and a few other families and may appear even in the wood rays and phloem of these groups. Cystoliths are not common or important features of cell structure.

All these changes are in nature either those of direct chemical transformation or molecular replacement; or those of apparent change brought about by infiltration, or by deposition of large amounts of new, different substances, within the original wall, so adulterating the latter that the first substance is obscured. Pseudochanges are secured by the addition of new material in lamellae over the old wall, completely burying the latter. Such modifications may accompany the maturation of the wall in cell ontogeny, or may take place long after cell maturity, for example, when sapwood becomes heartwood, and when cortical photosynthetic

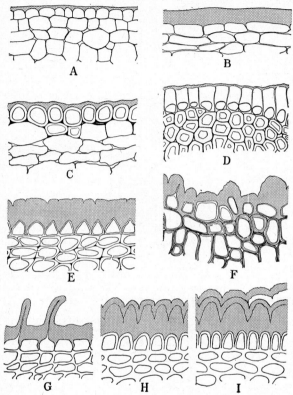

FIG. 33.—The cuticle. Sections of the epidermis and adjacent tissues, showing various degrees of thickness and of extent of the cuticle between epidermal cells. (For thin cuticles see Figs. 172*B*; 175*A,C*; very thin cuticles are not illustrated.) *A, Citrus sinensis* (orange), fruit; *B, Malus pumila* (var. Ben Davis), fruit; *C, Dracaena Goldieana*, stem; *D, Dasylirion serratifolium*, leaf; *E, Acer pennsylvanicum*, stem, the older, outer layer cracking and disintegrating; *F, Smilax rotundifolia*, stem; *G, Vaccinium corymbosum*, stem; *H* and *I, Cornus circinata*, stem, showing two stages, the outer, older layers being replaced by new ones formed below.

parenchyma cells or phloem parenchyma cells become sclereids after a long period of activity in their first condition.

THE CUTICLE

The surface of the aerial parts of vascular plants is covered by a layer of cutin, the *cuticle*. The cuticle covers the epidermis closely, following the contour of all cells (Fig. 33). Cutin is impermeable to water and the cuticle contributes to the control of water loss from the underlying cells. On young parts the cuticle is thin and pliable; as growth continues, it is stretched and added to in thickness. When growth of the tissues beneath is complete, the cuticle becomes firm and highly resistant to stretching and to breakdown by any agent. (Cuticles persist, little changed, even in coal.) In thickness and in character, the mature layer varies greatly (Chap. XIV); it is thin in plants of shade and moist habitats, generally thick in those of dry and sunny situations. Its thickness on the leaves and fruit of the same plant varies somewhat with position on the plant and from season to season with weather conditions. On ephemeral organs and aquatic plants it remains very thin and may not be readily detectable. Typical roots have no cuticle.

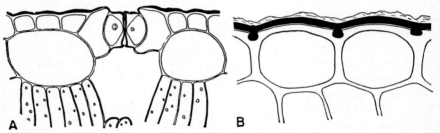

Fig. 34.—Cuticle of the leaf of banana (*Musa*). *A*, section of upper epidermis through stomatal region, showing extent of cuticle over guard cells; *B*, section through pulvinar band showing "pegs" of cuticle over anticlinal walls of epidermal cells, freedom of cuticle between pegs, and waxy layer on surface of cuticle. (*After Skutch.*)

The only breaks in the continuity of the cuticle are the openings of stomata and lenticels. The cuticle extends for varying distances into the stomatal apertures (Figs. 34*A*, 179*B*,*C*) and a thin layer of cutin may coat the stomatal chamber. The walls of large air chambers may have a thin cuticle, and small intercellular spaces between the epidermis and mesophyll cells may also be coated with cutin.

In origin, the cuticle is a secretion of the epidermal cells and perhaps also of underlying cells. The cutin, in apparently a liquid or semiliquid form, is passed through the outer wall in an unknown manner. On the surface it forms a continuous layer and becomes hard and tough. According to one view, droplets of cutin pass through exceedingly minute openings of the outer wall; according to another, they are formed in the deeper lying cells and migrate along the walls, reaching the surface through the anticlinal walls of the epidermis. It is clear that the cuticle is thickened from below, although in early stages—and perhaps at some other times—new material is apparently dissolved in the preexisting layer. An example of the manner in which the cuticle is repaired or added to is seen in young stems and leaves which in early stages are clothed with hairs that are later shed; the spaces left in the epidermis by the bases of the hairs—now breaks in the cuticle—are filled by deposits of cutin.

The surface of the cuticle is commonly smooth—always so in early stages—but when mature it may be roughened by cracks, by the sloughing of scales or minute particles from the surface, and by small nodules or slight ridges and irregularities (Fig. 33). Many so-called "cuticular pegs and ridges" are projections of the wall of the epidermal cells coated by the cuticle (Fig. 33). Such are the minute longitudinal flutings of the surface of stems and petioles of some herbs, for example, *Rumex*. The term "cuticular pegs" has been incorrectly applied also to the cutinized radial walls of epidermal cells which, when seen in sections, appear as inward projections of the cuticle.

The floral parts, especially petals and stamen hairs, of some plants are reported to have in early stages a wrinkled cuticle, the wrinkles appearing as minute folds when the deposition of cutin is rapid and growth of the epidermis in area is slow. The folds are said to form a "reserve surface," and later, when growth of the organ is very rapid, are stretched out or reoriented. Over the mature epidermis the only folds remaining are those whose axes lie in the direction of greatest growth of the cells beneath. The radiating striations of the cuticle of papillate epidermal cells of petals are said to represent this condition.

The cuticle commonly adheres closely to the surface of the epidermis but has been reported to lie free and loosely over the walls in some floral hairs and to be loose, but pegged down to the epidermis over pulvinar swellings, where the surface undergoes great changes in curvature, as in the leaf of banana (*Musa*) (Fig. 34). The separation of the cuticle from the epidermis by an air space is said to be responsible for the silvery appearance of some leaves.

The development of the cuticle, *cuticularization*, is accompanied, or followed in many plants by *cutinization* of the cells below. The outer walls of the epidermal cells are first modified. These may be increased greatly in thickness becoming many times as thick as the other walls, and as the cuticle itself (Fig. 33F, G). They may be so heavily impregnated with cutin that they are separable only with great difficulty from the layer of pure cutin, the cuticle. The radial and inner walls may similarly be cutinized and even the walls of the subjacent cells. Where the entire walls of the epidermal cells become cutinized, the cells may die, but frequently they remain alive, with normal pits and plasmodesmata. The term cuticle is unfortunately sometimes incorrectly applied in a functional sense to the entire outer "waterproofing" layer, the cuticle plus all cutinized cells. The layer which is stripped from leaves as "the cuticle" is usually the cuticle plus cutinized parts of underlying layers.

In some woody stems where the epidermis persists for more than one year—for example, species of *Acer, Cornus, Kerria*—the original cuticle is cracked or ruptured and is repaired by secretion of new material from beneath, or a complete new layer is formed (Fig. 33*H,I*).

Many fruits, such as plum, tomato, persimmon, and some varieties of apple (Fig. 33*B*), have, when ripe, a thick cuticle. This reaches full thickness only as the fruit attains mature size. The heavy layer aids in the conservation of water in the fruit after the supply of water ceases. Some of the keeping qualities of fruits are due to the presence of a heavy cuticle. Varieties of apples with thick cuticles keep better than those with thinner cuticles. Fruits grown in sunshine have heavier cuticles than those grown in the shade; and apples grown in a "dry year" have heavier cuticles than those borne on the same tree in a "wet year." In the drying of prunes the cuticle is chemically or mechanically perforated to hasten the drying.

Minute particles or rods of wax on the surface of the cuticle are responsible for the glaucous character of stems and leaves and constitute the bloom of fruits. Wax may occur in larger particles, as in the bayberry, *Myrica*, or may heavily coat the leaves, as in the wax palms.

References

The Cell—General

Lewis, F. T.: The typical shape of polyhedral cells in vegetable parenchyma and the restoration of that shape following cell division, *Proc. Amer. Acad. Arts Sci.*, **58**, 537–552, 1923.

———: A further study of the polyhedral shapes of cells, *Proc. Amer. Acad. Arts Sci.*, **61**, 1–34, 1925.

———: The shape of cells as a mathematical problem, *Amer. Scientist*, **34**, 359–369, 1946.

Marvin, J. W.: Cell shape and cell volume relations in the pith of *Eupatorium purpureum*, *Amer. Jour. Bot.*, **31**, 208–218, 1944.

Matzke, E. B.: Volume-shape relationships in lead shot and their bearing on cell shapes, *Amer. Jour. Bot.*, **26**, 288–295, 1939.

———: The three-dimensional shapes of bubbles in foams, *Proc. Nat. Acad. Sci.*, **31**, 281–289, 1945.

Meeuse, A. D. J.: A study of intercellular relationships among vegetable cells with special reference to "sliding growth" and to cell shape, *Rec. Trav. Bot. Neer.*, **38**, 18–140, 1942.

Neeff, F.: Über Zellumlagerung, Ein Beitrag zur experimentallen Anatomie, *Zeitschr. Bot.*, **6**, 465–547, 1914.

Priestley, J. H.: Cell growth and cell division in the shoot of the flowering plant, *New Phyt.*, **28**, 54–81, 1929.

Seifriz, W. (Ed.): "The Structure of Protoplasm," Ames, Iowa, 1942.

SHARP, L. W.: "Introduction to Cytology," 3d ed., New York, 1934.
———: "Fundamentals of Cytology," New York, 1943.
SINNOTT, E. W.: Structural problems at the meristem, *Bot. Gaz.*, **99**, 803–813, 1938.
THOMPSON, D'A. W.: "On Growth and Form," 2d ed., Cambridge, 1942.

THE PROTOPLAST

BAILEY, I. W.: The cambium and its derivative tissues, V. A reconnaissance of the vacuome in living cells, *Zeitschr. Zellforsch. Mikr. Anat.*, **10**, 651–682, 1930.
BAILEY, I. W., and C. ZIRKLE: The cambium and its derivative tissues, VI. The effects of hydrogen-ion concentration in vital staining, *Jour. Gen. Phys.*, **14**, 363–383, 1931.
FAULL, A. F.: Elaioplasts in *Iris:* a morphological study, *Journ. Arnold Arboretum*, **16**, 225–267, 1935.
GUILLIERMOND, A.: Observations vitales sur le chondriome des chromoplastides et le mode de formation des pigments xanthophylliens et carotiniens. *Rév. Gén. Bot.*, **31**, 372–413, 446–508, 532–603, 635–770, 1919.
———: "The Cytoplasm of the Plant Cell," Engl. transl. by L. R. Atkinson, Waltham, Mass., 1941.
HEITZ, E.: Untersuchungen über den Bau der Plastiden, I. Die gerichteten chlorophyllscheiben der Chloroplasten, *Planta*, **26**, 134–163, 1936.
LUBIMENKO, V.: Les pigmentes des plastes et leur transformation dans les tissues vivantes de la plante, *Rév. Gén. Bot.*, **39**, **40**, 1927, 1928.
MEYER, F. J.: Das trophische Parenchym, A. Assimilationsgewebe. *In* Linsbauer, K.: "Handbuch der Pflanzenanatomie," IV, 1923.
MOULTON, F. R. (Ed.): The cell and protoplasm, Publ. No. 14, *Amer. Assoc. Adv. Sci.*, 1940.
NEWCOMER. E. H.: Concerning the duality of the mitochondria and the validity of the osmiophilic platelets in plants, *Amer. Jour. Bot.*, **33**, 684–697, 1946.
RABINOWITCH, E. I.: "Photosynthesis and Related Processes," I, Chaps. XIV, XV, XVI, New York, 1945.
RANDOLPH, L. F.: Cytology of chlorophyll types of maize, *Bot. Gaz.*, **73**, 337–375, 1922.
SCHÜRHOF, P. N.: Die Plastiden, *In* Linsbauer, K.: "Handbuch der Pflanzenanatomie," I, 1924.
SINNOTT, E. W., and R. BLOCH: Division in vacuolate plant cells, *Amer. Jour. Bot.*, **28**, 225–232, 1941.
STRAUS, W.: Recherches sur les chromatophores, IV. Sur la structure des chromatophores de la carotte, *Helvetica Chim. Acta*, **25**, 1370–1383, 1943.
WEIER, E.: The structure of the chloroplast, *Bot. Rev.*, **4**, 497–530, 1938.
ZIRKLE, C.: The structure of the chloroplast in certain higher plants, *Amer. Jour. Bot.*, **13**, 301–320, 1926.
———: The growth and development of plastids in *Lunularia vulgaris*, *Elodea canadensis*, and *Zea mays*, *Amer. Jour. Bot.*, **14**, 429–445, 1927.
———: Vacuoles in primary meristems, *Zeitschr. Zellforsch. Mikr. Anat.*, **16**, 26–47, 1932.

PLASMODESMATA

CRAFTS, A. S.: A technic for demonstrating plasmodesmata, *Stain Tech.*, **6**, 127–129, 1931.
GARDINER, W., and A. W. HILL: The histology of the cell wall with special reference to the mode of connection of cells, *Phil. Trans. Roy. Soc. London*, **194B**, 83–125, 1901.

HUME, M.: On the presence of connecting threads in graft hybrids, *New Phyt.*, **12**, 216–225, 1913.
KUHLA, F.: Die Plasmaverbindungen bei *Viscum album*, *Bot. Zeit.*, **58**, 29–58, 1900.
LIVINGSTON, L. G.: The nature and distribution of plasmodesmata in the tobacco plant, *Amer. Jour. Bot.*, **22**, 75–87, 1935.
MEEUSE, A. D. J.: On the nature of plasmodesmata, *Protoplasma*, **35**, 143–151, 1941.
———: Plasmodesmata, *Bot. Rev.*, **7**, 249–262, 1941.
MÜHLDORF, A.: Das plasmatische Wesen der pflanzlichen Zellbrücken, *Beih. Bot. Centralbl.*, **56**, 171–364, 1937.
POIRAULT, G.: Recherches anatomiques sur les cryptogames vasculaires, *Ann. Sci. Nat. Bot.*, 7 sér., **18**, 113–256, 1893.
PRIESTLEY, J. H.: The physiology of cambial activity, II. The concept of "sliding growth." *New Phyt.*, **29**, 96–140, 1931.

ERGASTIC SUBSTANCES

BLACK, O. F.: Calcium oxalate in the dasheen, *Amer. Jour. Bot.*, **5**, 447–451, 1918.
FUCHS, P. C. A.: Untersuchungen über den Bau der Raphidenzelle, *Oester. Bot. Zeitschr.*, **48**, 324–332, 1898.
NETOLITZKY, F.: Die Kieselkörper, Die Kalksalze als Zellinhaltskörper, *In* Linsbauer, K.: "Handbuch der Pflanzenanatomie," III, 1929.
PFITZER, E.: Über die Einlagerung von Kalkoxalat-Krystallen in die pflanzliche Zellhaut, *Flora*, **1872**, 97–102, 129–136, 113–120, 1872.
SAFFORD, W. E.: The useful plants of the island of Guam, *Contr. U.S. Nat. Museum*, **9**, 67–71, 1905.
VAN DE SANDE-BAKHUYSEN, H. L.: The structure of starch grains from wheat grown under constant conditions, *Proc. Soc. Exp. Biol. Med.*, **23**, 302–305, 1926.

THE CELL WALL

AJELLO, L.: Cytology and cellular interrelations of cystolith formation in *Ficus elastica*, *Amer. Jour. Bot.*, **28**, 589–593, 1941.
ALDABA, V. C.: The structure and development of the cell wall in plants, I. Bast fibers of *Boehmeria* and *Linum*, *Amer. Jour. Bot.*, **14**, 16–24, 1927.
ANDERSON, D. B.: A microchemical study of the structure and development of flax fibers, *Amer. Jour. Bot.*, **14**, 187–211, 1927.
———: The structure of the walls of the higher plants, *Bot. Rev.*, **1**, 52–75, 1935.
ANDERSON, D. B., and T. KERR: Growth and structure of the cotton fiber, *Ind. Eng. Chem.*, **30**, 48–54, 1938.
ARZT, T.: Untersuchungen über das Vorkommen einer Kuticula in den Blättern dikotylen Pflanzen, *Ber. Deutsch. Bot. Ges.*, **51**, 471–500, 1933.
BAILEY, I. W.: The structure of the bordered pits of conifers and its bearing on the tension hypothesis of the ascent of sap in plants, *Bot. Gaz.*, **62**, 133–142, 1916.
———: Structure, development, and distribution of so-called rims or bars of Sanio, *Bot. Gaz.*, **67**, 449–468, 1919.
———: The formation of the cell plate in the cambium of higher plants, *Proc. Nat. Acad. Sci.*, **6**, 197–200, 1920.
———: The significance of the cambium in the study of certain physiological problems, *Jour. Gen. Physiol.*, **2**, 519–533, 1920.
———: Phragmospheres and binucleate cells, *Bot. Gaz.*, **70**, 469–471, 1920.
———: The cambium and its derivative tissues, III. A reconnaissance of cytological phenomena in the cambium, *Amer. Jour. Bot.*, **7**, 417–434, 1920.

―――: The cambium and its derivative tissues, IV, The increase in girth of the cambium, *Amer. Jour. Bot.*, **10**, 499–509, 1923.
―――: The cambium and its derivative tissues, VIII. The structure, distribution, and diagnostic significance of vestured pits in dicotyledons, *Jour. Arnold Arboretum*, **14**, 259–273, 1933.
―――: Cell wall structure of higher plants, *Ind. and Eng. Chem.*, **30**, 40–47, 1938.
―――: The microfibrillar and microcapillary structure of the cell wall, *Bull. Torrey Bot. Club*, **66** (4), 201–213, 1939.
―――: The wall of plant cells, *In* The cell and protoplasm, *Publ. Amer. Assoc. Adv. Sci.*, **14**, 31–45, 1940.
――― and E. E. BERKLEY: The significance of X-rays in studying the orientation of cellulose in the secondary walls of tracheids, *Amer. Jour. Bot.*, **29**, 231–241, 1942.
――― and A. F. FAULL: The cambium and its derivative tissues, IX. Structural variability in the redwood, *Sequoia sempervirens*, and its significance in the identification of fossil woods, *Jour. Arnold Arboretum*, **15**, 233–254, 1934.
――― and T. KERR: The visible structure of the secondary wall and its significance in physical and chemical investigations of tracheary cells and fibers, *Jour. Arnold Arboretum*, **16**, 273–300, 1935.
――― and ―――: The structural variability of the secondary wall as revealed by "lignin" residues, *Jour. Arnold Arboretum*, **18**, 261–272, 1937.
――― and M. R. VESTAL: The orientation of cellulose in the secondary walls of tracheary cells, *Jour. Arnold Arboretum*, **18**, 185–195, 1937.
BARANETZKI, J.: Épaississement des parois des éléments parenchymateux, *Ann. Sci. Nat. Bot.*, 7 sér., **4**, 134–201, 1886.
BEER, R., and A. ARBER: On the occurrence of multinucleate cells in vegetative tissues, *Proc. Roy. Soc. London*, B **91**, 1–17, 1919.
BELL, H. P.: The protective layers of the apple, *Can. Jour. Res.*, C **15**, 391–402, 1937.
BERKLEY, E. E.: Cellulose orientation, strength and cell wall development of cotton fibers, *Textile Res.*, **9**, 355–373, 1939.
BOEKE, J. E.: On the origin of the intercellulary channels and cavities in the rice root, *Ann. Jard. Bot. Buitenzorg*, **50**, 199–208, 1940.
BUSTON, H. W.: Observations on the nature, distribution and development of certain cell wall constituents of plants, *Biochem. Jour.*, **29**, 196–218, 1935.
COLIN, H., and A. CHAUDIN: Pectine et ciment intercellulaire, *Bull. Soc. Chem. Biol.*, **16**, 1333–1343, 1934.
COMMITTEE ON NOMENCLATURE, INTERNATIONAL ASSOCIATION OF WOOD ANATOMISTS: Glossary of terms used in describing wood, *Trop. Woods*, Yale University School of Forestry, No. **36**, 1933.
DADSWELL, H. E., and D. J. ELLIS: Contributions to the study of the cell wall, I Methods for demonstrating lignin distribution in wood, *Jour. Council Sci. Ind. Res.* (Aust.), **13**, 44–54, 1940.
FOSTER, A. S.: Structure and development of sclereids in the petiole of *Camellia japonica* L., *Bull. Torrey Bot. Club*, **71**, 302–326, 1944.
―――: Origin and development of sclereids in the foliage leaf of *Trochodendron aralioides* Sieb. & Zucc., *Amer. Jour. Bot.*, **32**, 456–468, 1945.
FREY, A.: Der heutige Stand der Micellartheorie, *Ber. Deutsch. Bot. Ges.*, **44**, 564–570, 1926.
―――: Über die Intermicellar-Räume der Zellmembranen, *Ber. Deutsch. Bot. Ges.*, **46**, 444–456, 1928.

FREY-WYSSLING, A.: The submicroscopic structure of cell walls, *Sci. Prog.*, **34,** 249–262, 1939.
FROST, F. H.: Histology of the wood of angiosperms, I. The nature of the pitting between tracheary and parenchymatous elements, *Bull. Torrey Bot. Club*, **56,** 259–264, 1929.
GÉNEAU DE LAMARLIÈRE, L.: Sur les membranes cutinisées de plantes aquatiques, *Rév. Gén. Bot.*, **18,** 289–295, 1906.
GOLDSTEIN, B.: A study of progressive cell plate formation, *Bull. Torrey Bot. Club*, **52,** 197–219, 1925.
GRIFFIN, G. J.: Bordered pits in Douglas fir; a study of the position of the torus in mountain and lowland specimens, *Jour. For.*, **17,** 813–822, 1919.
HOCK, C. W.: Microscopic structure of the cell wall, *In* Seifriz, W. (Ed.): "The Structure of Protoplasm," Ames, Iowa, 1942.
JÖNSSON, B.: Siebähnliche Poren in den trachealen Xylemelementen der Phanerogamen, hauptsächlich der Leguminosen, *Ber. Deutsch. Bot. Ges.*, **10,** 494–513, 1892.
JUNGERS, V.: Recherches sur les plasmodesmes chez les végétaux, I. *La Cellule*, **40,** 5–81, 1930.
KAMP, H.: Untersuchungen über Kuticularbau und kutikuläre Transpiration von Blättern, *Jahr. Wiss. Bot.*, **72,** 403–465, 1930.
KERR, T., and I. W. BAILEY: The cambium and its derivative tissues, X. Structure, optical properties and chemical composition of the so-called middle lamella, *Jour. Arnold Arboretum*, **15,** 327–349, 1934.
KOEHNE, E.: Ueber Zellhautfalten in der Epidermis von Blumenblättern und deren mechanische Function, *Ber. Deutsch. Bot. Ges.*, **2,** 24–29, 1884.
KÜNEMUND, A.: Die Entstehung verholzter Lamellen, untersucht besonders an *Salix alba*, *Bot. Arch.*, **34,** 462–521, 1932.
Lee, B., and J. H. PRIESTLEY: The plant cuticle, I. Its structure, distribution, and function, *Ann. Bot.*, **38,** 525–545, 1924.
MARTENS, P.: Recherches sur la cuticle, II. Dépouillement cuticulaire spontané sur les pétales de "*Tradescantia*," *Bull. Soc. Roy. Belgique*, **66,** 58–64, 1933.
———: Recherches sur la cuticle, III. Structure, origine et signification du relief cuticulaire, *Protoplasma*, **20,** 483–515, 1934.
———: Recherches sur la cuticle, IV. Le relief cuticulaire et la differentiation épidermique des organes floraux, *La Cellule*, **43,** 289–318, 1934.
———: Nouvelles recherches sur l'origine des espaces intercellulaires, *Beih. Bot. Zentralbl. Abt.*, A. **58,** 349–364, 1938.
MEEUSE, A. D. J.: Development and growth of the sclerenchyma fibers and some remarks on the development of the tracheids in some monocotyledons, *Rec. Trav. Bot. Neer.*, **35,** 288–321, 1938.
———: A study of intercellular relationships among vegetable cells with special reference to "sliding growth" and to cell shape, *Rec. Trav. Bot. Neer.*, **38,** 18–140, 1942.
MOOG, H.: Ueber die spiraligen Verdickungsleisten der Tracheen und Tracheiden unter besonderer Berücksichtigung ihrer Auszichbarkeit, *Beih. Bot. Centralbl.*, Abt. I. **42,** 186–228, 1925.
PHILLIPS, E. W. J.: Movement of the pit membrane in coniferous woods with special reference to preservative treatments, *Forestry*, **7,** 109–120, 1933.
PRESTON, R. D.: The organization of the cell wall of the conifer tracheid, *Phil. Trans. Roy. Soc.*, B **224,** 131–174, 1934.

PRIESTLEY, J. H.: The cuticle in angiosperms, *Bot. Rev.*, **9**, 593–616, 1943.

—— and L. I. SCOTT: Studies on the physiology of cambial activity, II. The concept of sliding growth, *New Phyt.*, **29**, 96–140, 1930.

—— and ——: The formation of a new cell wall at cell division, *Proc. Leeds Phil. and Lit. Soc. Sci. Sec.*, **3**, 532–545, 1939.

RECORD, S. J.: Cystoliths in wood, *Trop. Woods*, Yale University School of Forestry No. 3, 10–12, 1925.

——: Spiral tracheids and fiber-tracheids, *Trop. Woods*, Yale University School of Forestry, No. 3, 12–16, 1925.

——: "Identification of the Timbers of Temperate North America," New York, 1934.

RENDLE, B. J.: Gelatinous wood fibres, *Trop. Woods*, Yale University School of Forestry, No. **52**, 11–19, 1937.

S., D. H. (Notice of Book.) Dr. G. Krabbe: "Das gleitende Wachstum bei der Gewebebildung der Gefässpflanzen," Berlin, 1886. *Ann. Bot.* **2**, 127–136, 1888.

SCARTH, G. W., R. D. GIBBS, and J. D. SPIER: The structure of the cell-wall and the local distribution of the chemical constituents, *Trans. Roy. Soc. Can.*, **5**, 269–288, 1929.

SCOTT, F. M.: Cystoliths and plasmodesmata in *Beloperone*, *Ficus*, and *Boehmeria*, *Bot. Gaz.*, **107**, 372–378, 1946.

SINNOTT, E. W., and R. BLOCH: Changes in intercellular relationships during the growth and differentiation of living plant tissues, *Amer. Jour. Bot.*, **26**, 625–634, 1939.

—— and ——: Development of the fibrous net in the fruit of various races of *Luffa cylindrica*, *Bot. Gaz.*, **105**, 90–99, 1943.

SKUTCH, A. F.: Anatomy of leaf of Banana, *Musa sapientum* L. var. Hort. Gros Michel, *Bot. Gaz.*, **84**, 337–391, 1927.

SPONSLER, O. L.: The molecular structure of the cell wall of fibers, *Amer. Jour. Bot.*, **15**, 525–536, 1928.

TUPPER-CARY, R. M., and J. H. PRIESTLEY: The composition of the cell wall at the apical meristem of stem and root, *Proc. Roy. Soc.*, B **95**, 109–131, 1923.

VAN WISSELINGH, C.: Die Zellmembran. *In* Linsbauer, K.: "Handbuch der Pflanzenanatomie," III, 1925.

WAREHAM, R. T.: "Phragmospheres" and the "multinucleate phase" in stem development, *Amer. Jour. Bot.*, **23**, 591–597, 1936.

WERGIN, W.: Über den Feinbau der Zellwande höherer Pflanzen, *Biol. Zentralbl.*, **63**, 350–369, 1943.

WRIGHT, J. G.: The pit-closing membrane in the wood of the lower gymnosperms, *Trans. Roy. Soc. Can.*, Sec. V, Biol. Sci. III, **22**, 63–95, 1928.

ZIEGENSPECK, K.: Über das Ergusswachstum des Kutins bei *Aloe*-Arten, *Bot. Arch.*, **21**, 1–8, 1928.

CHAPTER III

MERISTEMS

Growth in an organism consists fundamentally of increase in the volume of the protoplasmic body. Closely associated with this increase is differentiation of the cells that represent this increase. Differentiation is usually thought of as a phase of growth. In some of the more primitive plants and in some parts or stages of more advanced plants, growth may take place without the partition of the protoplasmic body into cellular units. In cellular plants, however, any considerable amount of growth is usually manifested by new-cell formation and the subsequent enlargement and modification of the derivative cells. In plants with little or no tissue specialization, cell division may occur throughout the plant body or in any part thereof, increasing in this way the mass of the plant body. In plants with specialized tissues, on the other hand, the formation of new cells is chiefly localized in certain more or less definitely organized regions known as *meristems*.

The terms *meristem* and *meristematic*, as applied to developing cells and tissues, are somewhat loose. Writers differ greatly in the use of them. Rigid definitions cannot be made and no sharp line can be drawn in description between tissues that constitute meristems and those that are merely in some measure meristematic, or between these and permanent tissues. Growth and differentiation constitute a continuing process, and correlation of structure with stages of this process is a matter of convenience. In this treatment the term "meristem" is applied to *regions of more or less continuous cell and tissue initiation;* the adjective "meristematic" is used to indicate *resemblance in an important way to a meristem,* but not necessarily as *consisting of* or *constituting meristem;* that is, it is applied to those cells, tissues, and regions that have characteristics of developing structures—especially cell division—but do not themselves strictly constitute meristems. For example: the apices of stems and the cambium are regions of tissue initiation; developing xylem and phloem are meristematic tissues, for they form some new cells and are immature, but they are not permanent or semipermanent initiating regions. Again, cells in mature tissue, such as the primary cortex of stems, may divide. Such cells are meristematic, but neither they nor the tissues of which they are a part constitute a meristem. During the recently renewed interest in meristems, the use of these terms has, unfortunately, been broadened and confused.

The cells of meristem differ from those of mature tissues in that commonly they have abundant cytoplasm with vacuoles small or lacking, large nuclei, thin walls, and no intercellular spaces. But initials of apical meristems and the cambium may have vacuoles of good size, the radial walls of cambium are thick (Fig. 88), and minute spaces may be present among the derivatives of apical initials.

The term "embryonic" is sometimes used as synonymous with meristematic in describing young tissues. Such use is confusing; the term is best applied to tissues of the embryo only.

Meristematic and Permanent Cells and Tissues.—From the standpoint of stage of development, cells and tissues are *meristematic* or *permanent*. Meristematic cells and tissues are those in which new-cell formation is going on and differentiation is incomplete. When such cells and tissues have become fully differentiated and mature, they are said to be *permanent*. Permanent cells are not necessarily permanent in the usual sense of the word, since they may change both in form and in function after a longer or shorter period of existence as completely differentiated, or mature, cells. For example, epidermal cells or typical cortical cells may form a cork cambium months after those parts of the stem in which they lie have become mature; similarly, photosynthetic cells of the cortex and parenchyma cells of the older phloem may become stone cells. Cells may change their nature markedly a long time, even years, after becoming mature; they may again become active in division, forming other cells and even meristems. Any living cell is commonly believed to be potentially meristematic. Cells of elaborate shape and extremely thick walls are possible exceptions. Even though no cells may be strictly permanent, a distinction between meristematic and permanent cells and tissues is of value in an understanding of the origin and development of growing regions.

CLASSIFICATION OF MERISTEMS

Meristems are classified on several bases: stage of development, structure, position in plant, origin, function, topography. The bases of these classifications are not mutually exclusive, and the definitions cannot be rigid. A discussion of the more important types of meristems follows.

Meristems Based on Stage or Method of Development

Promeristem.—The region of new growth in a plant body where the *foundation* of new organs or parts of organs is initiated constitutes *promeristem*. (The terms "primordial meristem," "Urmeristem," and "embryonic meristem" have also been applied to this initiating region.) Structurally this region consists of the initials and their immediate derivatives, all young cells with diameters much alike; walls thin, with early stages of pits; cytoplasm active, vacuolate or nonvacuolate; nuclei large; and intercellular spaces absent or minute. The promeristem of a given organ or region is of rather limited extent, varying in amount in different plants, in different organs, and under differing growth conditions. It is of course not definitely set off from the somewhat older meristematic tissue into which it merges. As soon as cells of the promeristem begin to change in size, shape, and character of wall and cytoplasm, setting off the beginnings of tissue differentiation, they are no longer a part of typical promeristem; they have passed beyond that earliest stage. For example, there is at the tip of an organ a meristem of some length; only the youngest part of this, a small apical portion, is promeristem. The remainder of the meristem represents the early stages of the tissues formed by the promeristem. No term exists for this partly developed region in which segregation of tissues is beginning, but cell division continues freely.

Mass, Plate, and Rib Meristems.—On the basis of plane of division, *mass*, *plate*, and *rib meristems* have been distinguished as "growth forms" of meristem. In mass meristem, growth is by three-plane or all-plane division and produces increase in mass; plate meristem is by divisions—chiefly in two planes—so that there is plate-like increase in area; rib meristem, by continuing divisions (anticlinal) in only one plane, produces rows or columns of cells, functioning chiefly in the increase of organs in length. Examples of mass meristem are early stages of many embryos; developing sporangia; the endosperm of many plants; young pith and cortex of some plants. Plate meristem (one-layered) forms epidermis, and (two- to several-layered) is prominent in leaf development where in early stages divisions in the plane of the leaf blade and at right angles to that plane build up great increase in blade area with little increase in thickness. Rib meristem plays a prominent part in the development of young roots and of the pith and cortex of young stems. The term "rib" meristem is in frequent use at present but is unfortunately often applied not to regions of cell initiation but to those of cell differentiation where the cells stand in long rows. In general, these "meristems" are merely parts of meristematic tissue-masses distinguished by planes of cell division. Recently the term *file meristem* has to some extent replaced rib meristem; this is a more accurately descriptive term. It is obvious that these types can be distinguished only when the cells are chiefly of typical parenchyma-cell form; meristems with cells of prosenchymatous form cannot be thus classified. Cells that are beginning to elongate constitute the first stages of *procambium* (Chap. V), of *collenchyma* (Chap. IV), or of strands of fibers.

Meristems Based on History of Initiating Cells

Primary and Secondary Meristems.—On the basis of type of tissue in which origin occurs, meristems are classified as *primary* and *secondary*. *Primary meristems* are those that build up the fundamental (primary) part of the plant and consist in part of promeristem. (In some uses, promeristems are considered distinct from primary meristems.) In primary meristems, promeristem is always the earliest stage, and transition stages to mature tissues constitute the remainder of the meristem. The possession of promeristem continuously from an early embryonic origin is characteristic of primary meristems; no stage wherein all or some of the cells have become permanent, or modified from the meristematic, has entered their history. The chief primary meristems are the apices of stems and roots and the primorida of leaves and similar appendages.

Exceptions to the promeristem continuity of primary meristems from early embryonic tissue are found in adventitious buds and roots and in some types of wound tissue. (Many so-called "adventitious buds" arise from buried dormant bud intials, not *de novo* in permanent tissue.) The originating meristems of true adventitious organs arise secondarily in more or less nearly permanent tissues, but they are, because of their structure and behavior, primary meristems. Once established, they may persist indefinitely.

Secondary meristems are set apart from primary meristems in that they always arise in permanent tissues; in their history there is interposed a stage of permanent tissue or at least of partial development toward permanency. They have no typical promeristem, though their initiating layers may to some extent resemble this tissue. *Secondary meristems* are so called because they arise as new meristems in tissue which is not meristematic. The cork cambium is an example of secondary meristem; it is formed from mature cells—cortical, epidermal, or phloem cells or from any mature cells that are not too highly specialized.

The primary meristems build up the early and, for a time, structurally and functionally complete plant body. The secondary meristems later add to that body, forming supplementary tissues that functionally replace or reinforce the early formed tissues or serve in protection and repair of wounded regions.

The cambium, one of the most important meristems, does not fall definitely in either group. It arises from apical meristem of which it is a late and specialized stage. (Accessory cambia may be secondary.) The tissues formed by the cambium are secondary, as are all those formed by secondary meristems, whereas other primary meristems form only primary tissues.

The classification of meristems as primary and secondary has sometimes been considered of little value but it is especially helpful in an understanding of the manner in which the complex mature plant body is attained and modified with continuing growth.

Meristems Based on Position in Plant Body

On the basis of position in the plant body, meristems are usually classified as *apical,* "*intercalary,*" and *lateral* (Fig. 35). Apical meristems

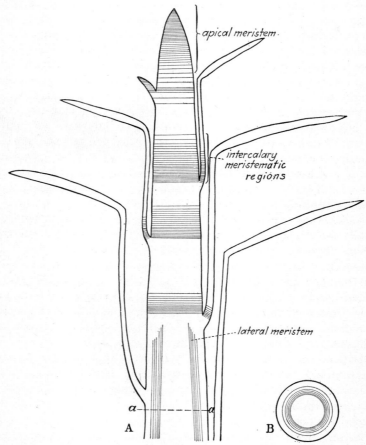

Fig. 35.—Diagram to show position of meristems. The closely lined areas are youngest; the unlined areas are mature or slowly growing. *A*, longitudinal view; *B*, cross section of *A*, at level *a–a*.

are those which lie at the apices of the axis and of the appendages and are commonly called growing points. "Intercalary meristems" are those which lie between regions of permanent tissue, as, for example, at the base of the leaves of many monocotyledons. Lateral meristems, as the name implies, are situated laterally in an organ. The cambium and the cork cambium are lateral meristems.

Apical Meristems.—Apical meristems, or growing points, occur universally at the tips of the roots and stems, and often of the leaves, of vascular plants. The activity of these meristems brings about increase in the length of these organs, laying down the primary body of the plant. Initiation of growth is by one or more cells situated at the tip of the organ which maintain their individuality and position and are known as *apical initials* or *apical cells*. These cells may be strictly terminal or terminal and subterminal.

Apical Cells.—Among vascular plants, solitary apical cells (except those of leaves) occur in the horsetails, most of the ferns, and a few other pteridophytes. In other vascular plants a group or groups of apical, or apical and subapical cells constitute the initiating body. The behavior of these cells in tissue formation has not been extensively and critically studied until recent years and is even yet only partly understood. Since the initials apparently differ little, if at all, from their recently formed daughter cells and since all of these cells may continue to divide freely, it is most difficult to determine the number and limits of the initials. The number of initials is probably fairly constant for a given organ and a given species, but it is probable that at least in some plants, the number of initials and, to some extent, the form of the group vary from time to time even in the same organ. Meristems are highly plastic. Variations are related to vigor of growth, to seasonal conditions, and to the morphology of the organ developing. It is now known that in seed plants there are many types of apical-meristem differentiation. For each major plant group there is apparently a fairly characteristic plan of development, but this plan may not be consistently followed in detail even within an individual plant (see Tunica-corpus Theory).

Initials, where solitary, appear to persist indefinitely; where there are several or many, they may function for a time and then be replaced by new cells. During periods of rapid growth, additional cells may serve more or less temporarily as initials.

Types of Apical Cells.—Solitary apical cells of several types are distinguished by shape and by the number of sides from which new cells are cut off. The two most common are the lenticular, or two-sided cell (not found in vascular plants), and the pyramidal, or three-sided cell (Figs. 37*A*, 41). The former is strictly a three-sided cell in shape and the latter a four-sided cell, but new cells are formed only from two and three sides respectively, hence terms have been given which apply to the activity of the cell rather than to its shape. The side from which no new cells are formed lies in the direction of growth. In roots, cells are cut off on this face also (Fig. 41). The size of the apical cell is reduced only temporarily by the formation of daughter cells, and its position is maintained indefinitely.

The behavior of solitary apical cells in segmentation, and of the cells which are cut off by the apical cells, is of importance in the study of the morphology, both vegetative and reproductive, of the bryophytes. In these plants, groups of cells in certain positions and of certain origin always form the same tissues or organs. That this must also be true in general of vascular plants was formerly believed, but such is clearly not the case. The behavior of a given segment of an apical meristem may have little morphological significance in the groups of plants above the bryophytes.

"Intercalary Meristems."—So-called "intercalary meristems" are merely portions of apical meristems that have become separated from the apex during development by layers of more mature or permanent tissues and left behind as the apical meristem moves on in growth (Fig. 35). The more mature layers are the nodal regions and the "intercalary meristems" are therefore internodal. At early stages the internode is uniformly or nearly uniformly meristematic; later, some part of it matures more rapidly than the rest, and a definite sequence in development within the internode is set up and maintained. The region of youngest cells is in most plants at the base of the internode, but it may be in the middle or at the top. Where the youngest region is basal, the order of maturing in the tissues of the internode is of course acropetal; where it is at the top, the order is basipetal; and where it is median, the sequence is both upward and downward.

It has been generally believed that "intercalary meristems" are separated from the mother meristem at so early a stage that they retain a layer of the promeristem of the apical region and that this promeristem continues, after its segregation, to function as a promeristem, initiating complete new segments of the organ. But it has been shown that this is not true in some grasses, and it is perhaps not so in any plants. Although these regions consist partly of promeristem cells, vascular strands with protoxylem and protophloem cells in various stages of maturation run through them. With such structure they are not typical meristems but meristematic regions and should be so called.

The best known intercalary meristematic regions are those of the stems of grasses and other monocots and of horsetails, where they are basal, and those of some mints, where they lie just below the node. In some peduncles, such as those of *Taraxacum* and *Plantago*, they are said to occur at the top. Leaves of many monocotyledons (grasses, *Iris*) and some other plants, such as *Pinus*, have basal meristematic regions. Intercalary meristematic regions ultimately disappear; they become wholly transformed into permanent tissues.

Lateral Meristems.—Lateral meristems are composed of initials that divide chiefly in one plane (periclinally) and increase the diameter of an organ. They add to existing tissues or build new tissues. The cambium and the cork cambium are meristems of this type. The former has been called "the lateral meristem," but phellogen layers also belong here. The briefly functioning "marginal meristems" of some leaves have been called lateral but fall in none of the above groups.

Meristems Based on Function

In physiological anatomy a classification of meristems based chiefly on functional significance is used. The outermost layer of the young growing region, from which the epidermis develops, is called the *protoderm;* the elongating, tapering cells of this region, the *procambium;* the remainder of the meristematic tissue, *fundamental*, or *ground-tissue* meristem. In this usage the term "procambium" is not synonymous with *procambium* (Chap. V) as this term is used morphologically to indicate early stages of vascular tissue; the physiological term covers not only cells from which vascular tissues develop but others also.

THEORIES OF STRUCTURAL DEVELOPMENT AND DIFFERENTIATION

The progress of differentiation in meristems brings about in cellular organization a structural segregation, or zonation, with regions distinguishable by (*a*) the number and position of the initiating cells, (*b*) the planes of division and consequent arrangement of cells, (*c*) the size, shape, and content of the cells, and (*d*) the rate of maturation of the cells. Several theories dealing with the methods of origin of the patterns formed by this zonation and their histological and morphological significance have been proposed.

The Apical Cell Theory.—Solitary apical cells occur in many of the algae, in the bryophytes, and, among vascular plants, in the Psilotaceae, the horsetails, most of the ferns, and some species of *Selaginella*. Early studies of the apical meristems of bryophytes and of the pteridophytes with a single apical cell demonstrated morphological significance in the segmentation of the apex. It was believed that the same conditions, although difficult to demonstrate in seed plants, held for all higher plants, and the *apical cell theory* was proposed as the basis for an understanding of the method of growth and morphology in all groups. But controversy arose over the interpretation of the complex apices of gymnosperms, and it became evident that the theory was not applicable to seed plants.

The Histogen Theory.—As a basis for the interpretation of the growing points of seed plants, the *histogen theory* replaced the older theory. Under this theory the more or less distinct major regions of the stem and root apex were called *histogens,* that is, tissue builders, in the sense of builders of specific parts of an organ. The histogens were the *plerome,* a central core; the *dermatogen,* a uniseriate, external layer; the *periblem,*[1] the region between the plerome and the dermatogen (Fig. 36). The histogens, or their individual layers, were believed developed by separate initials. This theory, in contrast to the apical cell theory, placed the origin of axis apices in a group of initials. It differed from the earlier theory also in morphological interpretations or implications—the plerome formed the pith and primary vascular tissues; the dermatogen, the epidermis; the periblem, the cortex.

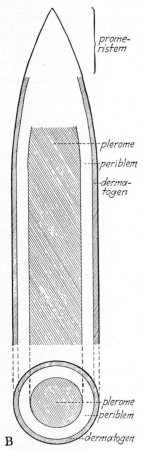

Fig. 36.—Diagram of axis tip to show "histogen" regions. *A*, longitudinal section; *B,* transverse section.

The viewpoint and the terminology of the histogen theory have long dominated anatomical interpretation of tissues, regions, and even organs in vascular plants. But the distinction of these histogens in an apex cannot be made in some plants, and in others the regions have no morphological significance. The plerome may form only the pith or the entire central cylinder and part of the cortex; the periblem may form the cortex and the outer stelar tissues or part of the cortex. Even in homologous axes on the same plant the histogens may form different parts.

The terms clearly have no morphological value, and, because their prominence has rested largely on such value, they are becoming less used. They are found occasionally in histological descriptions of stems, commonly in those of roots, but serve chiefly to indicate regions in a rather loose or topographical sense.

[1] These terms were originally applied to the *zones* of meristematic tissue in early stages of development from initials but they have been occasionally applied to the initials only.

MERISTEMS

The Tunica-corpus Theory.—Interest in the development of seed-plant axes reawakened about twenty years ago with the formulation of the *tunica-corpus theory* and increasing attention has been given to the ontogeny of the stem tips of vascular plants. According to the tunica-corpus theory (which has been applied only to the leafy stem, or shoot) different rates and method of growth in the apex set apart two regions of unlike structure and appearance—a central core, the *corpus*, and an outer enveloping layer, around and above the corpus, the *tunica*. In the corpus the cells are large, with arrangement and planes of cell division irregular, and increase is in volume; in the tunica the cells are usually smaller than

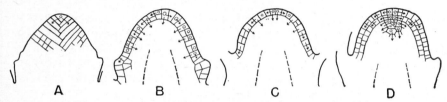

FIG. 37.—Diagrams to show position and planes of division of stem-apex initials. *A*, initial solitary, with oblique anticlinal divisions only; *B*, initials many, superficial, with their divisions and those of the dermatogen both anticlinal and periclinal; *C*, initials several, superficial, with their divisions both anticlinal and periclinal and those of the dermatogen largely anticlinal; *D*, initials in three tiers, the two outer, with divisions anticlinal only, forming a two-layered tunica and the innermost, with divisions in all planes, forming a corpus. (Initials indicated by outlined nuclei.)

those of the corpus and lie in layers or sheets, with divisions strictly, or chiefly, anticlinal, and growth is primarily in area. These regions vary greatly in the distinctness of their limits and in their form and relative size; their boundaries are often not clear-cut; the corpus may be massive or slender; the tunica, many-, few-, or one-layered. The important elements in such an elaboration of regional development are the position and number of initials and the planes of division in these cells and their derivatives.

The number of initials is few to several or many. Rarely, in small, very slender apices, such as those of grass seedlings, there may be only one or two in the tunica and about two in the corpus. The determination of initials as self-maintaining and persistent cells is often impossible because they differ little or not at all from their recently formed daughter cells. Certainly it is unlikely that the initials remain constant in number, form, and sequence of division.

Where all the initials are on the surface, their divisions are both anticlinal and periclinal, and no clearly defined tunica and corpus are formed (Figs. 37*A*,*B*,*C*, 38*A*,*B*, 39*A*,*B*). Where the initials are in a cluster or mound, the outer ones, if their divisions are strictly anticlinal, form a definite tunica, and the inner ones, with divisions in more than one plane, form a corpus (Figs. 37D, 38D, 39C, D). Under these conditions tunica and corpus have completely independent origins and are fairly distinct. Between these extremes lie many intermediate conditions in which distinction of tunica and corpus is weak or doubtful.

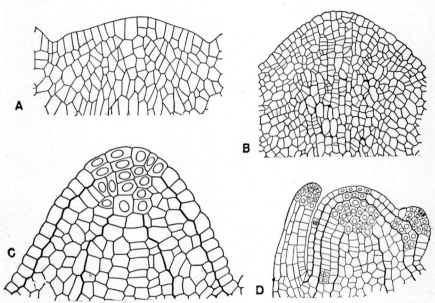

Fig. 38.—Structure of stem apices as seen in longitudinal section. *A, Lycopodium Selago*, apex built up by superficial initials which divide both anticlinally and periclinally. *B, Zamia umbrosa*, apex built up by central group of superficial initials which divide both anticlinally and periclinally. *C, Sequoia sempervirens*, apical initials divide both anticlinally and periclinally; subapical initials divide in all planes (initials indicated by outlined nuclei). *D, Sinocalamus Beecheyana*, tunica two-layered, its initials in two layers, with divisions, with rare exceptions, anticlinal; initials of corpus many, in a mass below tunica initials, their divisions in all planes (initials of apex and of leaf apices and bud indicated by outlined nuclei). The heavier lines in *C* and *D* indicate not thicker cell walls but separate zones, limiting in *C*, a surface layer, the dermatogen, and a central core that forms the pith; in *D*, the two tunica layers, and the central part of the corpus, formed by periclinal divisions of the lowest corpus initials. (*A*, after Haertel; *B*, after Johnson; *C*, after Cross; *D*, after Hsü, modified.)

Below the initials or below the direct derivatives of these cells (the two groups constituting promeristem), elaboration of cells in size, form, and arrangement initiates tissue specialization and the building of the framework of the primary body (Chap. V). The various regions, or stages, merge on their margins and overlap longitudinally. Major interruptions in the progressively more advanced stages back from the apex are brought in by the appearance of nodal areas and lateral structures.

Types of Stem Apices.—In vascular plants, zonation in stem-apex differentiation follows more or less definite patterns that appear to be characteristic of the major groups. These patterns show increasing com-

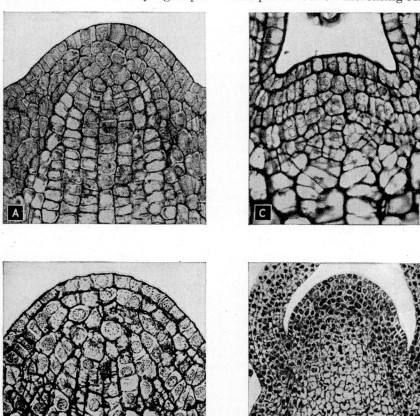

Fig. 39.—Structure of stem apices as seen in longitudinal section. *A, Abies concolor*, initials of the surface layer divide both anticlinally and periclinally, subapical initials in all planes; *B, Sequoia sempervirens*, initials of the surface layer divide chiefly anticlinally, subapical initials divide in all planes; *C, Vinca rosea*, tunica of three layers and corpus present, initials of tunica dividing only anticlinally, those of the corpus dividing in all planes; *D, Phlox Drummondii*, floral apex. (*A, after Korody; B, after Cross; C, after Cross and Johnson; D, after Miller and Wetmore.*)

plexity from the lower to the higher groups and seem to represent a series in specialization from simplicity to complexity. In lower groups the initials form a uniseriate surface layer, and there is no evidence of differentiation into tunica and corpus; in the highest groups the initials form a two-storied cluster of which the superficial members build a tunica—which, in its greatest specialization, is probably uniseriate—and the inner members form the corpus. The majority of seed plants are intermediate in apex structure between these types.

The Primitive Type of Stem Apex.—Among the pteridophytes, *Lycopodium*, *Isoëtes*, and some species of *Selaginella*, and among primitive gymnosperms, the cycads, have simple apices with surface initials and no distinction of tunica and corpus.

Lycopodium (Fig. 38*A*) serves as an example of this primitive type. The initiating layer is a weakly defined, uniseriate surface area which divides freely both anticlinally and periclinally. No definable initials can be distinguished; all the cells of the layer are morphologically alike. The anticlinal divisions increase the area of the surface layer; the periclinal divisions form an inner core.

The Stem Apex with Weak Tunica-corpus Segregation.—Beginnings in distinction of tunica and corpus are perhaps present in some of the lower conifers. In the Pinaceae (*Abies*, *Pinus*), the initials form a terminal uniseriate group. From these, by both periclinal and anticlinal divisions, a central core and an enveloping uniseriate layer are formed (Figs. 37*C*, 39*A*). The latter suggests a tunica in appearance, but it has frequent periclinal divisions even on the flanks of the apex, and there is no clear line of separation between the tissues formed by the two regions.

Some higher conifers have a somewhat more specialized apex (Figs. 38*C*, 39*B*). As in the Pinaceae, the initials are a small group of surface cells in one tier, with divisions both anticlinal and periclinal. These divisions form, respectively, a dermatogen-like layer and a central mass, but in the outer layer the periclinal divisions are rare or absent, except close to the apex. The part played by the outer layer in building the body of the stem is restricted to the cells at the very apex. The outer layer early becomes tunica-like in its restriction of planes of division.

Of the conifer genera thus far critically studied, only *Cryptomeria* and *Taxodium* (apices of permanent shoots) have a "dermatogen" in which there are no, or almost no, periclinal divisions. The apices of conifers have the structural appearance of tunica-corpus segregation but in those so far studied there is only one tier of initials and therefore no independent meristematic regions.

The conifers show much variety in apex structure. Even within a species (*Taxodium distichum*, *Sequoia sempervirens*) there are marked differences, correlated in part with morphology of the stem concerned, vigor of growth, and other conditions.

The Stem Apex with Definite Tunica and Corpus.—In the angiosperms the segregation of apical-meristem zones is usually more definite than in lower groups; there are two sets of initials, one above the other, which give rise to tunica and corpus that are in large measure or wholly independent (Figs. 37D, 38D, 39C). The tunica has no or only rare periclinal divisions and ranges in thickness from several layers to one, with two or three layers probably most frequent. The larger numbers of tunica layers occur more frequently in the dicotyledons. A single-layered tunica such as that of grasses (*Avena, Triticum*) probably represents the most specialized condition, but even in this, occasional periclinal divisions may occur, as in *Zea*. The corpus varies from a large complex type to a slender, simple type. The number of layers in the tunica may vary even in an individual plant. The limits of the two groups of initials and of the tunica and corpus is often definite but may be difficult or impossible to determine and in some genera (Cactaceae) the apex shows no distinction of tunica and corpus, resembling the apex of the primitive gymnosperms. *Sinocalamus Beecheyana* (a bamboo) (Fig. 38D) and *Vinca rosea* (Fig. 39C) serve as examples. In the former the tunica is two-layered (sometimes one- or three-layered); in the latter it is four-layered.

Discussion, Tunica-corpus Theory.—The tunica-corpus theory has served well in the establishment of an understanding of the complex, diverse, and varying meristematic patterns of the stem tips of seed plants. Providing a basis and a method, it has stimulated intensive study, and detailed information has been obtained for a considerable number of plants. The position, number, and behavior of the initiating cells in seed-plant stems is now in some measure known, and early stages in the development of the primary body of the shoot are much better understood. Although, as with the histogens of the histogen theory, a distinction of tunica and corpus has little or no morphological significance, it is of topographical value in studies of detailed development. Caution is necessary lest morphological interpretations accompany the use of the theory: the boundaries of tunica and corpus are often obscure; where the limits are fairly distinct, the two regions may be inconstant in structure and function, varying with seasonal conditions, vigor of growth, age of plant, position on plant, and morphology of the stem tip concerned. For example, in some plants the innermost tunica layer of a primary branch becomes an outermost part of the corpus on secondary branches. The number of tunica layers varies even in apices of the same type at the same time; it is commonly less on the less vigorous and the lateral branches than on strong, leading stems. The differences found, even on the same plant, in the thickness and distinctness of the tunica at different times and under different morphological conditions—as in main and lateral branches, deciduous and persistent twigs, indeterminate vegetative tips and the determinate apices of flowers and thorns—are related directly to growth status and to the morphological nature of the developing organ. A flexible basis is necessary for all descriptions of meristems. Tunica and corpus should be recognized as dynamic and fluid, not functionally or morphologically distinct or constant regions.

The lateral organs of the stem—leaves, branches, and floral organs—arise near the apex and studies of tunica and corpus have added greatly to a knowledge of the origin and early development of these organs.

The Floral Apex.—The structure of the floral apex (Fig. 39*D*) differs in no fundamental way from that of the vegetative stem. Such differences as can be distinguished are those inherent in the determinate nature of the axis, the telescoping of internodes, and the crowding of appendages.

It has been claimed that the floral apex differs markedly from the vegetative apex; that apical initials build up the "central core" in the vegetative axis, but that both apical and flanking tissues build up the floral apex; that the tunica of the floral apex has no definable initiating zone and that periclinal divisions occur at any depth within it. Such differences may occur, varying greatly in degree with the form and structure of the flower concerned, but they are not morphological differences. They are associated directly with the function of an apex that is developing into a flower—elongation ceases; the apical initials of the earlier vegetative growth lose their identity, and growth activity is restricted to the peripheral region where many appendages are arising at compacted nodal levels; the numerous periclinal divisions at all depths are associated with the broadening of the receptacle and with the origin of the floral appendages which cover it. Distinction of outer and inner zones varies in the floral apex as it does in the vegetative apex. The number of layers in the tunica of the floral apex is more in some plants, less in others, than in the vegetative apex. In ontogeny a floral apex develops by the gradual or abrupt transformation of a vegetative apex.

The Root Apex.—The apical meristematic region of the root, lacking developing appendages and segregation into nodal and internodal regions, is simpler in gross structure than that of the stem but is complicated at the tip by the *root cap* and different methods of formation of this structure. The root cap is a terminal portion of the root tip, more or less definitely set off from the tissues beneath, covering and protecting the initiating apex of the root proper. It is formed by the same initials as those that build the root proper in all types of root tip except those of the monocotyledons where it has independent origin. Its formation is necessarily discussed with that of the root apex; its structure and occurrence are considered in Chap. X.

The apical meristem is short compared with that of the stem; specialization into zones is concentrated, with the early stages lying within the terminal millimeter beneath the cap (Figs. 41, 42). Growth proceeds in cap and root proper in opposite directions, with sequence in development toward the tip in the cap and away from the tip in the root itself. In number, the initials range from one to many (Fig. 40). Where the initials are more than one, they are arranged in one to four fairly distinct, uniseriate groups (Fig. 40B,C,D). In each group there are one to several initials. Since the initials are the central members of uniseriate layers and can be distinguished from their recent derivatives only by

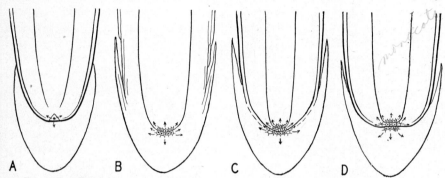

Fig. 40.—Diagrams of root-apex types. A, initial solitary, cap distinct, but not independent in origin; B, initials in two groups, cap not structurally distinct, formed by same initials as periblem and dermatogen; C, initials in three groups, cap not distinct, formed by same initials as dermatogen; D, initials in three groups, cap distinct in structure and independent in origin.

position and restriction in division to anticlinal planes, the number can often be determined only approximately. Furthermore, the number of initials in a group seems to vary with diameter and rapidity of growth of the root. In slender roots it may be reduced to one—as in some grasses—but the zonation remains clear. Lines of zonal segregation are most obscure in root tips of large diameter. Where there is more than one group, the groups lie adjacent to one another on the longitudinal axis of the root (Fig. 40).

Each of these groups quickly develops one or more growth zones which are usually more clearly marked than are similar zones in the stem apex (Figs. 41, 42). In many plants these zones appear to represent "the histogens," and interpretations of the root apex have long continued under the histogen theory. Although the terms dermatogen, periblem, and plerome are no longer in general use in descriptions of stem ontogeny, they have been continued for convenience to indicate general zones in studies of root development. A fourth histogen, the *calyptrogen*, is added where the cap has an independent origin.

In published descriptions there is much confusion as to method of origin of the zones and no agreement as to the number of types of development. The distinctness of the zones varies, and forms intermediate between the types have been described. There seem, however, to be basic patterns for the major plant groups. This pattern is determined by the number of initials, the number of groups of these initials, the zones formed by each group, the morphological nature of the cap, and the degree of independence of the cap.

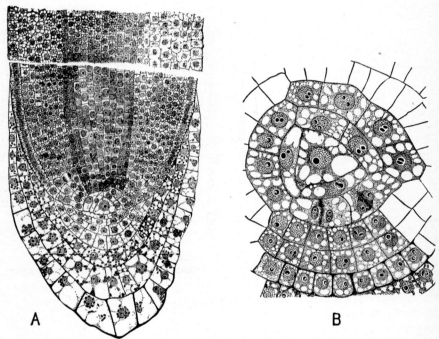

Fig. 41.—Root tip of *Pteris*. *A*, median longitudinal section, showing apical cell with four "cutting faces" (one not in plane of section), the development of tissues from this cell, and plerome, periblem, and dermatogen differentiated; *B*, transverse section through apical cell, showing the three faces from which cells are cut off to form the root proper. (*After Hof.*)

The vascular cryptogams that have a solitary apical cell in the stem— the horsetails, most of the ferns, some species of *Selaginella*—have a similar solitary apical cell in the root (Figs. 40*A*, 41). This one cell forms the entire root and the cap which is usually sharply distinct structurally. The origin of the cap from the large apical cell is clear.

In many gymnosperms there are two groups of initials (Fig. 40*B*). The inner forms the plerome; the outer forms the periblem and the cap. No line can be drawn between these two regions; the cap appears as a distal proliferation of the periblem. A dermatogen is not set off at the very apex, as in all other groups, but is formed from layers of the periblem a little way back from the apex where the base of the cap is separated from the periblem.

In the angiosperms there are three, rarely four, groups of initials. In the dicotyledons the distal group forms the cap and the dermatogen; the median group, the periblem; the innermost, the plerome (Fig. 40C).

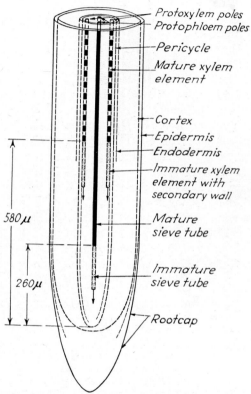

Fig. 42.—Diagram of structure of a diarch tobacco root tip, showing position of stages of development of first vascular elements. (*After Esau.*)

The outstanding characteristic of root-apex development in this group is the common origin of cap and dermatogen. These two protective parts of the root develop from the same initials and the cap can be looked upon morphologically as a specialized development of the epidermis.

The monocotyledons, like the dicotyledons, have three groups of initials which form four zones, but the outermost, independently, forms the cap, and that next beneath, the dermatogen and periblem. The outstanding characteristic of this type is the independence of the cap in origin and structure. Further, the two zones that are formed by one group of initials (dermatogen and periblem) are different from those (cap and dermatogen) similarly formed in the dicotyledons. Although the group of initials from which the dermatogen and periblem arise is usually uniseriate (Fig. 40D), it apparently in some genera may be two or even several layers thick.

Rarely all the zones arise from separate groups of initials. Only a few aquatic monocotyledons (*Pistia, Hydrocharis*) are known to have this four-group condition.

Variations of the types have been described as transitional and there has been much disagreement as to the number of types and the validity of the bases used in determination of them. The root apex is in need of broad and critical comparative study such as that given the shoot apex under the tunica-corpus theory.

Relation of Types of Root-tip Development.—Evolutionary progress has perhaps been from a single center of growth to a group of three or four centers, with progressive increase in independence of the zones. The cap seems to have had different methods of origin. In the lower groups it has a common origin with all other parts of the root. In the gymnosperms it is a specialized, distally developed part of the periblem. The dermatogen, under this condition, is delayed in development and arises by specialization of periblem layers below the surface. In the dicotyledons the cap appears to be a specialized development of the epidermis. In the monocotyledons, the cap has attained independence in origin.

References

ARTSCHWAGER, E.: Anatomy of the vegetative organs of the sugar cane, *Jour. Agr. Res.*, **30**, 197–221, 1925.

BALL, E.: The development of the shoot apex and of the primary thickening meristem in *Phoenix canariensis* Chaub., with comparisons to *Washingtonia filifera* Wats. and *Trachycarpus excelsa* Wendl, *Amer. Jour. Bot.*, **28**, 820–832, 1941.

BOKE, N. H.: Zonation in the shoot apices of *Trichocereus spachianus* and *Opuntia cylindrica*, *Amer. Jour. Bot.*, **38**, 656–664, 1941.

BOND, T. E. T.: Studies in vegetative growth and anatomy of the tea plant (*Camellia thea* Link) with special reference to the phloem, II. Further analysis of the flushing behaviour, *Ann. Bot.*, **9**, 183–216, 1945.

BROOKS, R. M.: Comparative histogenesis of vegetative and floral apices in *Amygdalus communis* with special reference to the carpel, *Hilgardia*, **13**, 249–299, 1940.

BRUMFIELD, R.: Cell lineage studies in root meristems by means of chromosome rearrangements induced by X-rays, *Amer. Jour. Bot.*, **30**, 101–110, 1943.

CROSS, G. L.: The structure and development of the apical meristem in the shoots of *Taxodium distichum*, *Bull. Torrey Bot. Club*, **66**, 431–452, 1939.

———: The shoot apices of *Athrotaxis* and *Taiwania*, *Bull. Torrey Bot. Club*, **70**, 335–348, 1943.

———: A comparison of the shoot apices of the Sequoias, *Amer. Jour. Bot.*, **30**, 130–142, 1943.

ENGARD, C. J.: Organogenesis in *Rubus*, *Univ. Hawaii Res. Publ.*, **21**, 1–234, 1944.
ESAU, K.: Ontogeny in the vascular bundle in *Zea mays*, *Hilgardia*, **15**, 327–356, 1943.
———: Phloem anatomy of tobacco affected with curly top and mosaic, *Hilgardia*, **13**, 437–490, 1941.
FLAHAULT, C.: Recherches sur l'accroissement de la racine chez les phanérogams, *Ann. Sci. Nat. Bot.*, 6 sér., **6**, 1–168, 1878.
FOSTER, A. S.: Structure and growth of the shoot apex in *Ginkgo biloba*, *Bull. Torrey Bot. Club*, **65**, 531–556, 1938.
———: Problems of structure, growth and evolution in the shoot apex of seed plants, *Bot. Rev.*, **5**, 454–470, 1939.
———: Further studies on zonal structure and growth of the shoot apex of *Cycas revoluta* Thunb., *Amer. Jour. Bot.*, **27**, 487–501, 1940.
———: Comparative studies on the structure of the shoot apex in seed plants, *Bull. Torrey Bot. Club*, **68**, 339–350, 1941.
GRÉGOIRE, V.: La morphogénèse et l'autonomie morphologique de l'appareil floral. I. Le carpelle, *La Cellule*, **47**, 287–452, 1938.
HÄRTEL, K.: Studien an Vegetationspunkt einheimischer Lycopodien, *Beitr. Biol. Pflanzen*, **25**, 126–168, 1938.
HELM, J.: Das Erstärkungswachstum der Palmen und einiger anderer Monokotylen zugleich ein Beitrag zur Frage des Erstärkungswachstums der Monokotylen überhaupt, *Planta*, **26**, 319–364, 1936.
HSÜ, J.: Structure and growth of the shoot apex of *Sinocalamus Beecheyana* McClure, *Amer. Jour. Bot.*, **31**, 404–411, 1944.
JANCZEWSKI, E. de: Recherches dur l'accroissement terminal des racines dans les phanérogames, *Ann. Sci. Nat. Bot.*, 5 sér., **20**, 162–201, 1874.
JOHANSEN, D. A.: A proposed new botanical term, *Chron. Bot.*, **6**, 440, 1941.
JOHNSON, M. A.: Structure of the shoot apex in *Zamia*, *Bot. Gaz.*, **101**, 189–203, 1939.
———: Zonal structure of the shoot apex in *Encephalartos*, *Bowenia*, and *Macrozamia*, *Bot. Gaz*, **106**, 26–33, 1944.
KEMP, M.: Morphological and ontogenetic studies on *Torreya californica* Torr. I. The vegetative apex of the megasporangiate tree, *Amer. Jour. Bot.*, **30**, 504–517, 1943.
KLEIM, F.: Vegetationspunkt und Blattanlage bei *Avena sativa*, *Beitr. Biol. Pflanzen*, **24**, 281–310, 1937.
KOCH, L.: Ueber Bau und Wachstum der Sprossspitze der Phanerogamen, I Die Gymnospermen, *Jahrb. Wiss. Bot.*, **22**, 491–680, 1891.
KORODY, E.: Studien an Spross-Vegetationspunkt von *Abies concolor*, *Picea excelsa* und *Pinus montana*, *Beitr. Biol. Pflanzen*, **25**, 23–59, 1938.
MILLER, H. A., and R. H. WETMORE: Studies in the developmental anatomy of *Phlox Drummondii* Hook., III. The apices of the mature plant, *Amer. Jour. Bot.*, **33**, 1–10, 1946.
NEEFF, F.: Über Zellumlagerung. Ein Beitrag zur experimentellen Anatomie, *Zeitschr. Bot.*, **6**, 465–547, 1914.
NEWMAN, I. V.: Studies in the Australian Acacias. VI. The meristematic activity of the floral apex of *Acacia longifolia* and *A. suaveolens* as a histogenetic study of the ontogeny of the carpel, *Proc. Linn. Soc. N.S.W.*, **61**, 56–88, 1936.
REEVE, R. M.: Comparative ontogeny of the inflorescence and the axillary vegetative shoot in *Garrya elliptica*, *Amer. Jour. Bot.*, **30**, 608–619, 1943.
RÜDIGER, W.: Die Sprossvegetationspunkte einiger Monocotylen, *Beitr. Biol. Pflanzen*, **26**, 401–443, 1939.

SATINA, S., A. F. BLAKESLEE, and A. G. AVERY: Demonstration of the three germ layers in the shoot apex of *Datura* by means of induced polyploidy in periclinal chimaeras, *Amer. Jour. Bot.*, **27**, 895–905, 1940.

SCHMALFUSS, K.: Untersuchungen über die interkalare Wachstumszone an Glumifloren und dikotylen Blütenschaften, *Flora*, **124**, 333–366, 1930.

SCHMIDT, A.: Histologische Studien an phanerogamen Vegetationspunkten. *Bot. Arch.*, **8**, 345–404, 1924.

SCHÜEPP, O.: Meristeme, *In* Linsbauer, K.: "Handbuch der Pflanzenanatomie," IV, Berlin, 1926.

SHARMAN, B. C.: Developmental anatomy of the shoot of *Zea mays* L., *Ann. Bot.* n.s., **6**, 245–282, 1942.

SINNOTT, E. W.: Structural problems at the meristem, *Bot. Gaz.*, **99**, 803–813, 1938.

SOUÈGES, R.: "La Différentiation," III, La différentiation organique, Paris, 1936.

STERLING, C.: Growth and vascular development in the shoot apex of *Sequoia sempervirens* (Lamb.) Endl., I. Structure and growth of the shoot apex, *Amer. Jour. Bot.*, **32**, 118–126, 1945.

STRASBURGER, E.: "Die Coniferen und die Gnetaceen, eine morphologische Studie," Jena, 1872.

STRUCKMEYER, B. E.: Structure of stems in relation to differentiation and abortion of blossom buds, *Bot. Gaz.*, **103**, 182–191, 1941.

TIEGS, E.: Beiträge zur Kenntnis der Entstehung und des Wachstums der Wurzelhauben einiger Leguminosen, *Jahr. Wiss. Bot.*, **52**, 622–646, 1913.

TREUB, M.: "Le Méristèm Primitif de la Racine dans les Monocotylédones." Leiden, 1876.

VON GUTTENBERG, H.: Der primäre Bau der Angiospermenwurzeln, *In* Linsbauer K.: "Handbuch der Pflanzenanatomie," VIII, Berlin, 1940.

———: Der primäre Bau der Gymnospermenwurzeln. *In* Linsbauer, K.: "Handbuch der Pflanzenanatomie," VIII, Berlin, 1941.

WAGNER, N.: Über die Entwicklungsmechanik der Wurzelhaube und des Wurzelrippenmeristems, *Planta*, **30**, 21–66, 1939.

WARDLAW, C. W.: The shoot apex in pteridophytes, *Biol. Rev. Cambridge Phil. Soc.*, **20**, 100–114, 1945.

ZIMMERMANN, W. A.: Histologische Studien am Vegetationspunkt von *Hypericum uralum*, *Jahrb. Wiss. Bot.*, **68**, 289–344, 1928.

CHAPTER IV

TISSUES AND TISSUE SYSTEMS

Among many lower plants there is little differentiation in kinds of cells; only simple vegetative and reproductive cells are present. Groups or masses of these cells that are alike in origin, structure, and function form *tissues*. The plant body consists of "vegetative tissue" and "reproductive tissue." In the higher plants the body is complex in cellular structure, made up of many kinds of cells, of most varied form and function and of different origin. For these plants a definition of "tissue" that is less rigid than one that requires similarity of origin, structure, and function is necessary. Morphologically, a *tissue* is a continuous, organized mass of cells, alike in origin and in principal function. Within a tissue there may be great diversity of cellular form and function, but the cells that make up a tissue must be contiguous (not dispersed among other cells) and must form a structural part of the plant.

Classification of Tissues.—Tissues are classified on several bases—position in the plant body, kind of constituent cells, function, method or place of origin, stage of development. Each classification is made primarily or wholly on one of these bases. The classifications can be divided into those of descriptive anatomy and morphology and those of physiological anatomy. Continuity of cells in the tissue is essential from the morphological viewpoint, whereas it is not important from the physiological viewpoint where function alone binds cells together as a tissue. This major difference in basis can be made clear by examples. Of the morphological tissue wood, or *xylem*, only part—the actual conducting cells—forms the "wood" of physiological anatomy, the supporting cells belong to a different tissue, the storage cells to still another. Morphologically the epidermis forms a single tissue, but in physiological treatments the majority of the cells of the epidermis are grouped with the periderm to form "dermal tissues," and the guard and accessory cells of the stomata form a part of the "aerating tissues." Scattered and isolated cells and groups of cells form many of the tissues of physiological antomy.

TISSUE TYPES BASED ON STAGE OF DEVELOPMENT

Meristematic and Permanent Tissues.—Tissues are distinguished as "meristematic" and "permanent" on the same basis as are cells.

Meristematic tissues are immature tissues in which growth is taking place; permanent tissues are those in which growth has ceased, at least temporarily. Permanent tissues—as a whole or in part—may again become meristematic.

TISSUE TYPES BASED ON KIND OF CONSTITUENT CELLS

Simple and Complex Tissues.—From the standpoint of number of kinds of cells making up a tissue, tissues are *simple* or *complex:* they are simple when homogeneous, consisting of one kind of cell; complex when heterogeneous, consisting of more than one kind of cell. This classification aids in detailed descriptions of tissues. There are very few simple tissues in plants. The common ones are *parenchyma, collenchyma,* and *sclerenchyma*. These terms are applied also to cells; for example, a cell may be called "a parenchyma cell." Such a cell is either a unit of the simple tissue, parenchyma, or a cell with the characters of the cells of parenchyma but a member of a complex tissue. Parenchyma and sclerenchyma cells are common constituents of complex tissues; collenchyma cells do not occur with other kinds of cells. The adjectives "parenchymatous" and "sclerenchymatous" indicate cells which possess some of the characters of the cells of parenchyma and of sclerenchyma, respectively, but which do not belong definitely in those tissues. There are parenchymatous fibers, sclerenchymatous cork cells, etc.

The conducting tissues, *xylem* and *phloem*, are the prominent complex tissue types. They need separate discussion here because of their great complexity and variety and because of their structural and functional prominence. Other complex tissues can be interpreted as combinations of parenchyma and sclerenchyma or modifications of these tissues. Xylem and phloem are considered in this chapter as to general structure and function; they are treated further in other chapters.

Parenchyma.—Simple vegetative tissue, that is, tissue which is usually not complex or elaborate in structure or form, like that which makes up the body mass of lower plants and the nonspecialized portions in more complex plants, is called *parenchyma*. Parenchyma is a rather loosely used term in that it is applied to all generally unspecialized and fairly simple tissues which are concerned largely with the ordinary vegetative activities of a plant. Parenchyma is, obviously, phylogenetically the primitive tissue since the lower plants have undoubtedly given rise to the higher plants through specialization and since the single type or the few types of cells found in the lower plants have become by specialization the many and elaborate types of the higher plants. Further, all meristematic tissue is unspecialized. Hence, it is parenchyma-like—in fact, is often called parenchyma—and it can be said that, ontogenetically also, parenchyma is the primitive tissue.

The general characters of parenchyma cells are diameters essentially equal, walls thin, protoplast present, and a capability for cell division even when the cells are permanent cells (Figs. 43, 53). Exceptions occur in all these characters. Parenchyma makes up large parts of various

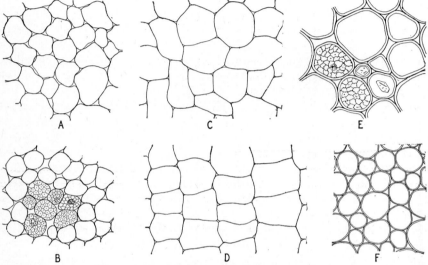

Fig. 43.—Parenchyma. *A*, from pith of rhizome of *Polypodium vulgare*; *B*, from cortex of root of *Asclepias incarnata*, the cells filled with starch grains; *C*, *D*, from pith of *Zea*, transverse and longitudinal respectively; *E*, thick-walled, lignified cells from pith of twig of *Castanea dentata*, the cells containing starch grains or crystals; *F*, thick-walled cells from pith of *Clematis virginiana*.

organs in many plants. Pith, the mesophyll of leaves, and the pulp of fruits consist chiefly of parenchyma; the cortex and pericycle are often wholly or in large part parenchyma, and parenchyma cells occur freely in xylem and phloem.

In early studies of plant structure all tissues were divided on the basis of general form and function into *parenchyma* and *prosenchyma*. The latter was distinguished from parenchyma chiefly by its elongate, pointed, thick-walled cells and its specialized functions of support, protection, and conduction. But entirely different types of tissue are present in "prosenchyma" and the term became obsolete. For convenience, cells and tissues are still frequently described as "prosenchymatous" in contrast with "parenchymatous."

Collenchyma.—In many stems, support in early stages is given in large part by a soft but strong tissue known as *collenchyma*. Important characteristics of this tissue are its early development and its adaptability to changes in the rapidly growing organ, especially those of increase in length. When it becomes functional, no other strongly supporting tissues have appeared. It consists of elongate cells, various in shape, with unevenly thickened walls, rectangular, oblique, or tapering ends, and persistent protoplasts. The cells overlap and interlock in varying degrees, forming strands similar to those of fibers. The walls consist of cellulose

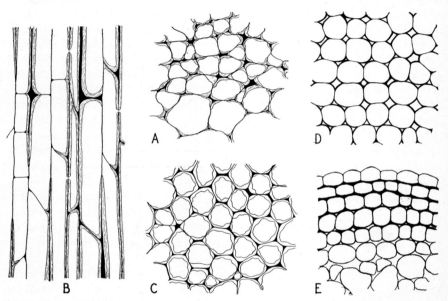

Fig. 44.—Collenchyma. *A, B*, transverse and longitudinal sections from stem of *Solanum tuberosum*; *C, D, E*, transverse sections from stem of *Abutilon Theophrasti*, stem of *Asclepias syriaca*, and petiole of *Asarum canadense*, respectively.

and pectin and have a high water content. (The presence of pectin probably accounts for the high water-absorbing capacity.) They are plastic, extensible, and readily adapted to rapid growth. The strands are at first of small diameter but are added to, as growth continues, from surrounding meristematic tissue. Cells on the borders of the strands may be transitional in structure, passing into the parenchyma type.

The areas of greater wall thickness are in the form of longitudinal strips (Fig. 44*B*) which occupy the corners of the cells (Fig. 44*A,C*), cover the tangential walls (Fig. 44*E*), or are confined to those parts of the walls that abut on intercellular spaces (Fig. 44*D*). Where spaces are surrounded by thickened wall strips, hollow rod-like structures are formed, and the collenchyma appears to consist of thick-walled cells among thin-walled cells (Fig. 44*D*). The three types of collenchyma are apparently related to cell arrangement: where the cells are irregularly arranged, the first type is formed; where the cells lie in tangential rows, the second; where intercellular spaces are present, the third. These have been called respectively, "angular," "lamellar," and "tubular" collenchyma. A distinction of these types is hardly necessary since all may occur in a single small strand and merge into one another (Fig. 154). The commonest condition, considered the typical, is that with thickenings at the corners.

The pits are simple, large or small, with apertures rounded or slit-like. Under special treatment, the wall can be seen to be many-layered, each layer extending completely around the cell but thicker where the wall is thickest. The layers are alternately rich in cellulose and rich in pectin. The thickenings appear early in the development of the cells while increase in cell size is still going on and provide an example of maintenance and strong increase in wall thickness while increase in area is continuing. The cytoplasm is prominent and may contain chloroplasts, although photosynthesis is not commonly a function of this tissue. The contents of some cells may differ from those of others in the same strand, as in *Rumex*, where some cells are tannin-bearing.

Ontogenetically collenchyma develops from elongate, procambium-like cells that appear very early in the differentiating meristem. Small intercellular spaces are present among these cells, but they disappear in angular and lamellar types as the cells enlarge, either closed by the enlarging cells or filled by intercellular substance.

Collenchyma is a tissue, first of all, of temporary support, and in some plants it serves only in this way, becoming crushed and even absorbed as secondary-tissue development crowds the outer primary tissues against the epidermis or the first periderm layer. This crushing obtains in many herbaceous plants which have considerable secondary thickening, such as *Melilotus, Chenopodium, Solidago*. The stems of typical woody plants rarely possess collenchyma, and roots are without it. Permanent collenchyma remains in a normal functional condition in mature plant parts. Petioles, such as those of *Solanum* and *Sambucus*, which are soft and not particularly woody, are supported largely by permanent collenchyma. Soft, herbaceous stems, such as those of *Impatiens* and *Pilea*, are also supported to a considerable extent in this way, and even the strongly woody stems of some herbs for example, species of *Aster, Nepeta, Pelargonium, Rumex*, and *Lactuca* possess small amounts of it.

Collenchyma occurs chiefly in the outer parts of stems, petioles, and leaf midribs. Its supporting value is increased by its peripheral position where it often constitutes the bulk of the ridges and angles of the organs. Some peduncles, for example, those of *Taraxacum*, and many

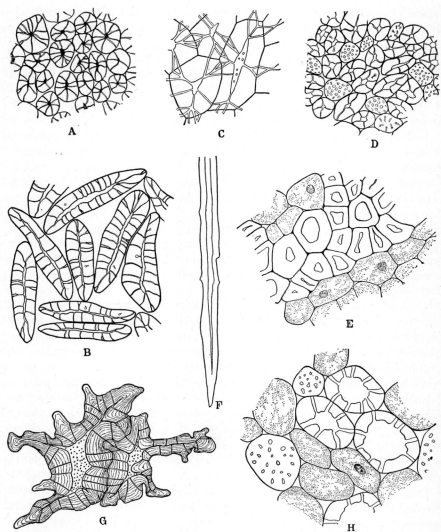

Fig. 45.—Sclerenchyma. *A, B*, transverse and longitudinal sections of elongate sclereids from endocarp of *Cocos nucifera* (coconut shell); *C*, sclereids from endocarp of *Crataegus* showing the pits fused in groups; *D*, sclereids ("grit cells") from the fleshy pericarp of *Pyrus communis*; *E*, fibers of the primary phloem region of *Cannabis* in transverse section; *F*, longitudinal section of part of one fiber shown in *E*; *G*, irregular sclereids from the phloem of *Tsuga*; *H*, sclereids and parenchyma cells from the stem cortex of *Dracaena fragrans*.

pedicels owe their support almost wholly to it. In the stems and leaves of herbs this tissue may form bundle caps and similar isolated strands, as in celery and beet.

Sclerenchyma.—Another type of supporting tissue, which is also in large measure protecting, is *sclerenchyma*. The cells of this tissue, in contrast with those of collenchyma have hard, usually lignified walls with a low percentage of water. At maturity they usually have no protoplasts. The walls are uniformly and strongly thickened. In shape and size, sclerenchyma cells are most diverse, but two general types are recognized—*fibers* and *sclereids*. The distinction of these forms of sclerenchyma is one of convenience in description; it has no morphological significance. Intermediate forms are many (Fig. 45B), and both types may occur in the same tissue, even intermingled, and serve the same general function.

Fibers.—Fibers are elongate sclerenchyma cells, usually with pointed ends (Figs. 47, 54). Chemically, the walls are usually lignified, although there are fibers with walls largely of cellulose and others with gelatinous walls. The pits of fibers are always small, round or slit-like in outline and, in mature cells, unless a protoplast is present, are doubtless functionless. Pits may be numerous but are commonly rather few in number, and in fiber types with excessively thick walls they may be few or present only as vestigial structures. The lumen of fibers is small; it often is a mere channel through the center of the fiber, and this opening may be blocked in spots (Fig. 54B) so that at certain levels in the fiber no lumen exists, only a line or spot representing, in cross section, its former position. In the development of fibers the protoplast often becomes multinucleate. In most kinds of fibers, however, the protoplast disappears as the cell reaches maturity, and the permanent cell is dead and empty. Fibers that retain their protoplasts and other types of fibers are discussed in more detail under the tissues in which they occur (*see* Xylem, Phloem, and Cortex).

Classification of Fibers.—If the usual and loose use of the term "fiber" —such as is covered by the above definition—is accepted, fibers may occur in nearly all parts of the plant. They are most generally found and are most abundant in the cortex, pericycle, phloem, and xylem. Morphologically, there are two distinct types. The fibers of the cortex, pericycle, and phloem possess simple pits and are thus different from those of the xylem which have bordered pits (although these pits may be so reduced as to be essentially simple) since xylem fibers are, morphologically, reduced tracheids. Fibers are sometimes divided into classes: "bast fibers" and wood fibers. These groups are essentially the same as the two just discussed, but the terms are poor since the word "bast" is involved and has, unfortunately, several uses. In its most common use, the term "bast" is synonymous with phloem or refers to fibers of the secondary phloem. It is more fully discussed on page 111. "Bast fibers" in the above classification include fibers of the cortex and pericycle as well as of the phloem. Fibers can best be designated by means of the tissue or region in which they occur, as cortical fibers, pericyclic fibers, phloem fibers, wood fibers, etc.

Fibers occur singly or in small groups scattered among other cells. Usually they form strands or sheets of tissue extending longitudinally for considerable distances. Their value as strengthening tissue is largely due to their arrangement in these long masses and to the overlapping and interlocking of the cells. They serve also to give general firmness to tissues.

Fibers develop in two ways. In those only a few millimeters long—for example, those of Manila hemp, *Agave, Sansevieria*—all parts of the cell are always at the same stage of development at the same time. As in most cells, the secondary wall is deposited simultaneously throughout the cell when full size is attained. In much longer fibers, for example, those of flax and hemp, the cell elongates apically, keeping pace with the growth of surrounding cells, and the secondary wall develops in part of the cell while the apex is still growing.

The term "fiber" is popularly applied to various plant structures that are not morphologically fibers: hairs (cotton); strands of cortical or phloem fibers (flax, hemp); foliar vascular bundles with their sclerenchyma sheaths or caps, or the caps alone (Manila hemp, New Zealand flax, sisal); strands of collenchyma (celery); wood cells generally (paper pulp); fragments of leaves and woody tissues (various drugs); bits of "seed" coats (wheat flour).

Sclereids.—In contrast with fibers, sclereids have diameters essentially alike. Some sclereids are elongate (Fig. 45*B*), and sclerenchyma types, especially in seed and fruit coats, may fall definitely in neither class. Sclereids vary greatly in type—shape, thickness of wall, form and number of pits, relation to surrounding cells—and various terms have been applied to the different types. Loosely, all cells of this type are sometimes called *sclerotic cells*, especially when they occur in soft tissues. The term "stone cells" is often used synonymously for sclereids but is better restricted in application to unbranched sclereids without uniform or extreme form. Resemblance in shape to stones often well describes such cells. Stone cells may be solitary, may lie loosely together, or, when angular, may be closely packed and even interlocked. Stone cells in fleshy fruits are sometimes called *grit cells*. The gritty parts of the flesh of pears and quinces are clusters of stone cells (Fig. 45*D*).

Stone cells have also been called *brachysclereids*. Sclereids more extreme in form have been distinguished as: *macrosclereids* (rod cells), more or less columnar in shape, constituting the "palisade layer" of many seeds and fruits and occurring in some xerophytic leaves and stem cortices; *osteosclereids* (bone cells) of bone-like or barrel shape, constituting hypodermal layers in many seeds and fruits and frequent in xerophytic leaves; *astrosclereids* (stellate cells), with extreme lobes or arms, in leaves and stems of xerophytic organs; *trichosclereids* ("internal hairs"), branched sclereids with lobes projecting, like hairs, into intercellular spaces, in leaves and stems of hydrophytes. The last two types are structurally closely alike. A classification of sclereids on shape and structure is of little value. The character of "branching" or "non-branching pit cavities," which has been used, in part, in distinguishing sclereid types, is largely dependent upon the thickness of the cell wall. Sclereids may occur almost anywhere in the plant body but are most abundant in the cortex, in the phloem, and in fruits and seeds. Like fibers, they occur singly, in groups of a few cells, or in large masses. The hard parts of seeds, nuts, and hard fruits generally are made up largely of stone cells of various types. Where stone cells are scattered, they merely give firmness, as in leaves and the flesh of fruits. In masses, they give hardness and mechanical protection, as in many kinds of bark and in the shells of nuts.

The wall of sclereids is very thick and strongly lignified. Occasionally, it is suberized or cutinized. The pits are very small, with round apertures, and their cavities have the form of more or less branching canals because of the fact that, as the area of the cell wall is reduced on the inside by the increased thickening, the pits are brought together (Fig. 45A,C). Two or even several pits thus fuse to form one structure which has only one aperture in each cell but has as many arms as there were original pits. Sclereids are usually dead cells. The shriveled remains of protoplasm and inclusions of the protoplast, such as tannin and mucilage, may be present.

THE IMPORTANT COMPLEX TISSUES

The structural and functional prominence of the vascular system renders its tissues of great importance as tissue types. These vascular tissues are, in fact, the only complex tissues which need separate discussion; all others may be interpreted as combinations of parenchyma and sclerenchyma and of modifications of these tissues. The following treatment of vascular tissues is of general nature only since a description of xylem and phloem as primary and secondary tissues and as specially modified in various organs will be found in other chapters. The development of xylem and phloem is discussed in part in Chap. VI.

Xylem

The Tracheid.—The fundamental cell type in xylem is the *tracheid*. The tracheid is an elongate cell with tapering ends (Fig. 46) which, when mature, is nonliving, that is, without a protoplast. The walls are hard but not thick and are usually lignified. In cross section the tracheid is typically angular, though more or less rounded forms occur. The tracheids of secondary xylem, owing to their method of arrangement, have fewer sides than those of primary xylem and are often more sharply angular. The ends of the tracheids do not taper uniformly to the tip in all planes, but the tapering is confined largely to the radial plane and usually to one side of the cell only. The end of a tracheid of the secondary wood is more or less chisel-like. The tapering is seen in tangential sections of the tracheid; radial sections do not show tapering, the end of the cell in such sections being rectangular or somewhat rounded. The pits are abundant and of the bordered type, though they vary in size, in outline, and in distribution over the walls. The lumen of a tracheid is large and free of contents of any kind. The tracheid is apparently well adapted structurally to its functions which are water conduction, primarily, and support, secondarily. It is a long, empty, firm-walled tube extending parallel with the long axis of the organ. It is in communication with contiguous tracheids—as well as with other types of cells, living and nonliving—by means of numerous, well-developed pits. These thin areas permit ready diffusion into adjacent cells. The arrangement of tracheids is always such that contiguous cells overlap at least over the tapering portions, and over these areas of the wall, the pits are often abundant. Channels for longitudinal conduction are thus provided through a series of lumina which form a more or less direct line or an anastomosing system.

The Continuity of the Walls of the Tracheid.—The individual lumina are shut off from one another by thick walls and, in the pits, by the closing membranes. The closing membranes, however, except in the central region, the torus, are very delicate and, at least in some of the conifers, are perforated. In *Larix* and *Sequoia*, for example, the perforations, though microscopically minute, are so numerous that the torus is suspended by a meshwork of strands (Fig. 25*A*). The openings in perforated pit membranes are not readily seen unless the membrane is well stained, but their presence is demonstrable by the passing of solid particles such as those of carbon in India ink under pressure from one tracheid to another through the pits.

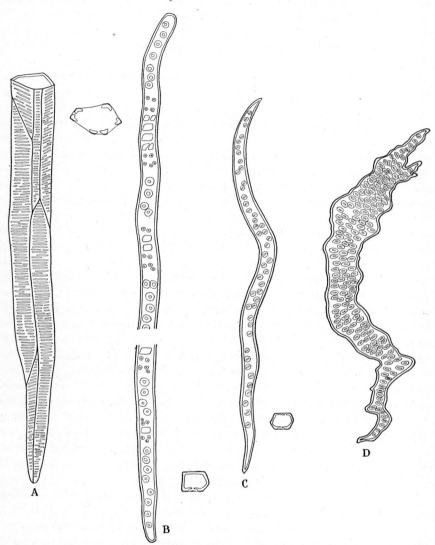

Fig. 46.—Tracheids. *A*, from *Woodwardia virginica*, one-sixth of cell shown; *B*, from *Pinus Strobus*, one-third of cell shown; *C*, *D*, from *Quercus alba*, *C*, a normal, and *D*, a flattened and distorted tracheid from the spring wood. (The illustrations of tracheids are drawn to scale and to the same scale as are those of fibers and vessels.)

Pits of Tracheid Walls.—The position of pits in the wall of the tracheid and the size and shape of pits depend upon the position and nature of the contiguous cells. The various larger plant groups have more or less constant types of bordered pits which have characteristic shape, torus, and extent of border. The ferns and the clubmosses have transversely elongated pits with narrow borders and little or no torus. The pits lie close together, covering the wall, giving it a ladder-like appearance (Fig. 46A), and the cells are called *scalariform tracheids*, or, better, *scalariform-pitted tracheids*. The former term is undesirable in this connection since it is also used with other meanings and where definite pits do not occur (protoxylem). In other gymnosperms and most angiosperms the bordered pits are chiefly rounded with wide borders (Fig. 46B, C); those of the angiosperms are mostly much the smaller. The best development of the torus is in the gymnosperms where the bordered pit probably reaches its highest development (Figs. 24, 25). The closing membrane of these pits is so constructed that its position in the pit cavity can be readily changed from a median position to a lateral one with the torus closely appressed to the aperture of the pit (Figs. 22C, 25B, C). A valve-like action is thus secured, the pit being freely open when the torus is in the median position, diffusion—or in perforated closing membranes, direct passage—taking place around the torus through the peripheral part of the closing membrane. When one of the pit mouths is closed by the placing of the torus against it, direct passage is largely or wholly shut off, and diffusion must take place through the thicker and denser torus. Thus, apparently by changes in the position of the closing membranes of bordered pits, there may be some control over the passage of fluids in xylem. It is significant that the type of pit in which this control of the passage of fluids is present is characteristic of water-conducting cells and does not occur elsewhere; also that in the structurally reduced pits of fiber-tracheids and fibers the pit membranes have lost the capability for movement. The bordered pit is further discussed in Chaps. II and VII. In simple pits no structural complexity occurs.

Function of the Tracheid.—The tracheid is structurally adapted, both in lumen and in wall, to the function of conduction. The thick and firm walls of tracheids also aid in support, and, where there are no fibers or other supporting cells, the tracheids play a prominent part in the support of an organ. For this the overlapping and interlocking of the cells and their union into strands and cylinders are as important as are thick walls.

Tracheids alone perhaps made up the xylem of very ancient plants, but in living forms wood is a complex tissue, with parenchymatous wood rays even in the simplest forms of secondary xylem; and primary xylem (as a morphological tissue unit) always contains parenchyma cells. The more complex types of xylem may contain several kinds of cells—tracheids, *fibers* of one or more kinds, *vessels* of one or two types, parenchyma cells (known as *xylem parenchyma* or *wood parenchyma*), and *wood-ray parenchyma*. Further discussion of the constitution of xylem will be found in Chap. VII.

The tracheid clearly serves both as a conducting and as a supporting cell. But evolutionary development in xylem has resulted in a specialization of this once simple tissue in such a way that the original functions of the tissue have become segregated in distinct cell types—support to fibers of various kinds and conduction (of an apparently more efficient type) to vessels. The new function of food storage has been acquired, and this is carried on by wood parenchyma. The wood-ray parenchyma is also concerned with food storage; the ray is, however, probably primarily concerned with lateral conduction. In complex xylem in which the definitely dissociated functions of support and conduction are given over to fibers and vessels respectively, tracheids are usually not found. But in the wood of *Quercus* and some other genera all three kinds of cells are present.

Fibers and Fiber-tracheids.—In the phylogenetic development of the fiber, the thickness of the wall has been increased and the diameter of the lumen correspondingly decreased; the length of the cell, especially of the tapering end, in most types decreased, and the number and size of the pits reduced. (Reduction of the pit in size is discussed in detail in Chap. II.) Where reduction has reached a stage where the lumen is so narrow—often nearly occluded—and the pits so small that it seems there can be but little or no conduction, typical fibers are formed. Between such cells and normal tracheids all intergrading forms occur. These intermediate types, which cannot be called either typical tracheids or typical fibers, are designated *fiber-tracheids*. Tracheids have pits in size and type like those in vessels of the same tissue; in fiber-tracheids, the pits are smaller than those in vessels, with reduced or vestigial borders and with the inner aperture narrow and extending beyond the limits of the chamber. A line between tracheids and fiber-tracheids and between fiber-tracheids and fibers cannot, of course, be drawn. Extreme specialization results in the formation of a type of fiber with very thick walls and with pits so reduced as to be essentially simple (Fig. 47*C,F,G*). Such fibers are called *libriform wood fibers* because of their similarity to phloem fibers (phloem or phloem fibers having been known to early students as "liber"). Libriform wood fibers occur abundantly in woody dicotyledons, chiefly in the more specialized families, such as the Leguminosae. Gelatinous layers occur in the walls of the fiber-tracheids and fibers of many genera in

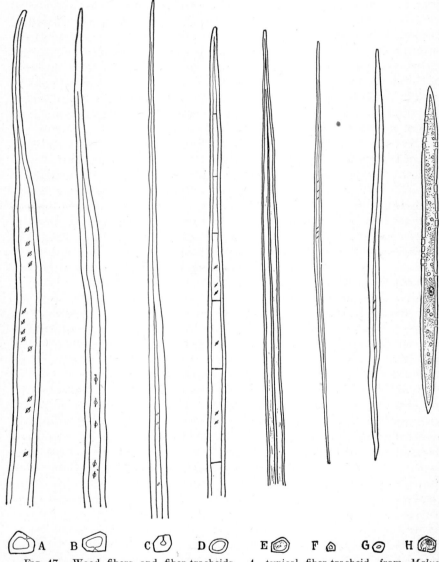

Fig. 47.—Wood fibers and fiber-tracheids. *A*, typical fiber-tracheid, from *Malus pumila*, two-thirds shown; *B*, typical fiber-tracheid, from *Liriodendron Tulipifera*, two-thirds shown; *C*, libriform fiber, from *Quercus alba*, one-fourth shown; *D*, septate fiber-tracheid, from *Swietenia Mahagani*, one-half shown; *E*, gelatinous fiber, from *Quercus rubra*, about one-half shown; *F*, libriform fiber, from *Carya ovata;* *G*, libriform fiber, from *Guaiacum sanctum* (lignum vitae); *H*, "fusiform wood parenchyma cell" (sometimes called "substitute fiber"), from first annual ring of *Sassafras variifolium* (like a fiber only in shape and in direct origin from cambium derivative; without the divisions usual in wood parenchyma).

different families and occasionally in tracheids. Cells with these layers are called *gelatinous tracheids*, *fiber-tracheids*, and *fibers*. The so-called "*substitute wood-fiber*" (Fig. 47H) is not a fiber but a parenchyma cell with fiber-like form. It is discussed under wood parenchyma.

In some fiber-tracheids the protoplast persists after the secondary cell wall is mature and may divide so that two or more protoplasts, separated by thin, transverse partitions are enclosed within the original wall. Such fiber-tracheids are known as *septate fiber-tracheids* (Fig. 47D). They are, in fact, not individual cells but rows of cells; the transverse plates are true walls, and each chamber has a protoplast with nucleus. In a mass of fiber-tracheids in which septation is occurring, all or only part of the cells may divide. The protoplasts of these cells, divided or undivided, may persist for some time. The "septa" are true walls, formed after cell division, but they lack secondary layers. A septate fiber can readily be distinguished from a row of wood-parenchyma cells by the absence of a secondary wall on the septa and the failure of the septum to extend, at the line of union with the mother-cell wall, beyond the wall surface of the mother cell. The septa, further, remain unlignified. Septate fiber-tracheids occur in many plants, especially in shrubs and the more "woody" herbs, in vines, and in tropical trees.

Vessels.—In the evolutionary development of the tracheid the diameter of the cell has increased and the wall has become perforated by large openings. These specializations permit direct transmission of water from cell to cell. In the more primitive types of vessel elements, the general form of the tracheid is retained and increase in diameter is not great; in the most advanced types, increase in diameter is enormous and the cell becomes drum-shaped (Fig. 48L). The tracheid is considerably longer than the cambium cell from which it is derived; the primitive vessel only slightly longer; the most advanced type retains the length of the cambium cell or is slightly shorter, with a diameter greater than its length. In the series from least to greatest specialization the ends of the cells change in shape; the angle made by the tapering end wall becomes greater and greater until the end wall is at right angles to the side walls (Fig. 48L). Some of the intermediate forms have tail-like tips beyond the end wall where, under elongation, a part of the end of the cell has penetrated between adjacent cells (Fig. 48A,G,I). Wall thickness remains about as in the tracheid or is less, though vessels with very thick walls are frequently found, as in *Carya*, *Fraxinus*, and *Diospyros*. The pits are often more numerous and smaller than are those of tracheids and may cover the wall closely. When abundant, they may be scattered or arranged according to some definite pattern. The distribution of pits is, of course, controlled primarily by the nature and position of the contigu-

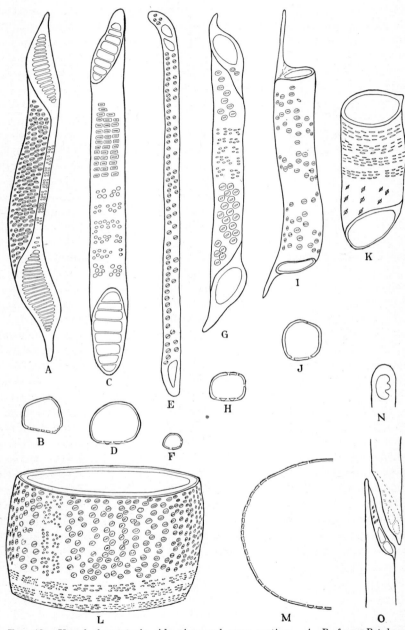

FIG. 48.—Vessel elements in side view and cross section. *A, B*, from *Betula alba*; *C, D*, from *Liriodendron*; *E, F*, from *Lobelia cardinalis*; *G, H*, from *Quercus alba*; *I, J*, from *Malus pumila*; *K*, from *Acer Negundo*; *L, M*, from *Quercus alba*; *N*, end of vessel element from *Lobelia* showing perforation indicating derivation of porous vessel from scalariform; *O*, ends of vessel elements from *Lobelia*, showing method of union of elements in a series.

ous cells. Thus, if another vessel is contiguous to the vessel in question, the wall is heavily pitted over that part of its surface in contact with the other vessel, whereas only a few small pits or none exist in the area lying against a fiber. The pit-pairs between a vessel and tracheids, wood-parenchyma cells, and wood-ray cells are similarly controlled in position, number, and type by the pitting normal to these cells.

The Term "Vessel."—Since the perforations of the cell occur usually in the end walls, the development of end walls transverse to the long axis of the cell brings a series of cells into a definite tube-like system, which provides for transmission in a more or less nearly straight line. Such a condition is in contrast with the indirect lines of conduction in a group of tracheids. A tube-like series of cells thus formed has long been known as a *vessel*, or *trachea*.

Both these terms have unfortunately been applied in two ways: to the series or system of cells, and to individual cells with perforate walls that serve in direct water conduction. The former has the support of priority and long usage and should be continued. The unit cells making up such a series are called *vessel elements*, *vessel members*, or *vessel units*. (The term "vessel segment," which has been in frequent use is inappropriate and should be discarded because it implies that the series is a unit which has been divided to form the cells.) Some difficulties arise in the application of these terms under all conditions. In plants in which the water-conducting cells are tracheid-like in shape, union of the cells forms a mesh-system, and vessels of the tube type are not present. The terms "vessel element" and "vessel member" can be applied here to the individual cells. Vessels are present in the sense of continuous series of fused, perforate cells although individual tubes cannot be distinguished.

Types of Vessel Perforations.—The openings in vessel-element walls, known as *perforations*, are restricted to the end walls except in certain slender, tapering types where definite end walls cannot be distinguished, and they are said to be present on the side walls (Fig. 48A,E). Most vessel elements are perforated in two areas, one at each end, but three and even four such areas occur in some vessel elements. Where there are three or four, connections are made with other vessels or with branches of a meshwork system of vessel elements. The more or less definitely delimited or outlined area of the wall in which perforation occurs is the *perforation plate*. Commonly this is an end wall. The portion of the plate remaining after perforation has occurred is the *perforation rim;* the strips of cell wall between scalariform perforations are the *perforation bars*.

Perforation plates are described as having a *simple perforation* if there is but one opening; *multiple perforations* if there are two or more openings. Multiple perforations are grouped as *scalariform*, where the openings are more or less elongate and parallel; and *reticulate*, where the arrangement of the openings makes a mesh-like structure. The common types are the simple and scalariform. The reticulate type is infrequent and not clearly separable from the scalariform. Small vessel elements occasionally have one type of perforation at one end, another at the other. Simple perforations are usually round but range in form in slender elements to narrowly elliptical. In annular and spiral vessels the perforation is of either type, but most frequently is simple. Commonly, end walls that are transverse have simple perforations, and those that are oblique have scalariform openings, but there are many exceptions. Phylogenetically the scalariform type is primitive and forms transitional to the simple are frequent.

Vessels are characteristic of the angiosperms; in only a few are they lacking—the Winteraceae, Trochodendraceae, and Tetracentraceae (which are primitively vesselless families) and in some xerophytic, parasitic, and aquatic genera where they have been lost in reduction. In many monocotyledons they are absent from the stems and leaves; in others from either the stems or the leaves. They are present in some species of *Selaginella;* among the ferns, in two species of *Pteridium;* among the gymnosperms, in the Gnetales. In each of these groups the vessel has clearly arisen independently, in the angiosperms probably more than once.

Vessel Length and Width.—Vessels extend for distances that vary with the kind of plant, the type of xylem, the type of vessel element, the location in the organ, and apparently with the rate of growth of the organ. The limits of an individual vessel are difficult to determine. In climbing plants and in trees with ring-porous wood and simple-perforate elements they may be several, perhaps many meters long. But they are often less than one meter and frequently only a few centimeters long. In a single tree they are apparently progressively longer from near the pith outward. It is highly doubtful that they extend "from root tip to stem tip" in any plant. Where branching and interlocking of elements occur, no question of length arises. Even the usually direct-series type may branch, and determination of unit length becomes uncertain. The terminal elements have but one perforation and taper at the blind end.

Although vessels characteristically are wide, the more slender forms may be even narrower than are typical tracheids. From this diameter they range to a maximum of somewhat over one millimeter. The wider vessels are characteristic of certain herbs, such as *Zea*, many woody vines and lianas, and of some trees, such as, *Castanea, Quercus*, and *Fraxinus*.

Ontogeny of the Vessel.—Vessels are formed from series of xylem mother cells—procambium cells or cambium derivatives—by the fusion of the cells end to end during the last stages of development. This fusion involves the loss of the end walls, or parts of those walls, so that the lumina of the series of cells are freely open into one another, and, with the walls, form a long tube (Figs. 49, 50).

From the meristematic stage the vessel elements enlarge rapidly, increasing greatly in diameter. Those with scalariform perforations and

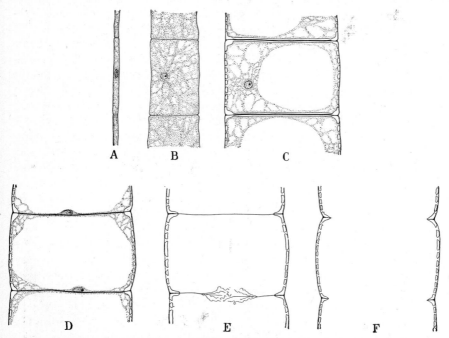

FIG. 49.—The ontogeny of the vessel element in *Robinia Pseudo-Acacia*. *A*, the cambium initial; *B*, the cell much enlarged; *C*, the cell still further enlarged, the secondary wall well developed, except on the perforation areas where the primary is thickened, and the pits present; *D*, the cytoplasm restricted to the periphery, the nucleus adjacent to the wall where dissolution is occurring, the secondary wall removed from the pore areas; *E*, the cytoplasm lost, the very thin end walls disintegrating; *F*, the mature, perforated, empty cell.

the more elongate, simply perforate types may increase in length somewhat, the tips forming 'tails' which penetrate between surrounding cells. Those that become of great diameter, and especially those developing from stratified cambium initials, do not elongate and may even become somewhat shorter.

During the rapid growth in cell size, the primary cell wall, although greatly and rapidly increased in extent, remains constant in thickness except in the areas which later disintegrate in the formation of the perforations. These areas become thicker and marginally limited (Figs. 49*C*, 50*B*). In section they are lens-shaped (in many herbs) or plate-like

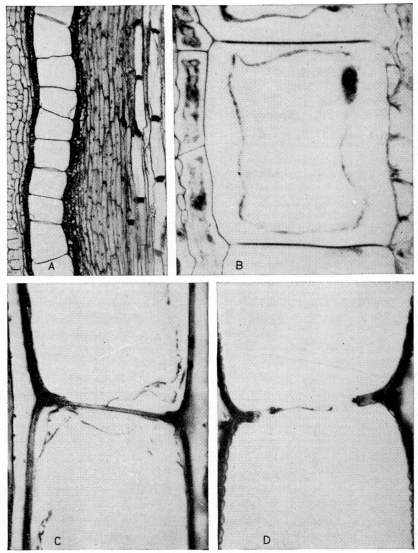

Fig. 50.—Ontogeny of vessel elements. *A, B,* in *Cucurbita; C, D,* in *Zea. A,* longitudinal section of part of vascular bundle, showing on the left, a series of vessel elements approaching maturity with intact end walls; on the right, phloem with conspicuous sieve tubes. *B,* longitudinal section of a vessel element at about full size, with protoplast, and perforation plate (part of end walls that will be dissolved) thickened. (The intercellular spaces between parenchyma cells at the extreme left have been formed by the tearing apart of these cells during expansion of the vessel element.) *C, D,* last stages in the maturation of the vessel element: the lateral walls and the perforation rim with secondary wall, the primary wall of the perforation plate disintegrating, only remnants of the protoplasts remaining. (*After Esau and Hewitt.*)

(in many woody plants) and can often be seen to be three-layered, consisting of the primary walls of the two adjacent cells and the middle lamella. The primary walls are chiefly of cellulose, with doubtless some pectic substances; the middle lamella is largely pectic. Multinucleate stages have been reported in the development of primary vessel elements in the Euphorbiaceae, but the uninucleate condition is apparently characteristic of all types of vessel elements.

Throughout the enlargement of the cell the cytoplasm remains abundant and active. As maturity is reached, it begins a slow disintegration. In some woody plants the nucleus becomes small and greatly flattened and lies in scant cytoplasm against the wall where perforation is about to occur (Fig. 49D); in other plants it maintains a more or less nearly central position in the cell.

After the primary wall is mature, and in some woody plants the secondary wall is partly, perhaps even fully formed, the perforation of the end wall and loss of the protoplast begin. The wall in the perforation area becomes thinner, and, as the cytoplasm gradually goes to pieces, it also disintegrates—in some plants simultaneously throughout, in others beginning in the center. The disappearance of this piece of wall is in some plants at first by a thinning, suggesting a dissolving, followed by a breaking up into several delicate layers (Fig. 49E). Statements that these walls are ruptured early in vessel development under tension as the cells rapidly increase in diameter and that the broken walls retract to form the rim are based on inaccurate observation.

Maturation does not proceed simultaneously in all members of a vessel series but progresses from one end to the other. The statement that a vessel matures simultaneously from the base of a tree trunk to its tip is not borne out by careful study. The cambium, which forms the rows of vessel-element initials, itself progresses in activity from one region to another in various directions.

The progress of perforation of vessel elements seems to be closely similar in the protoxylem and metaxylem of herbs and the secondary xylem of trees. It apparently has not been studied in detail elsewhere.

The perforation rim closely resembles the rest of the cell wall. It varies in width with vessel type from a broad band to a ridge so narrow as to be hardly evident. It always has the same characters as the side walls—the same thickness and lignification and the same sculpturing: pit-pairs, bars, or spirals. In annular cells, a terminal ring of secondary wall at the somewhat constricted end of the vessel element forms an inconspicuous rim. In slender, loose-spiral elements, a terminal turn of the spiral may form the rim; in cells with more tightly coiled spirals, the rim has usually two or more turns of the spiral which are interrupted by the perforation. The obscure cell ends and rim in annular vessel elements—which are usually very long because of the excessive stretching to which these first-formed xylem cells are subjected—are easily overlooked, and a series of cells may be interpreted as one.

That the perforations of vessel elements represent enlarged and fused bordered pits from which the closing membranes have disappeared is clear. All transitional stages are frequently found; occasionally, a single end wall shows typical, normal, bordered pits, similar pits without membranes, and semifused groups of two or more pit-shaped openings. This fusion is, of course, not ontogenetic but phylogenetic.

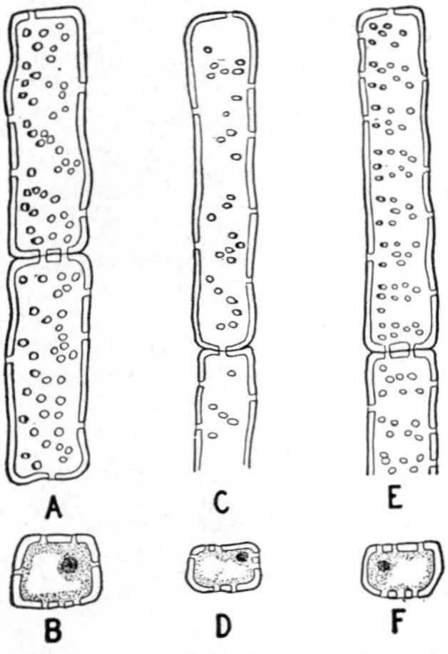

In simply perforated vessel elements the proportion of the end wall occupied by the perforation is greatest where the wall is transverse; under this condition the wall is reduced to a narrow perforation rim and in extreme cases may be almost or quite lacking. The limit of the vessel members is always made clear, however, by the division of the lateral wall and frequently also by the median bulging of the walls of the individual cells. A given species or genus may possess one type of vessel exclusively, or both types may be present in the same tissue. Where both types occur, the vessels with the larger elements are usually simply perforate and those with the smaller ones scalariform.

The vessels of the secondary wood in many plants possess spiral thickenings of the secondary wall (Fig. 17). The functional significance of this structure is unknown.

Fig. 51.—Wood parenchyma cells in longitudinal and transverse section (the protoplasts omitted in the longitudinal). *A*, *B*, from *Quercus alba*; *C*, *D*, from *Malus pumila*; *E*, *F*, from *Carya ovata*. (The scale of this figure is twice that of the figures of tracheids, fibers, and vessels.)

Wood Parenchyma.—Parenchyma cells are a common constituent of the xylem of most plants. In secondary xylem they occur as vertical series of more or less elongated cells placed end to end (Fig. 51), known as *wood* or *xylem parenchyma*, and radial transverse series which form part or all of the wood rays and are known as *wood-* or *xylem-ray parenchyma*. Wood parenchyma varies in amount in secondary wood from none, as in the wood of certain conifers, such as *Pinus*, *Taxus*, and *Araucaria*, to a considerable proportion, as in many dicotyledons, and may even constitute nearly all the vertically arranged tissue, as in some herbs and spongy-wooded trees. The primary wood of all except some highly specialized plants has always a considerable amount, especially in the first-formed parts. The parenchyma of this tissue consists wholly of vertically arranged cells. Parenchyma cells of xylem, unlike tracheids, vessel elements, and most kinds of fibers, remain alive as long as the tissue in which they lie is functioning in conduction. Xylem parenchyma cells may be thin-walled or thick-walled; those of secondary wood often have thick, more or less strongly lignified walls. In function, wood parenchyma serves for food storage and is probably also associated with conduction, either directly or indirectly. Those parenchyma cells which constitute the xylem rays are considered in the discussion of secondary xylem.

Function of Xylem.—That the xylem as a whole is the water-conducting tissue and the chief supporting structure of vascular plants is without question. Secondary functions in wood parenchyma and xylem-ray parenchyma are apparent, such as the storage of food (starch, oils, and other substances) and of materials of such nature as crystals, gums, and resins. Water conduction undoubtedly takes place in the lumina of tracheids and vessels and possibly also in the walls, but an explanation of the rise of sap through these cells has not yet been found. Whether physical processes directly and alone are concerned, or whether the activities of living cells are also involved, has long been in dispute. Histological evidence strongly suggests that the presence of living cells is related to the upward conduction of water by tracheids and vessels. Every water-conducting cell is in contact in some part of its wall surface with one or more living cells, and abundant pits are present in this contact area. In highly specialized wood, where the tissue consists largely of fiber-tracheids and fibers, and where the water-conducting cells are relatively few, each vessel is sheathed with parenchyma cells, and no parenchyma cells occur among the nonconducting fibers. In xylem that lacks wood parenchyma all tracheids are heavily pitted with ray parenchyma.

Phloem

In xylem and phloem, evolutionary specialization has progressed along somewhat similar lines in that increased efficiency in conducting structures has resulted from the arrangement of cells in longitudinal series with closer functional intercellular relationship. In xylem a series of tracheids, structurally and functionally united, has become a vessel; in phloem a series of cells similarly united, forms a *sieve tube*.

In xylem the fundamental structural and functional cell type is the tracheid; similarly, in phloem the basic cell type is the *sieve element*.[1] Of

---

[1] "Sieve element" is used as a term of convenience to cover both sieve cell and sieve-tube element.

this fundamental phloem cell there are two forms: a simple, more primitive one, the *sieve cell* of gymnosperms and lower forms where series of united cells do not exist; a specialized one, the unit of a series, the *sieve-tube element* or *sieve-tube unit*. (The term "sieve-tube segment," which is occasionally used, should be discarded because it implies that this cell is formed by segmentation.) Unfortunately, in common usage the term "sieve tube" is applied not only to the series of united cells and to the individual members of this series but to cells that are not united in series. The term "sieve tube" was first applied to the series and is descriptive of this structure. It should be retained in this usage and so restricted as far as possible. For convenience, the loose, double use will probably continue; the context will in most cases indicate the exact meaning, but certainly the terms "sieve-tube element" and "sieve-tube unit" for the members of a sieve tube are preferable and not ambiguous. In the study of sieve elements the sieve-tube element has received greater attention than the sieve cell, and most of the descriptions of the structure of the sieve element are based on the study of the sieve-tube element.

Phloem, like xylem, is a complex tissue. It may be made up (*a*) of sieve cells and phloem parenchyma only, as in the pteridophytes and many gymnosperms, (*b*) of sieve cells, parenchyma, and phloem fibers, as in some gymnosperms, or (*c*) of two or more of the following cell types: sieve tubes, companion cells, phloem parenchyma of one or more kinds, phloem fibers, sclereids, and various kinds of secretory cells, as in angiosperms.

The Sieve Cell and Sieve-tube Element.—Sieve cells and sieve-tube elements are morphologically equivalent and are alike in fundamental structure and in function. They differ in that the perforations of the walls of the sieve cell and their cytoplasmic strands are all alike whereas those of the sieve-tube element are of two degrees of specialization, and in that sieve cells are not arranged in series as are sieve-tube elements. These sieve cells and sieve-tube elements are elongate living cells with a thin cellulose wall. The protoplast has a large central vacuole and a thin peripheral layer of cytoplasm. No nucleus is present when the cell is mature. The cytoplasm contains leucoplasts which in some plants accumulate starch or similar substances. In some dicotyledons the vacuole contains slimy materials (of proteinaceous nature) which may be distributed throughout the cell sap or accumulate in masses in various places. The *slime plugs* seen in sections in some plants (Figs. 109*C*, 110*C*) are doubtless artifacts, aggregations of slime caused by injury to the tissues. The wall apparently is composed of primary layers only. In some genera, for example, *Liriodendron* (Fig. 108*A*), it may become thick at some stages, perhaps serving then as a place of food storage.

Sieve Areas and Sieve Plates.—Sieve cells and sieve-tube elements are unique as living, functioning cells in the absence of a nucleus and the presence in the walls of fine pores through which extend strands of cytoplasm that resemble plasmodesmata but are commonly of much greater diameter and, unlike plasmodesmata, are sheathed with callus. The connecting strands of cytoplasm of sieve elements are clustered in areas of the wall known as *sieve areas*. These areas, which are unspecialized in gymnosperms and pteridophytes, become highly specialized in angiosperms by enlargement of the strands and by sharper limitation. Such elaborated areas occupy more or less definitely restricted parts of the cell wall—usually of the end walls—which are known as *sieve plates*. Two types of sieve plates are distinguished—*simple*, with one sieve area; *compound*, with several sieve areas. In simple sieve plates the pores and strands tend to be very large, and the plate occupies all or nearly all of the end wall, which is commonly transverse. In compound sieve plates the pores tend to be smaller than those of simple sieve plates, and the plates occupy only part of the usually oblique end wall.

The sieve areas are scattered over the side and end walls or are in part or wholly restricted to certain walls, as to the radial and end walls in sieve tubes with long-tapering ends and to the end walls where the cell tapers more abruptly or has a transverse end wall (Fig. 52). The number of sieve areas on the side walls varies greatly: there are usually few or none where the end wall is transverse or nearly so (Fig. 52P, S); where the end wall is long-tapering, the side wall—as well as the end wall from which it is hardly to be distinguished—may be completely covered by closely set sieve areas (Fig. 52A,E). The number and position of sieve areas is to a large extent controlled by the position and arrangement of the surrounding sieve tubes. Sieve-tube elements may be somewhat lobed or forked and may even have three definite ends. Such cells stand at points of branching of a sieve tube. In phloem with tubes of the more primitive type such branching may be frequent, the tubes forming a loose mesh system rather than linear unbranched or rarely branched series, as in the highest type. Where well-marked sieve areas are lacking on the side walls, vestigial sieve plates, known as *lattices*, are often present (Fig. 52H, S). These resemble typical sieve areas and plates of various types but are indefinite in outline and often ghost-like; the perforations are exceedingly minute, being usually no larger than normal plasmodesmata (Fig. 52K) and often apparently lacking. Differences in the size of the pores in sieve plates and lattices of woody plants is shown by the following measurements: in *Juglans nigra* size of pores in the sieve plate is 1.8 to 3.5 micra, that in the lattices, 0.5 to 0.6 micra; in *Populus deltoides*, the sizes are, respectively, 3.5 to 5.5 micra and 0.5 to 0.6 micra; in *Salix*

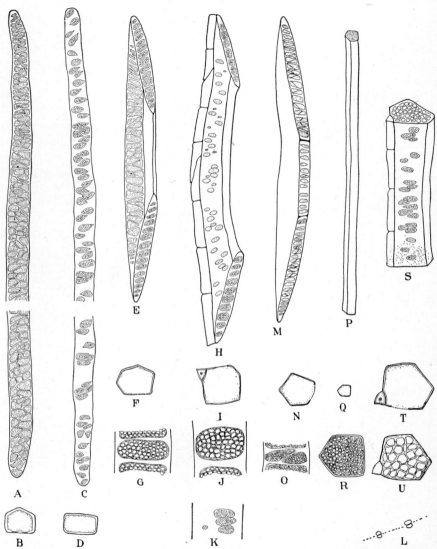

Fig. 52.—Sieve cells and sieve-tube elements in side view and cross section, with detail structure of sieve plates. *A,B,* from *Pteridium,* only one-fourth of cell shown; *C,D,* from *Tsuga canadensis,* only one-third of cell shown; *E,F,G,* from *Juglans nigra, G,* part of sieve plate in detail; *H,I,J,K,L,* from *Liriodendron*—*H,I,* with companion cells attached, *J,K,* detail of sieve plate and of lattice respectively, *L,* sieve plate in section; *M,N,O,* from *Malus pumila, O,* detail of sieve plate; *P,Q,R,* from *Solanum tuberosum, R,* detail of sieve plate; *S,T,U,* from *Robinia Pseudo-Acacia,* with companion cells attached, *U,* detail of sieve plate. (The sieve tubes are drawn on the same scale as tracheids, fibers, and vessels in Figs. 46, 47, and 48. All enlargements of sieve plates and lattices are on the same scale.)

nigra, 2.0 to 3.0 micra and 0.4 to 0.5 micra. Stages intermediate between typical sieve areas and lattices may be found. Lattices are common in sieve tubes of intermediate type; they are usually absent in most herbaceous plants. The term "*sieve field*," sometimes applied to lattices, has become confused in use and is probably unnecessary.

Sieve Tubes.—Union of the sieve-tube elements to form a *sieve tube* is secured structurally by modification in shape and arrangement so as to form a linear tube and functionally by the greater specialization of the connecting strands in the end walls of the elements. In evolutionary development the sieve-tube elements have become increased in diameter and shortened. They have a range in shape similar to that of the tracheid-vessel series: in the most primitive type the ends are long-tapering, and an end wall as distinct from a side wall is hardly apparent (Fig. 52A,C); commonly the well-defined end wall is oblique (Fig. 52E,H,M) or, in the most specialized forms, transverse (Fig. 52P,S). Also in evolutionary development of sieve tubes their sieve areas decrease in number. The primitive type of sieve tube has many sieve areas on a plate occupying a long, oblique end wall, and these areas closely resemble the many areas of the side walls; the most advanced type has one area in a plate occupying nearly all of a transverse end wall, and areas on the side walls are scarce or absent. The diameter of the connecting cytoplasmic strands in this highest type is much greater than that of other types. The sieve-tube type is not constant in families or sometimes even in genera. The primitive type is found in some families considered advanced, such as the Caprifoliaceae, and the most advanced type occurs in such primitive families as the Moraceae and Ulmaceae. Evolutionary advance in sieve-tube type has apparently taken place within families (Fagaceae, Rosaceae, Leguminosae) and within genera (*Fraxinus, Prunus*). Although the sieve tubes of vines and herbs are usually of the highest type, all types occur in woody and in herbaceous forms.

Ontogeny of the Sieve Elements.—The mother cells of sieve elements vary in shape from short cylindric to elongate, slender, and tapering. Primary pit-fields are numerous on the walls, especially in secondary phloem, as on young tracheid walls. As these cells differentiate, they elongate; the cytoplasm becomes highly vacuolate and streams actively; the wall thickens; sieve areas develop from the pit-fields, and the cytoplasmic strands become prominent and increase in size; callus develops around the connecting strands. The relation of pit-fields to sieve areas in ontogeny is not altogether clear. Apparently one pit-field may form one or more sieve areas, or several pit-fields form one area where the sieve plate is simple. In plates of this simple type when the connecting strands are very large, one or more pit-fields perhaps take part in the formation of the pore, each strand representing one greatly enlarged plasmodesmata thread or the fused threads of one pit-field. In the development of callus, rings are first formed about the strands at the apertures of the pores; increasing deposition forms cylinders that sheath the strands.

As the sieve-tube element reaches maturity in size, the wall becomes thinner, the nucleus disintegrates, the connecting strands continue to increase in diameter, the streaming of the cytoplasm ceases, the peripheral layer of cytoplasm becomes extremely thin, the boundary between the cytoplasm and the vacuole becoms indistinct, and the semipermeable properties of the cell appear to be lost. At this stage the functioning period of the cell as a conducting structure apparently begins. In all plants conduction by sieve tubes probably lasts only a short time—from a few days in the early-formed primary phloem to a year or perhaps more in the secondary phloem of woody plants.

During the functioning life of the sieve tube, the callus increases in amount, lengthening the cylinders about the strands. Callus is deposited also on the wall around and between the strands forming with the cylinder a cushion-like mass over the sieve area. In the late stages of thickening in this cushion, the strands become attenuate, and some or all of them may be broken off. This condition seems to accompany the death of the protoplast. The callus cushion is for this reason known as the *definitive callus*. In some woody plants, where sieve tubes seem to function during a second growing season—perhaps even longer—the protoplast does not die, and the definitive callus is dissolved as activity in conduction is renewed. Under these conditions the sieve strands must remain unbroken. Little is known in detail concerning the structure of sieve tubes that appear to function through more than one growing season. Callus cushions form only weakly or not all over the sieve areas of lattices. This, with the presence of few weak connecting strands, or none, is evidence of the vestigial nature of these structures. Lattices form a transition stage in lateral wall structure between the numerous sieve areas of the primitive sieve-tube element and the absence or great reduction in number of the highest type.

Obliteration of Sieve Tubes.—After the loss of the protoplast of the sieve-tube element, the empty cell wall, at this stage as thin as in early stages of development, is crushed or collapses under the pressure and tension of the surrounding tissues which are brought about by increase in diameter or length of the organ of which they are a part. In many herbaceous plants the callus cushions and cylinders are still present when crushing takes place; in secondary phloem of woody plants they have disappeared. The crushing of the dead sieve tubes and their companion cells commonly becomes so complete that these cells are soon represented only by strands or sheets of structureless material, and this material may be soon absorbed. In most plants, monocotyledons probably excepted, the surrounding living cells crowd into the space formerly occupied by the crushed cells, and evidence of the earlier structure of this tissue is difficult to find. The crushing and absorption of sieve and companion cells is known as *obliteration*. Obliteration is further discussed in Chap. VIII.

Sieve Cells of Gymnosperms.—Sieve cells have not been studied so intensively as sieve-tube elements, and the structure of their sieve areas is more difficult to ascertain. The connecting strands are said to be arranged in small groups, and the callus cylinders appear to surround these groups rather than individual strands as in angiosperms. The history of development, functioning period, and obliteration is much the same as that of sieve-tube elements. It is possible that the sieve cells of vascular cryptogams have a much longer functioning period—even several years—than those of seed plants.

Companion Cells.—The *companion cell* is a specialized type of parenchyma cell which is closely associated in origin, position, and function with sieve-tube elements. These cells occur only in the angiosperms but in these plants accompany most sieve-tube elements. Perhaps protophloem sometimes lacks them, and they appear to be scarce in the primary and early secondary phloem of some woody plants. In highly specialized phloem, such as that of many monocotyledons, they are abundant, together with sieve tubes making up the entire tissue.

Companion cells are formed by longitudinal, or oblique-longitudinal, division of the mother cell of the sieve-tube element before specialization of this cell begins. One daughter cell may become a companion cell and the other a sieve-tube element; or the latter may divide further, forming more companion cells. Transverse divisions in the companion-cell initial may form a row of companion cells so that one to several may accompany each element. A companion cell or a row of a few such cells formed by transverse division of a single companion-cell initial may extend the full length of the sieve-tube element. In a species the number of companion cells accompanying a sieve-tube element is fairly constant. Solitary, long companion cells are common in primary phloem and herbaceous plants; short and numerous companion cells appear to be characteristic of the secondary phloem of woody plants.

Companion cells have abundant granular cytoplasm and a prominent nucleus which is retained through the life of the cell. They do not contain starch at any time. They live only so long as the sieve-tube element with which they are associated and they are crushed with those cells.

When seen in transverse section the companion cell is usually a small, triangular, rounded, or rectangular cell beside a sieve-tube element (Fig. 52*I,T*). Often it seems to lie as if within the limits of the sieve-tube element. If it does not, it may be impossible to determine with which particular sieve tube it is associated for it may be in contact with more than one. In secondary phloem of many gymnosperms the marginal cells of the rays, called *albuminous cells*, differ markedly from the other ray cells. They are discussed in Chap. VIII.

Phloem Parenchyma.—Phloem typically contains parenchyma cells of other types than companion cells. Various terms have been given to these on a basis of shape and probable function, but these terms are confused in usage and the differences hardly merit distinction. In shape these cells range from elongate and tapering to broadly cylindrical, sub-

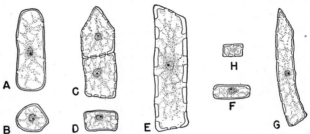

Fig. 53.—Phloem parenchyma, longitudinal and transverse sections. *A, B,* from *Salix nigra; C, D,* from *Robinia Pseudo-Acacia; E, F,* from *Liriodendron; G, H,* from *Malus pumila.*

spherical, and polyhedral (Fig. 53). The elongate cell, while young, may divide forming a row of cells that retain in their mature form and position evidence of this origin. In content also these cells show great variety: crystals, tanniferous substances, mucilage, latex, etc. Most of the parenchyma cells are filled with starch or oil during dormant periods. They remain alive until cut off from the inner living tissues by periderm formation. Those of secondary phloem in age may be transformed into sclereids, as in *Quercus.*

Parenchyma may be lacking in phloem—the tissue consisting of sieve tubes and companion cells only, as in the vascular bundles of many monocotyledons.

The parenchyma of the phloem rays is discussed under Secondary Phloem (Chap. VIII).

Phloem Fibers and Sclereids.—Sclerenchymatous cells are rare or absent in phloem of living pteridophytes, and they are absent from this tissue in some gymnosperms and angiosperms. But in many seed plants, fibers form a prominent part of both the primary and the secondary phloem (Fig. 54). The fibers of the primary phloem—or part of these cells—have in some plants been mistaken for pericyclic fibers (Chap. V). Phloem fibers differ from xylem fibers in that the pits, which have small, linear, or rounded apertures, are always simple. The walls are lignified. In development the long-tapering ends become interlocked and strong strands are formed. The fibers often form tangential sheets or cylinders enclosing the inner tissues. These are obviously of structural importance as layers protective to the soft cambium region within, and also to some extent as longitudinal strengthening tissues. In some herbs and occasional woody plants, such as *Dirca palustris*, they apparently are of more importance in supporting the stem than is the xylem cylinder.

Fibers of the protophloem, forming the outermost cells of the primary phloem, are prominent features of many stems, both woody and herbaceous. In the early stages of stem development these fibers may contribute largely to the support of the stem. They are arranged in various ways: as continuous, uniform, or irregular bands; as scattered, isolated strands; as clusters "capping" the primary phloem strands. Such fibers are usually lignified, as are those of *Cannabis* (hemp), but may be of cellulose, as in *Linum* (flax). These primary phloem fibers are often similar to fibers of the cortex and to those of the secondary phloem. All or any of them with other fibers and sometimes vascular bundles constitute commercial "bast."

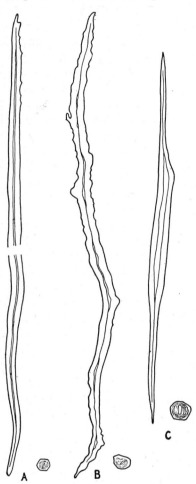

FIG. 54.—Phloem fibers, longitudinal and transverse sections. *A*, from *Salix nigra*, only two-thirds of cell shown; *B*, from *Malus pumila*; *C*, from *Robinia Pseudo-Acacia*. (The scale is twice that of the figure of sieve elements, Fig. 52.)

The Term "Bast."—Because of the strength of strands of phloem fibers, these have long been used in the manufacture of cords and ropes and in the weaving of matting and cloth. Fibrous tissue used in this way has been known since early days as *bast*, or *bass*. The term was originally applied to any fibers obtained from the outer part of a plant, though a large part of such material came from the secondary phloem, as in the basswood, *Tilia*. When the secondary phloem was recognized as distinct from the cortex, the term "bast" was applied to this phloem, which was the common source of fibers. And in this sense, that is, as synonymous with phloem—as wood is synonymous with xylem— "bast" is still in frequent use. With the term used in this morphological sense, the fibers of the phloem become *bast fibers*. But the term "bast fibers," or simply "bast," is also applied to any fibers from the outer parts of a plant. This is topographical, not histological or morphological use; such fibers may be a part of the cortex or pericycle. Secondary phloem is also frequently "divided into hard phloem, or bast, and soft phloem"; and "bark" is divided into "bast and living phloem." The term "bast" is used to such an extent without accurate botanical meaning that it should be discontinued as a technical term. Furthermore, it is superfluous, since the terms "phloem," "phloem fibers," "pericyclic fibers," and "cortical fibers" cover accurately all its uses.

Primary phloem occasionally contains sclereids, and the older secondary phloem of many trees contains abundant cells of this type. These cells develop from parenchyma cells as the tissue ages and the sieve tubes cease to function.

Pitting in Cells of the Phloem.—In phloem the complex cellular relationship, the delicate walls of many of its cells, and the similarity of pits with plasmodesmata to sieve areas with cytoplasmic strands make the determination of pitting difficult. Statements concerning wall structure in some positions vary greatly, and much additional information is needed. The pitting of the thicker walled cells is clear: the pit-pairs are the simple pit-pairs of parenchyma and sclerenchyma, except that in some types of parenchyma the pits resemble sieve areas (Fig. 110D). Between sieve tube and parenchyma cell there is a sieve area on the side of the former and a pit on the side of the latter. Between the sieve tube and companion cell the wall is usually very thin and specialized areas are not seen; rarely, this wall is thick, and "pits" have been reported. It has been generally believed that companion cells are not pitted with parenchyma, but pit-pairs have been described in some plants.

Function of Phloem.—The chief function of phloem is the conduction of elaborated foodstuffs, both proteins and carbohydrates. The sieve elements are believed to be the cells concerned in this conduction with the companion cells, or the albuminous cells, related in some way to their activities. The sclerenchymatous tissues—fibers and sclereids—serve in some measure in support of organs and protection of soft, underlying tissues. Many parenchyma cells are starch storage cells at certain periods; others perhaps serve in conduction of some substances; some are crystal storage regions.

Phloem is further discussed under "Primary Body" (Chap. V) and "Secondary Phloem" (Chap. VIII).

The Term "Phloem."—The term "phloem" is sometimes used—as is the term "xylem"—to indicate only the conducting cells of a complex tissue whose chief function is conduction. In this usage only the sieve elements of the tissue usually called phloem are indicated as "phloem." Occasionally all the "soft cells" are labelled "phloem," the sclerenchyma being excluded as "bast." The restriction of "phloem" to sieve tubes and companion cells is especially applied to protophloem. But histologically and morphologically the entire, continuous tissue is the phloem, and the term is so used in this book. For example: protophloem does not consist only of the sieve-tube elements that constitute its first-maturing cells but includes the much later maturing fibers that surround them.

"Transfusion Tissue."—A peculiar type of conducting tissue, which consists largely of short tracheids with thin, cellulose walls and bordered pits or reticulate or scalariform thickenings, often accompanies typical vascular tissue in the leaves of the gymnosperms. These cells are tracheid-like in the character of their pitting and in the absence of a protoplast but otherwise suggest elongate parenchyma cells. They lie adjacent to typical xylem at the sides of the bundle and may partly or even completely surround it. Since these cells seem to serve as conducting tissue, connecting the veins and the mesophyll of the leaves, taking the place of the usual very small branches of the veins, they form together what is called *transfusion tissue*. Though the function of this tissue is still uncertain, it undoubtedly represents modified vascular tissue.

TISSUE SYSTEMS

All the tissues of a plant which perform the same general function, regardless of position or continuity in the body, may be considered to form, together, a *tissue system*. In this sense the term is wholly a physiological one. There are found in physiological treatments of anatomy such tissue systems as the "mechanical system," the "absorbing system," and the "storage system." The various parts of most of such systems, however, are bound together only by function; they have little or no structural or morphological unity. Continuity and similarity of nature or of origin may be lacking.

From a morphological viewpoint the grouping of tissues into tissue systems is sometimes convenient. In the morphological sense, a tissue system must be a complex of cells extending continuously throughout the plant body or over a considerable portion thereof. It may be so simple as to consist of only one type of cell or one type of tissue or may consist of two or more types of tissues. Very few systems in any way structurally distinct can be distinguished. The older students of anatomy distinguished an *epidermal,* or *tegumentary, system;* a *fundamental,* or *ground, system;* and a *vascular system.* The distinction of these systems is not often made in present-day anatomy, doubtless because the systems have, in use, been morphologically too inclusive or indefinite or have been nonmorphological. The epidermal system has sometimes included a hypodermis and other cortical tissues—even periderm; the fundamental system, the cortex, pericycle, and pith. Recognition of these systems has value in a strictly topographical sense and is especially useful in the interpretation of meristematic tissues. In young organs, the epidermal system consists of the dermatogen (or protoderm) alone; the vascular system consists of the procambium and the earliest formed xylem and phloem elements; the fundamental system consists of ground meristem—all the remaining tissues, which at this stage show little or no differentiation. In the mature primary body these form, respectively, the epidermis, the vascular tissues, and the cortex, pericycle, pith, and mesophyll.

The epidermis and vascular system are tissue systems of such uniformity and continuity of structure and of such constancy of function that they constitute important gross structural features of the plant body. The terms "epidermal system"—if used to cover epidermis only—and "vascular system" are convenient and valuable. "Ground tissue system" covers the heterogeneous remaining parts.

Secretory Tissue

All cells directly concerned with the secretion of gums, resins, essential oils, nectar, and similar substances are together frequently referred to as "secretory tissue." Such a classification is purely a physiological one, as secretory cells and tissues often do not have common origin or morphological continuity. Secretory cells frequently are isolated from other similar cells, embedded in pith, xylem, phloem, cortex, or any region. On the other hand, secretory cells may be aggregated forming a tissue in the strict morphological sense. Not infrequently such cells constitute a definite, organized secretory structure or gland.

Secretory Cells.—Secretory cells are of two general types: those in which the secretion formed is exuded from the secreting cell, as in glandular hairs and secretory surfaces, such as nectaries and the epithelium of resin and oil canals; and those in which the secretion formed is stored within the secretory cell. The term *execretory cells* is often applied to the first type. This type of cell is usually characterized by a protoplast with richly granular cytoplasm and a conspicuous nucleus (Figs. 55*A,B,C*,

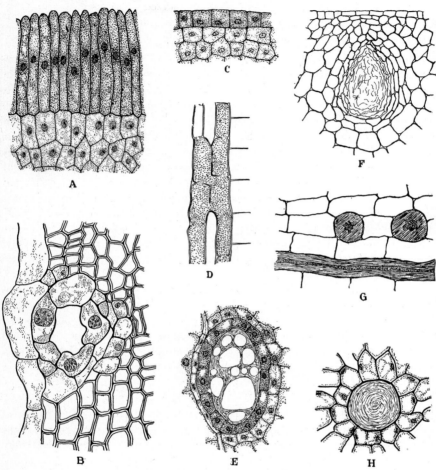

Fig. 55.—Secretory tissue. *A*, section of surface of nectary of *Euphorbia pulcherrima*; *B*, cross section of resin canal of *Pinus Strobus*; *C*, section of floral nectary of *Malus pumila*; *D*, latex vessel from *Tragopogon*, absorption of cross-walls in progress; *E*, cross section of oil canal of young fruit of *Angelica atropurpurea*; *F*, section of lysigenous oil cavity of rind of *Citrus sinensis*; *G*, latex cells from cortex of *Euphorbia splendens*; *H*, secretory cell of bud scale of *Liriodendron*. (*D, after Scott.*)

E, 56*B,C,D*); the secretory cell is usually large, with inconspicuous cytoplasm and large lumen filled with the secretion (Fig. 55*H*). This type of cell contains in various plants many different substances, such as essential oils in *Liriodendron, Sassafras,* and *Zingiber,* and mucilage in many ferns. Glandular hairs often show elaborate specialization related to various functions, as, for example, the stinging hairs of the nettles, *Urtica* (Fig. 56A). Such specialized hairs are often multicellular.

Glands.—Secretory cells are often organized into special secreting structures commonly known as *glands*. The term "gland" is also used more loosely to indicate secreting structures of any kind, including those

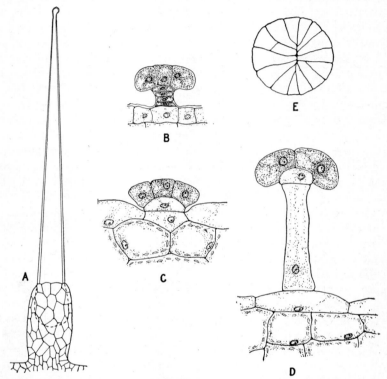

Fig. 56.—Secretory hairs. *A*, stinging hair from *Urtica gracilis*; *B*, glandular hair from ovary of *Gaylussacia baccata*; *C,D*, two types of glandular hairs from leaf of *Pinguicula*; *E*, top view of *D*. (The scale of *A* is much less than that of the others.)

of all stages of complexity of organization from the solitary secretory cell to the more elaborate structures. Glands are various in function, but among the more common types are those which secrete digestive enzymes, called *digestive glands*, and those which secrete nectar, known as *nectaries* (Fig. 55A, C). Other types are *hydathodes, resin ducts, oil ducts, laticiferous ducts* or *glands*. Glands or ducts may have central cavities where the secretion is stored or transferred. The cavities are either schizogenous (Chap. II), like the resin ducts of *Pinus* (Fig. 55B), or lysigenous, like the oil sacs in the rind of citrus fruits (Fig. 55F).

Digestive Glands.—In the majority of plants, enzyme secretion is not confined to specialized cells or tissues but is a characteristic of most living cells. In certain so-called "insectivorous" and "carnivorous" plants there are special glands which secrete protein-digesting enzymes; these enzymes act upon insects or other organisms so that the products of digestion can be absorbed by the plant. In *Drosera*, the secretory tissue is at the tips of the leaf "hairs" or "tentacles," structures which also serve to imprison the insects. Here, in addition to the digestive enzymes, there are secreted viscid substances which hold the insects. In such plants as *Nepenthes* and *Sarracenia*, which normally have pitcher-like traps partly filled with liquid, the glands are sessile and secrete the enzymes into the liquid from which the products of digestion are absorbed. In some other genera, for example, *Dionaea* and *Pinguicula* (Fig. 56C,D), the glands are inactive except when stimulated by contact with animal matter. Less specialized glandular tissue of this type is found in the embryos of certain seeds, but such tissue is usually not clearly differentiated.

Nectaries.—Many entomophilous plants produce nectar which attracts insects. This substance is secreted by specialized cells either on the floral parts themselves or, more rarely, on bracts or other structures outside the flower. Usually, the secretion of nectar is from specialized epidermal cells which cover certain regions of the flower, rather than from elaborate organs adapted to secretion alone. Definite and elaborate structures do, however, occur in certain families, for example, the Euphorbiaceae. In the less specialized nectaries the secreting cells are superficial upon the floral parts and in most plants closely resemble the other epidermal cells of the region but lack a cuticle (Fig. 55C). Sometimes they are set off from the surrounding epidermal cells by a somewhat more columnar or papillose shape and by denser cytoplasm (Fig. 55A). The nectar is exuded through the wall and exposed upon the outer or nectariferous surface. The secreting cells of stigmatic surfaces are of the same nature as those of nectaries but are often not clearly set off from normal epidermal cells.

The *septal nectaries* or *glands* of many monocotyledonous flowers are pockets in the septal walls of syncarpous ovaries where the fusion of the carpel walls is incomplete and the epidermal cells are glandular. These glands may be simple, slit-like cavities or deep pockets with canal-like passageways to the surface of the ovary.

Hydathodes.—Many plants possess structurally modified regions where water is exuded under conditions of low transpiration and abundant soil moisture. These are known as *hydathodes* or sometimes as *water pores* or *water stomata*. Morphologically, they are considered to be enlarged stomata which serve for water secretion. Structurally, they may resemble stomata closely but often they show elaborate structural specialization. Such structures do not "secrete" the fluid but merely provide and control the openings through which it escapes. Hydathodes occur commonly at the tips of leaves as in grasses, at the apices of serrations on the margins of leaves, and in other positions. They are found mostly on plants of humid climates.

Resin, Oil, and Gum Ducts.—In the gymnosperms generally and in many angiosperm families, resins, oils, gums, and other substances are secreted and conducted in ducts. In some plants, as in *Pinus*, these ducts or canals may form extensive systems extending both vertically and horizontally. In other plants the ducts may be local in occurrence and limited in extent, as in the fruits of the Umbelliferae. In *Pinus* and closely related genera, the resin ducts are schizogenous in nature, and when mature, have the structure of a tube with an epithelial lining (Fig. 55*B,E*). The oil ducts of the Umbelliferae (Fig. 55*E*) are of the same general nature. The secretory cells lining these cavities are thin-walled parenchyma with dense protoplasm. In general, these cells are elongate with the long dimension extending parallel with the long dimension of the duct. The substances secreted are various in nature and in some plants—for example, the resin of *Pinus* and of *Agathis* (Kauri gum), and some essential oils—are of much economic importance.

Another type of gland is that found in the rind of citrus fruits (Fig. 55*F*). Here there is a lysigenous cavity filled with essential oil and other substances that have been formed by the disintegration of the cells and as definite secretions before the breaking down of the tissues. The origin of this secretion is not well understood. These glands are the source of the essential oils of lemon and orange.

Laticiferous Ducts.—*Latex* is found in a considerable number of angiosperm families. This substance appears as a white, yellow, or reddish, sometimes slightly viscous fluid which has been shown to be an emulsion of proteins, sugars, gums, alkaloids, enzymes, rubber, and other substances suspended in a matrix of watery fluid. Starch grains may be abundantly present. The function of latex is not well understood. It is apparently secreted by the cells in which it is contained, and is conducted by them throughout the plant body. The latex of some plants is of great importance, especially as a source of rubber (*Hevea, Ficus*, etc.) chicle (*Achras*), and papain (*Carica*), as well as for other substances. Laticiferous ducts are of two types—one known as *nonarticulate latex ducts* or *latex cells* and the other as *articulate latex ducts* or *latex vessels*. The function and the contents of the two kinds of ducts are essentially the same, but the morphological nature and development are different.

Nonarticulate Latex Ducts.—These ducts are individual cells that extend as ramifying structures for long distances through the plant body (Fig. 57). The walls are soft and often thick, and the cytoplasm contains great numbers of nuclei. These ducts arise as typical small meristematic cells among other promeristem cells. They quickly become elongate, and their tips keep pace in growth with the surrounding meri-

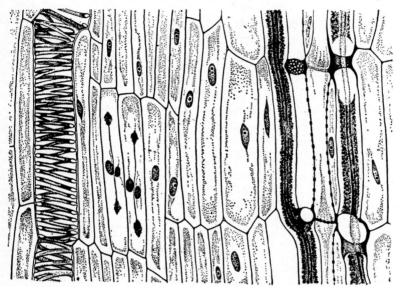

Fig. 57.—Radial section through growing cambium region in root of *Taraxacum kok-saghyz* (Russian dandelion). At the left, a mature, scalariform protoxylem vessel with simply perforated, transverse end walls; in the center, cambium and its recent derivatives, two with cell division incomplete, the phragmoplasts migrating to the end walls; at the right, mature phloem consisting of sieve tubes, articulate latex ducts, and storage parenchyma. (*After Artschwager and McGuire.*)

stem, penetrating among the new cells. They branch and extend through all the tissues of the plant, even in some genera (*Cryptostegia*) through secondary xylem. Although the branches of a cell come in contact with those of other cells, no anastomoses occur.

Two kinds of these nonarticulate ducts are described but the differences lie in the number per plant, the position of origin, and extent in the plant body rather than in structure or function. In one kind the cells arise only once, in small number, in the embryo. These few cells, extending with the meristems, ramify through the entire plant. The ducts of the Asclepiadaceae, most of the Apocynaceae and Euphorbiaceae, and those of some other families are of this kind. Ducts of the other kind arise repeatedly in the meristems and ramify into adjacent tissues but are restricted in their growth to one internode and attached leaf and branch. The ducts of the Urticaceae and those of some of the Apocynaceae (*Vinca*) are of this type.

Articulate Latex Ducts.—These ducts originate in the meristems from rows of cells by the absorption, complete or partial, of the separating walls early in the ontogeny of the cells. By branching and frequent anastomoses, a highly complex system is built up. A duct of this type resembles a xylem vessel in that it is made up of a series of cells united to form a tube by the dissolution of walls, but the latex tube is living and coenocytic. The Papaveraceae (poppy), Caricaceae (papaya), Compositae (dandelion), Musaceae (banana), and the genus *Hevea* (Brazilian rubber tree) have this type of laticiferous ducts.

References

(See also Chaps. V, VII, and VIII for references to vascular tissues.)

ABBE, L. B., and A. S. CRAFTS: Phloem of white pine and other coniferous species, *Bot. Gaz.*, **100**, 695–722, 1939.

ALEXANDROV, W. G. and K. J. ABESSADYE: Ueber die Struktur der Seitenwande der Siebröhren, *Planta*, **3**, 77–89, 1927.

ANDERSON, D. B.: A microchemical study of the structure and development of the flax fibers, *Amer. Jour. Bot.*, **14**, 187–211, 1927.

ARTSCHWAGER, E. F.: Anatomy of the vegetative organs of the sugar beet, *Jour. Agr. Res.*, **33**, 143–176, 1926.

——: Contribution to the morphology and anatomy of guayule (*Parthenium argentatum*), *U.S.D.A. Tech. Bull.* 842, 1–33, 1943.

——: Contribution to the morphology and anatomy of Cryptostegia (*Cryptostegia grandiflora*), *U.S.D.A. Tech. Bull.* 915, 1946.

——, and R. C. MCGUIRE: Contribution to the morphology and anatomy of the Russian dandelion (*Taraxacum kok-saghyz*), *U.S.D.A. Tech. Bull.* 843, 1943.

BAILEY, I. W.: The effect of the structure of wood upon its permeability, 1. *Amer. Ry. Assoc. Bull.* 174, 1915.

——: The structure of the bordered pits of conifers and its bearing on the tension hypothesis of the ascent of sap in plants, *Bot. Gaz.*, **62**, 133–142, 1916.

——: The problem of differentiating and classifying tracheids, fiber-tracheids, and libriform wood fibers, *Trop. Woods*, **45**, 18–23, 1936.

——: The development of vessels in angiosperms and its significance in morphological research, *Amer. Jour. Bot.*, **31**, 421–428, 1944.

BARANETZKI, J.: Épaississement des parois des éléments parenchymateux, *Ann. Sci. Nat. Bot.*, 7 sér., **4**, 134–201, 1886.

BEAUREGARD, H.: "Des Organes glandulaires des végétaux," 108 p., Paris, 1879.

BEHRENS, W. J.: Die Nectarien der Flüthen, *Flora*, **37**, 2–11, 17–27, 49–54, 81–90, 113–128, 145–153, 233–240, 241–247, 305–314, 369–375, 433–457, 1879.

BLASER, H. W.: Anatomy of *Cryptostegia grandiflora* with special reference to the latex system, *Amer. Jour. Bot.*, **32**, 135–141, 1945.

BONNIER, G.: "Les Nectaires," Bibliothèque de l'Académie de Médecine, Paris, 1879.

CHEADLE, V. I.: The occurrence and types of vessels in the various organs of the plant in the Monocotyledoneae, *Amer. Jour. Bot.*, **29**, 441–450, 1942.

——: The origin and certain trends of specialization of the vessel in the Monocotyledoneae, *Amer. Jour. Bot.*, **30**, 11–17, 1943.

———: Vessel specialization in the late metaxylem of the various organs in the Monocotyledoneae, *Amer. Jour. Bot.*, **30**, 484–490, 1943.
———: Specialization of vessels within the xylem of each organ in the Monocotyledoneae, *Amer. Jour. Bot.*, **31**, 81–92, 1944.
———, and N. B. WHITFORD: Observations on the phloem in the Monocotyledoneae, I. The occurrence and phylogenetic specialization in structure of the sieve tubes in the metaphloem, *Amer. Jour. Bot.*, **28**, 623–627, 1941.
DANGEARD, P. A.: Essai dur l'anatomie comparée du liber interne dans quelques familles de dicotyledones. *Le Botaniste*, **17**, 225–364, 1926.
ESAU, K.: Ontogeny of phloem in the sugar beet (*Beta vulgaris*), *Amer. Jour. Bot.*, **21**, 632–644, 1934.
———: Ontogeny and structure of collenchyma and of vascular tissues in celery petioles, *Hilgardia*, **10**, 431–467, 1936.
———: Vessel development in celery, *Hilgardia*, **10**, 479–488, 1936.
———: Ontogeny and structure of the phloem of tobacco, *Hilgardia*, **11**, 343–406, 1938.
———: The multinucleate condition in fibers of tobacco, *Hilgardia*, **11**, 427–434, 1938.
———: Development and structure of the phloem tissue, *Bot. Rev.*, **5**, 373–432, 1939.
——— and W. B. HEWITT: Structure of end walls in differentiating vessels, *Hilgardia*, **13**, 229–244, 1940.
FOSTER, A. S.: Structure and development of sclereids in the petiole of *Camellia japonica* L., *Bull. Torrey Bot. Club*, **71**, 302–326, 1944.
———: Origin and development of sclereids in the foliage of *Trochodendron aralioides* Sieb. & Zucc., *Amer. Jour. Bot.*, **32**, 456–468, 1945.
FREY-WYSSLING, A.: Saftergusses aus turgeszenten Kapillären, *Ber. Schweiz. Bot. Ges.*, **42**, 254–283, 1933.
FRITSCHÉ, E.: Recherches anatomiques sur le *Taraxacum vulgare* Schrk., *Arch. Inst. Bot. Liège.*, **5**, 1–24, 1914.
FROST, F. H.: Histology of the wood of angiosperms, I. The nature of the pitting between tracheary and parenchymatous elements, *Bull. Torrey Bot. Club*, **56**, 259–263, 1929.
———: Specialization in secondary xylem of dicotyledons, I. Origin of vessel, *Bot. Gaz.*, **89**, 67–94, 1930.
———: Specialization in secondary xylem of dicotyledons, II. Evolution of end wall of vessel segment, *Bot. Gaz.*, **90**, 198–212, 1930.
———: Specialization in secondary xylem of dicotyledons, III. Specialization of lateral wall of vessel segment, *Bot. Gaz.*, **91**, 88–96, 1931.
GAUCHER, L.: Du rôle des laticifères, *Ann. Sci. Nat. Bot.*, 8 sér., **12**, 241–260, 1900.
HANDLEY, W. R. C.: Some observations on the problem of vessel length determination in woody dicotyledons, *New Phyt.*, **35**, 456–471, 1936.
HILL, A. W.: The histology of the sieve tubes of angiosperms, *Ann. Bot.*, **22**, 245–290, 1908.
HUBER, B.: Das Siebröhresystem unserer Bäume und seine jahreszeitlichen Veränderungen, *Jahrb. Wiss. Bot.*, **88**, 176–242, 1939.
KROTKOV, G.: A review of literature on *Taraxacum kok-saghyz* Rod., *Bot. Rev.*, **11**, 417–461, 1945.
LEBLOIS, A.: Recherches sur l'origine et le développement des canaux sécréteurs et des poches sécrétrices, *Ann. Sci. Nat. Bot.*, 7 sér., **6**, 247–330, 1887.
LEEMANN, A.: Das Problem der Sekretzellen, *Planta*, **6**, 216–233, 1928.
LÉGER, L. J.: Recherches sur l'origine et les transformations des éléments libériens, *Mém. Soc. Linn. Normandie*, **19**, 49–182, 1897.

LEHMANN, C.: Studien über den Bau und die Entwicklungsgeschichte von Ölzellen, *Planta*, **1**, 343–373, 1925.
MACDANIELS, L. H.: The histology of the phloem in certain woody angiosperms, *Amer. Jour. Bot.*, **5**, 347–378, 1918.
MARTINET, J.: Organes de sécrétion des végétaux, *Ann. Sci. Nat. Bot.*, 5 sér., **14**, 91–232, 1872.
MAYBERRY, M. W.: Hydrocarbon secretions and internal secretory systems of the Carduaceae, Ambrosiaceae, and Cichoriaceae, *Bull. Univ. Kansas*, **37**, 71–112, 1936.
MÜLLER, C.: Ein Beitrag zur Kenntnis der Formen des Collenchyms, *Ber. Deut. Bot. Ges.*, **8**, 150–166, 1890.
PEIRCE, G. J.: Water and plant anatomy, *Proc. Calif. Acad. Sci.*, **25**, 215–220, 1944.
RENDLE, B. J.: Gelatinous wood fibers, *Trop. Woods*, **52**, 11–19, 1937.
SCHAFFSTEIN, G.: Untersuchungen an ungegliederten Milchröhren, *Beih. Bot. Centralbl.*, **49**, 197–220, 1932.
SCOTT, D. H.: The development of articulated laticiferous vessels, *Quart. Jour. Micro. Sci.*, new ser., **22**, 136–153, 1882.
SCOTT, F. M.: Differentiation in the spiral vessels of *Ricinus communis*, *Bot. Gaz.*, **99**, 69–79, 1937.
SPERLICH, A.: Das trophische Parenchym. B. Excretionsgewebe, *In* Linsbauer, K.: "Handbuch der Pflanzenanatomie," IV., 1939.
STRUCKMEYER, B. E., and R. H. ROBERTS: Phloem development and flowering, *Bot. Gaz.*, **100**, 600–606, 1939.
TETLEY, U.: The secretory system of the roots of the Compositae, *New Phyt.*, **24**, 138–161, 1925.
TOBLER, F.: Die mechanischen Elemente und das mechanische System, *In* Linsbauer, K.: "Handbuch der Pflanzenanatomie," IV, 1939.
VESTAL, P. A., and M. R. VESTAL: The formation of septa in the fiber-tracheids of *Hypericum Androsaemum* L., *Bot. Mus. Lfts., Harvard Univ.*, **8**, 169–188, 1940.
WILSON, C. L.: Lignification of mature phloem in herbaceous types, *Amer. Jour. Bot.*, 239–244, 1922.
WORSDELL, W. C.: On "transfusion-tissue": its origin and function in the leaves of gymnospermous plants, *Trans. Linn. Soc. Bot.*, 2 ser., **5**, 301–319, 1897.
ZIMMERMAN, J. G.: Über die extrafloralen Nektarien der Angiospermen, dissertation, Dresden, 1932.

CHAPTER V

THE PRIMARY BODY

THE PRIMARY TISSUES AND TISSUE SYSTEMS

The developing embryo of a vascular plant is early differentiated into an axis and appendages, and the axis is in a short time differentiated into stem and root. These first-formed parts of the plant body grow rapidly and (except for the interruption of the resting period of seeds) soon become mature. The root and stem remain meristematic at the tip, but the appendages, which are of limited growth, become mature throughout. Exceptions to the continued growth of the axis occur, for example, in flowers, thorns, and specialized roots; but normally the apical growth of the axis is unlimited, and by its activity the root and stem are increased in length, and the lateral structures, both the appendages of various ranks and the branches of the stem, are added. Branches of the root develop in a different way (Chap. X); adventitious roots and stems develop nearly always from new meristems formed in permanent or in traumatic tissue.

Apical meristems build up the new parts of the stem and root and form the appendages. These new parts of the body are structurally and functionally complete, at least temporarily, and constitute the *primary body* of the plant. *Secondary tissues*, constituting the *secondary body*, may be added later. These tissues are accessory only; they replace or reinforce structurally or functionally certain primary tissues. They increase the volume of existing primary types, as in xylem and phloem, or add a new kind of tissue which in function replaces a primary tissue of different kind, for example, cork.

The primary body is the fundamental morphological unit; the development of a secondary body does not change morphological structure but may in part conceal it. All fundamental tissues and body parts are present when primary growth is completed. These are, in the axis, the central cylinder with its xylem, phloem, pith, pericycle, and endodermis; the cortex; the epidermis; and, in the leaf, the corresponding regions and tissues. These parts show great variety of structure in different plants and in different parts of the same plant; for example, the vascular tissues of the central cylinder are arranged in one plant in a way entirely different from that in another plant, and at a node are differently placed than in the adjacent internodes. Such variations of grosser general structure are discussed later in the chapter; the present discussion is confined to the histological and anatomical features of the various primary tissues and regions.

Ontogeny of the Axis.[1]—At or near the apex of the axis lie the apical cells, the persistent initiating cells which develop the primordia of every region (Fig. 58). The apical cells form many cells closely alike in all ways, building up a region in which there is little or no evidence of tissue differentiation, the *promeristem* (Chap. III) (Fig. 58*a-a*). The cells of the promeristem differentiate as they become older, showing evidence in changes of size, shape, wall thickness, etc., of the various cell types into which they will develop. With this step the promeristem passes over into meristem in which a certain amount of cell differentiation is evident, and this meristem in turn passes over into permanent tissue. If a group of cells is viewed at different periods in its development, it will be found in the various stages mentioned. And, since the growth of the axis is apical and the various structural features of the axis are likewise continuously formed, there are found at various distances back from the apex, stages of development representative of the stages through which a given group of cells will pass as the cells become older. In other words, the development through which a given part of the axis passes, over a period of time from its beginning until it reaches maturity, may be found at any one time in different levels of the axis, progressively farther and farther from the apex. Thus, development as seen in spatial extent supplies the story of development in time. (Differences in nodal and internodal structure are not considered here.) It is thus evident that a series of cross sections of the axis taken in turn farther and farther away from the apex give the history of development of a cross-sectional area. Such a series is shown in Fig. 58, which also shows in longitudinal section progressive differentiation without break in continuity. In order that a comprehension of primary structure may be had, it is essential that this gradual differentiation in time be understood, as well as the fact that this may be seen expressed in space in the longitudinal extent of the growing axis tip.

The transition from meristem to permanent tissue does not take place simultaneously throughout a given level in the axis; there is overlapping of meristem upon mature tissues. For example, Fig. 58 *c-c* shows a stage where the pith and a part of the vascular tissues are mature. Other cells beside or around these are immature, being in the stage of procambium or even of promeristem. In a similar way, primary growth of a given section of the axis is not completed throughout the region in question

[1] Chap. III contains a more detailed discussion of the early stages in axis development.

THE PRIMARY BODY 125

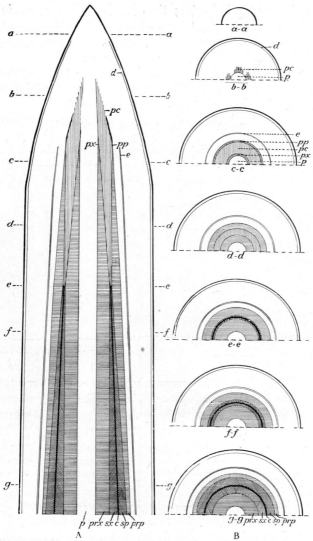

Fig. 58.—Diagrams showing development of the stem (leaf primordia, traces, gaps, and nodal differentiation omitted). *A*, longitudinal section of tip of elongating axis; *B*, *a-a* to *g-g*, cross sections of the axis at levels *a-a*, *b-b*, etc. At *a-a*, the entire axis is promeristem; at *b-b* the dermatogen (*d*), procambium (*pc*), and pith (*p*) are in early stages of development; at *c-c*, the procambium has increased in amount, forming a complete cylinder; the outermost and innermost procambium has become protophloem (*pp*) and protoxylem (*px*); the endodermis (*e*) is evident; at *d-d*, the procambium is less in amount, large portions of this tissue adjacent to the protophloem and protoxylem having become phloem and xylem; at *e-e*, the remaining layer of procambium has become the cambium, and has formed the first secondary phloem (*sp*) and secondary xylem (*sx*) cells; at *f-f*, the secondary tissues are increased in amount; the primary phloem (*prp*) is reduced by crushing; the cambium has moved outward; at *g-g*, further secondary growth has occurred.

before secondary growth begins; that is, there is here again an overlapping in development. It is impossible definitely to delimit by transverse planes the promeristem or the primary tips, but this can be done approximately.

Appendages consist chiefly of primary tissues. Leaves may be wholly primary but often the larger veins possess secondary tissues. The ontogeny of the leaf is discussed in Chap. XII.

PRIMARY VASCULAR TISSUE

The Procambium.—In the differentiation of axis and leaf from promeristematic tissue, the formation of a uniseriate external layer is usually the first evidence of the complex structure seen in the mature organ. The

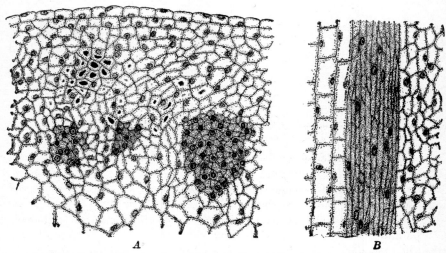

Fig. 59.—Procambium in cross and longitudinal section in stem tip of *Linum usitatissimum*. The procambium cells have smaller transverse diameter, are elongate, and have more dense cytoplasm. (At the stage illustrated here the cells of the pith are enlarging and no longer promeristematic.)

first cells to mature, however, in either leaf or axis, belong to the vascular tissue. In the promeristem, where all cells are essentially isodiametric and alike, continuing longitudinal divisions set apart in some areas strands of elongate, slender cells with dense cytoplasm. This meristematic tissue forms the primary phloem and xylem and is known as the *procambium*.[1] (The term "procambium" is used here to indicate the meristematic tissue that gives rise to the morphological vascular units, not as applied to any elongate meristematic cells that are cambium-like in form.) The form

---

[1] The terms "provascular tissue" and "provascular meristem" have recently been used in place of "procambium."

and arrangement of the procambium strands foreshadow the structure of the primary vascular skeleton. The first procambium appears as isolated strands very close to the apex in stem and root, commonly at a distance of only a few micra. It is continuous backward in the older tissues with older promeristem strands and mature vascular tissue, that is, development is acropetal.

The slender procambium strands increase in diameter by longitudinal cell division within themselves and by the addition of new cells upon their borders by transformation of adjacent promeristem cells. Other strands may appear a little later in other positions in the promeristem. Increase in size of the strands may be so great that some or all of the strands fuse,

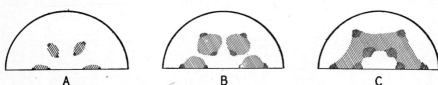

Fig. 60.—Diagrams showing stages in the development of a procambium cylinder from promeristem. *A*, the first procambium has appeared in the promeristem in the form of strands; *B*, the strands of procambium have increased in diameter, approaching one another, and the oldest portions (on the outside and the inside) have become phloem and xylem; *C*, the strands of procambium have united laterally, and more procambium has matured as phloem and xylem. In a later stage all procambium will have become vascular tissue, forming a primary vascular cylinder.

and a hollow cylinder or a solid central core may be formed. Ultimately the procambium skeleton so built up has the form of the vascular skeleton of the region. But at no one time is the complete procambium mass present because the oldest procambium cells quickly mature into phloem and xylem cells, often long before the last of the procambium cells appear.

As new cells are formed in a strand, cell length increases greatly, and the ends become more tapering. Increase in length is related to the rate and extent of increase in length of the region in which the cells lie. In this way the tissue keeps pace with the increase in the surrounding tissues with few or no transverse divisions. In later maturing procambium cells, the transverse diameter of the cells may also increase progressively. Both these changes are reflected in the size and form of the vascular cells formed from them: the first sieve tubes, tracheids, and vessels are short and exceedingly slender; the later matured cells increasingly longer and of somewhat greater transverse diameter. The difference in length is partly obscured by the stretching of first-formed cells.

As procambium develops, the diameter of the organ is increasing, and the promeristem cells are multiplying and enlarging. Spatial changes thus necessarily take place in the position of the strands in relation to each other and to the center of the organ. Further, within each strand, new-cell formation brings about spatial changes of the cells on the borders. In this way, as growth continues, the first phloem and the first xylem cells, which usually mature on the inner and outer margins of slender strands, are separated more and more widely as the strand becomes larger (Fig. 60).

The first-maturing cells in a young strand are phloem cells; the first xylem cells soon follow. The first mature phloem cells are close to the apex in both stem and root, at a distance of a few micra, frequently within one millimeter (Fig. 42).

The order of maturing, longitudinally, of the phloem cells in a strand is acropetal, like that of the procambium; that of the xylem is both

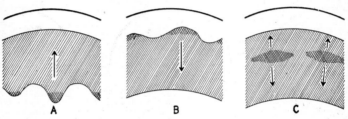

FIG. 61.—Diagrams showing order of development of primary vascular tissue transversely in an organ. *A*, centrifugal; *B*, centripetal; *C*, both centrifugal and centripetal. These diagrams may also illustrate types of xylem from the standpoint of order of development: *A*, endarch; *B*, exarch; *C*, mesarch.

acropetal and basipetal, from one or more points of beginning. Definiteness and uniformity of development transversely in a procambium strand are important characteristics of primary vascular tissue. The points of beginning and the order of cell maturation are constant for certain organs and for the major plant groups. The first xylem and phloem cells mature usually on the inner and outer margins of the strands and are separated radially by procambium in stems and leaves (Fig. 60) and tangentially by promeristem in roots. Order of development from the point of beginning is radial (as to the axis) and lateral in stems and leaves, radial only or chiefly in roots. After the first vascular cells mature, divisions in the central promeristem of the bundle tend to be in such planes that the later matured cells stand more or less in radial rows. This is especially true of herbs, vines, and woody plants of the more advanced families.

Centripetal and Centrifugal Growth.—The development radially in a procambial strand of mature vascular cells progressively from the point of beginning toward the center of the axis is known as *centripetal growth* (Fig. 61*B*), and that away from the center of the axis, *centrifugal growth* (Fig. 61*A*). The development of phloem is probably always centripetal, whereas that of xylem is sometimes centripetal and sometimes centrifugal.

Less commonly, the first-formed cells are situated, not at the edge of the procambium, but nearer the center, and development is then both toward and away from the center of the axis (Fig. 61C). In xylem when the development is toward the center of the axis, *centripetal xylem* is formed, and the xylem group or strand is said to be *exarch* (Fig. 61B); when it is *away from* the center of the axis, *centrifugal xylem* is formed and the xylem unit is said to be *endarch* (Fig. 61A). When development is such that both centripetal and centrifugal xylem are formed—even though the amount of one type be very small—the xylem is *mesarch* (Fig. 61C). (Development must always be considered as *relative to the center of the axis,* not to the center of the vascular bundle in question.) The terms "centripetal growth" and "centrifugal growth" refer only to sequence of appearance and of maturation of procambium cells in definite directions and do not imply growth in the sense of successive new-cell formation. The distinction of xylem as exarch, endarch, and mesarch is of morphological and phylogenetic importance because each type is characteristic of definite organs or parts of organs and is restricted more or less in its distribution in the larger plant groups. For example, the root is always exarch; the stem of seed plants is endarch; the axis of the club mosses is exarch; mesarch xylem is common in the ferns and infrequent elsewhere.

The First Vascular Elements: Protophloem and Protoxylem.—The first cells of the phloem to mature are known as the *protophloem;* those of the xylem, the *protoxylem.* These cells differ markedly from later formed cells of the same tissues in cell type; in size and shape, being very slender and long because stretched; and, in xylem cells, in the structure of the wall. Since they are formed very early in the ontogeny of the tissue in which they lie, they are subject to stresses due to increase in diameter and particularly to increase in length of the organ. To these stresses the surrounding meristematic cells accommodate themselves by new-cell formation and by increase in size, but the cells of the protophloem and protoxylem are mature and no longer subject to growth changes. In the region in which protophloem and protoxylem lie, the greatest change occurring is increase in length. The stresses brought about by this elongation tend to stretch the already matured cells. To such longitudinal stretching protoxylem and protophloem cells are, to a limited extent, adaptable. The cells are long and slender with walls which are thin and of cellulose, reinforced in protoxylem by bands of lignified secondary wall (Fig. 62). These bands, in the form of rings and spirals, prevent to some extent the collapse of the thin, plastic walls and the consequent closure of the lumen after stretching.

To those phloem and xylem cells which are thus capable of stretching, the terms "protophloem" and "protoxylem" are restricted. The parts of phloem and xylem which are not stretched by elongation of the region in which they lie are known as *metaphloem* and *metaxylem* respectively. Thus primary phloem consists of protophloem and metaphloem; primary

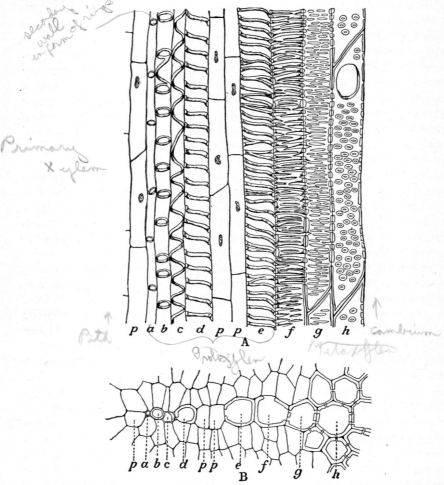

FIG. 62.—Protoxylem and metaxylem in transverse, *B*, and longitudinal, *A*, section in *Lobelia*. *a, b,* annular elements; *c, d, e,* spiral elements; *f,* scalariform element; *g,* scalariform-reticulate element; *h,* pitted vessel; *p,* parenchyma cell.

xylem, of protoxylem and metaxylem. In secondary phloem and secondary xylem no protophloem and protoxylem exist, for secondary tissues develop only after elongation ceases. A strand of xylem may consist wholly of protoxylem—the condition in most very small vascular bundles —or wholly of metaxylem, as in some slow-growing roots and rhizomes, but commonly both types of xylem are present, the proportions varying with the plant, the organ, and the rapidity of growth of the region in question. The phloem varies similarly.

It is obvious that in exarch xylem the metaxylem lies internal to the protoxylem (Fig. 61B, the protoxylem heavily shaded); in endarch xylem, external to the protoxylem (Fig. 61A); in mesarch xylem, both toward the inside and toward the outside of the protoxylem (Fig. 61C).

Primary Phloem.—The first vascular cells to mature in a given region are sieve elements of the phloem—sieve cells in gymnosperms and lower groups, sieve tubes in angiosperms. These cells are of the protophloem type. Their structure is difficult to determine because they closely resemble the procambium cells from which they are derived. Their diameter is very small; they live only a few days; they are soon greatly stretched, crushed between adjacent cells, and absorbed.

In phloem, as also in xylem, the first-formed cells are longest, because they develop while the region in which they lie is elongating most rapidly and they are therefore most stretched. Later formed cells are progressively shorter until metaphloem is formed. Similarly, the earliest of these cells have in general the smallest transverse diameter, and this diameter increases progressively in later formed cells. The change in diameter of these cells is related to the similar change in the procambium cells from which they mature. The first-formed phloem and xylem cells mature while all procambium cells are very slender.

The very first mature phloem cells of a region can be found only when the procambium is still in early stages. Later all evidence of their earlier presence has disappeared. Parenchyma cells and fibers mature later from the surrounding procambium cells and obliterate the spaces they occupied. The first protophloem probably consists only of sieve cells or sieve tubes distributed among procambium cells. Companion cells are rare or absent. Later, the procambium cells about these first-maturing cells differentiate as parenchyma cells and fibers. The fibers may be aggregated in clusters or masses which soon appear homogeneous because the early formed sieve tubes have disappeared, or they may be mingled with phloem parenchyma. These phloem fibers constitute what have been in many plants called pericyclic fibers. They, with the phloem parenchyma about them, make up a part, perhaps sometimes all, of the tissues commonly distinguished as pericycle. Mature conducting cells of protophloem are present in very young meristematic tissue—in stem and leaf tips as close to the apex as 0.3 mm; in root apices, about 1 mm.

The metaphloem is a complex tissue with well-developed cells of all types—sieve cells or sieve tubes, companion cells, parenchyma cells, and sclerenchyma in the form of fibers or sometimes sclereids. Parenchyma cells are added to the sieve cells or sieve-tube elements as soon as metaphloem begins to be formed. The various cell types are essentially the same as those of the secondary phloem of the same plant. The metaphloem, like the protophloem, may be early crushed. In woody plants and in herbs with well-developed secondary vascular tissues, the soft cells of the primary phloem are often completely destroyed within a short time after maturing by the development of secondary tissues beneath them. The remnants of the flattened cells may exist for some time or be quickly absorbed like those of the protophloem.

Primary phloem is of only temporary functional importance where secondary phloem is developed in considerable amount. Where little or no secondary phloem is formed, the primary phloem persists throughout the life of the organ and may be of the greatest importance physiologically. Structurally, such phloem is highly specialized, consisting usually only of sieve tubes of the highest type and their accompanying companion cells. Such is the condition in the phloem of most monocotyledons, and of such dicotyledons as *Cucurbita, Ranunculus,* and *Podophyllum.* In these genera the phloem of the bundles is in part secondary, but the primary and secondary phloem differ little histologically except that the protophloem may lack companion cells. In other herbaceous plants, such as *Solanum, Aster,* and *Lobelia,* where definite cylinders of secondary vascular tissue occur, but where the amount of secondary phloem is small, the primary phloem persists as functioning tissue through the life of the stem. In such forms the sieve tubes and companion cells often occur in small groups and are remarkable for their small size. An entire group may be of no greater diameter than one of the adjacent parenchyma cells (Fig. 138).

Primary Xylem.—Protophloem probably always consists only of one kind of cell, but protoxylem is a complex tissue made up of tracheids, vessels, and parenchyma cells. Fibers are probably always absent. Parenchyma cells surround the earliest formed tracheal cells and appear to mature with or soon after the conducting cells. (The term "protoxylem" is commonly applied only to the more or less isolated tracheids and vessels but is better used for the continuous tissue, including the parenchyma cells.)

The water-conducting cells of the protoxylem are the characteristic cells of this tissue because of the peculiar adaptation of their walls to the stretching which they normally undergo. The thin, plastic, primary walls of these empty cells are strengthened by the addition of a lignified secondary wall in the form of narrow rings and spiral bands. These bands apparently help to keep the conducting channels open during the elongation of the cells. The first-formed protoxylem cells have small amounts of secondary wall in the form of rings spaced at intervals along the cell (Fig. 62a,b). Such cells are called *annular cells*, or *annular elements*, or, more specifically, *annular tracheids* or *annular vessels*. (The term "element" is often used in speaking of cells of the protoxylem when the cells in question may be either tracheids or vessels.) Cells formed a little later than annular cells possess secondary walls in the form of spiral bands; such cells are called *spiral cells*, or *spiral elements* (Fig. 62c,d,e). The flatness of the coils and, to some extent, the number of the bands increase in the cells formed successively later and later. The proportionate amount of secondary wall thus increases in the successively formed cells. The first-formed cells are subjected to an excessive amount of elongation, those formed later and later to less and less. Structurally, the cells formed at any given period are adapted to the amounts of elongation occurring during their formation and functioning period. Annular cells can be stretched to a greater extent than spiral cells; cells with steep spiral coils more readily than those with flat coils; those with one spiral more freely than those with several. Where the structure of the secondary wall suggests a spiral lying in flat coils, the coils fused at the corners of the cell, or where a more definitely ladder-like appearance is produced by transverse bar-like thickenings running from corner to corner, the cell is known as *scalariform* (Fig. 62f). (The scalariform cell of protoxylem is best called a scalariform protoxylem cell, since the term "scalariform cell" is a loose descriptive term and may refer to a scalariform-pitted tracheid or to a scalariform-perforated vessel. A scalariform protoxylem cell may, of course, be also a scalariform vessel.)

The scalariform protoxylem cell is, obviously, capable of little if any stretching. Where the bands of secondary wall are more extensively and less uniformly tied together, forming a network of secondary thickening, a *reticulate cell* is formed (Fig. 62g). Where the secondary wall is still more extensive and the thin spots in the wall are definite and uniform in size and shape, the cell is *pitted* (Fig. 62h). Reticulate and pitted cells cannot be stretched. Primary xylem consisting of these cells is, therefore, *metaxylem*. Scalariform cells fall on the line between protoxylem and metaxylem. No sharp line separates any one type of cell from the next type, or protoxylem from metaxylem.

The water-conducting cells of protoxylem may be either tracheids or vessels. They are tracheids in pteridophytes and gymnosperms and in some of the less highly specialized angiosperms. In other flowering plants both kinds of cells may occur together, and in herbaceous forms and some woody forms the water-conducting cells of the protoxylem are largely vessels. Because protoxylem cells are made abnormally long by stretching, and the end walls are not readily seen, these cells, vessel-like in appearance, have been called "protoxylem vessels" regardless of their tracheidal or tracheal nature. This incorrect usage probably grew out of the idea of the seventeenth and eighteenth centuries that the animal-trachea-like appearance of spiral cells indicates a conducting tube of indefinite length. (The use of the term "vessel," or "trachea," in this sense is responsible for the opinion formerly held that "vessels occur in the gymnosperms only in the protoxylem.") It often cannot be readily determined whether a given cell of the protoxylem is a vessel or a tracheid; such a cell may then be called, for example, "an annular cell," or "an annular element"; it should not be called a vessel unless there is proof of the perforation of its end walls.

Proportion and Arrangement of Types of Protoxylem Elements.—In the protoxylem of a given vascular bundle, annular, spiral, and scalariform cells may all exist, and each in any proportion; or one or two of the types may be lacking. Spiral cells, especially those with closely coiled bands, constitute a large proportion of most protoxylem strands. Where growth of the axis is rapid, a high percentage of annular elements is found; where slow, very few or no such cells occur. As a strand of protoxylem forms, not only is the wall of the successively formed cells of different character, but the cells themselves each become somewhat larger than the preceding ones. The order of development may thus usually be determined, when several or all cells are mature, on the basis of size alone. Often there are variations and exceptions to the sequence in size, but the structure of the wall is constant in its relation to order of development.

Elongation of Protophloem and Protoxylem.—Though protophloem and protoxylem are structurally adapted to stretching, and the earliest formed cells especially so, the elongation to which they are subjected is often so great that their plastic capacity is surpassed and they are ruptured. The protophloem cells are rapidly absorbed after destruction, but the torn remains of the protoxylem tracheids and vessels persist permanently. Annular cells, being the first cells formed, are most frequently destroyed. Probably nearly all annular cells that are not destroyed are so distorted soon after their formation as to become nonfunctional. The stretching of these cells at first separates the supporting rings, and the thin, distended wall between them soon sags or collapses. The further pulling of the walls tilts the rings and turns them up on edge (Fig. 63B). Still further elongation fragments the cells, and isolated rings of secondary wall, often with pieces of the primary wall attached, are to be found where the cells have been destroyed (Fig. 63B). In spiral cells the spiral is pulled out so that it becomes steeply wound or is even straightened out (Fig. 63B). In such cells also the primary wall collapses

Protoxylem Lacunae.—Where a number of protoxylem cells lying together have been thus destroyed, a canal-like cavity, known as a *protoxylem lacuna*, may be formed (Fig. 63B). In some instances such an intercellular space is much enlarged by the pulling away of the surrounding parenchyma cells. Exaggerated protoxylem lacunae formed in this way are found in many herbaceous plants, and especially in the monocotyledons (Fig. 64C,D) and the horsetails (*Equisetum*). On the other

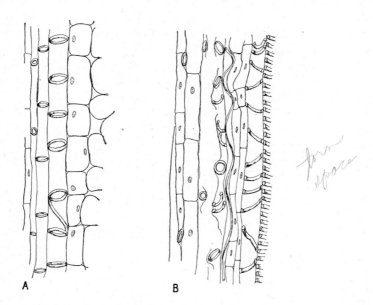

Fig. 63.—Protoxylem in longitudinal section. *A*, from fruit of *Arisaema*, the elements are all annular and there has been little stretching and no rupture; *B*, from stem of *Equisetum*, there has been extensive stretching of annular and spiral cells, with rupture of the former and the formation of a protoxylem lacuna. The space occupied by the first-formed cell has been occluded except where the rings lie.

hand, the growth of tissues surrounding the developing bundle at the time of the rupture of the protoxylem cells may be such that no lacuna is formed, but the surrounding parenchyma cells press in upon the flattening xylem cells and fill up the space occupied by them (Figs. 63B, left annular element; 64A,B). In annular or loosely spiral cells, parenchyma cells thus may be found pushing into the hollows formed by the sagging wall, or tylosis-like, into the lumen when the wall is ruptured. Rings from annular cells sometimes become pressed, upright, between parenchyma cells (Fig. 63B). Under such conditions the position of the first-formed protoxylem cells may be difficult to ascertain in transverse sections.

Ontogeny of Protoxylem Elements.—The stretching and tearing of the walls of the tracheary elements of protoxylem takes place in the thin, delicate primary wall. It has been claimed that the rings, spirals, and bars of the secondary wall are formed by the tearing apart of a previously formed, continuous secondary wall. Early studies demonstrated that in these tracheary elements of protoxylem the protoplast forms secondary

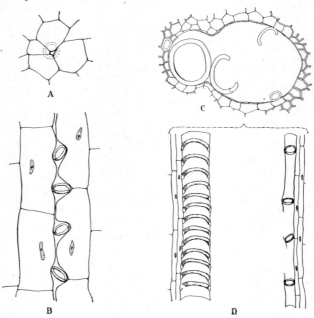

Fig. 64.—Protoxylem in transverse and longitudinal section, showing behavior of surrounding tissue. *A, B,* from *Lobelia,* showing a stretched annular element, the primary wall collapsed between the rings, the adjacent parenchyma cells pressing into the spaces, in cross section the space originally occupied by the cell nearly filled in; *C, D,* from *Zea,* the surrounding parenchyma pulled away from the protoxylem elements, leaving a large lacuna.

wall only in those areas where rings, spirals, etc., are later present. Recent work has confirmed the earlier interpretation. The presence of spiral bands stretched until straight is further evidence that stretching does not break up spirals and form rings.

Arrangement of Cells in Primary Vascular Tissues.—The cells of primary tissues of all kinds and in any organ may be arranged uniformly according to some pattern or may be without regularity of arrangement. The cells of secondary tissues, because of the method of their origin from meristems of the cambium type, tend to be arranged in rows or sheets. It is frequently stated that primary tissues are characterized by irregular, and secondary by regular arrangement, and this distinction is applied to vascular tissues especially, but the orderly arrangement of secondary tissues may be broken up by unequal development of the constituent cells. Also in the primary vascular tissues of many herbaceous and some woody angiosperms, the primary xylem and phloem, especially the protoxylem and metaphloem, stand in radial rows. This arrangement further suggests secondary origin because cells of the secondary tissues stand in rows continuous with them. Distinction of primary and secondary tissues can be made—except for protoxylem and part of the metaxylem—only on method of origin, by procambium or by cambium. Since the cambium itself is developed from the procambium as the formation of primary tissues ceases, only an approximate limit can be set between primary and secondary phloem and between primary and secondary xylem.[1]

Primary Xylem Types.—The type of primary xylem—exarch, endarch, or mesarch (Fig. 61)—is determined by the position of the protoxylem in relation to the metaxylem or, when the primary xylem consists only of protoxylem, by the determination of the position of the first-formed cells. The position of the first protoxylem cells is that of the protoxylem lacuna, if such is present or, where the tissue has closed in upon the torn or flattened cells, by the disturbance in normal cell arrangement and by the fragments of the destroyed cells. For the determination of xylem type, the study of longitudinal sections is sometimes necessary, since in

[1] It has been urged that distinction between primary and secondary tissues is of little or no value because in other tissues, as well as in vascular tissues, no clear segregation can be made; that, for example, internal phloem (in part) and small vascular bundles that form late in the ontogeny of organs arise not from promeristem but from permanent or semipermanent tissue. The development of primary tissues may continue indefinitely in any organ, especially with changes that do not involve increase of the fundamental body—such changes as the addition of vascular bundles with development of storage and fleshy tissues, secretory tissues, protective structures, fruit and seed coats. These new primary tissues are formed not as are typical secondary tissues—in radiating rows from persistent initials—but from short-lived procambium-like cells; they arise secondarily—the meristem is secondary—but the tissues themselves obviously constitute a part of the primary body. Other characters than that of method of development and arrangement can often be used to separate primary and secondary tissues. In the place where the question of limitation of primary and secondary nature most commonly arises—in the transition from primary to secondary xylem and phloem—the length of the elements can be used. Primary tracheary elements, especially in woody dicotyledons, are very much shorter than the similar elements of the adjacent secondary tissue. In vascular tissue the initiating cells of these tissues themselves differ much in length. Distinction of the primary (fundamental) plant body and the later formed secondary (accessory) body—and hence also of primary and secondary tissues—is necessary for an understanding of the changes secondary growth brings about in the primary body. Sharp lines of separation between these tissues are often not possible and are unnecessary.

cross sections annular cells may be with difficulty separated from spiral cells, and the first-formed annular cells may be obliterated at certain levels.

It cannot be too emphatically stated that the different types of protoxylem tracheids and vessels are formed at successively later and later stages in the ontogeny of the tissue. No cell, once mature, changes its character: an annular cell does not become a spiral cell, nor a spiral cell a scalariform cell. The increasing areas of secondary wall in the various cell types are found in cells successively formed.

The Term "Vascular Bundle."—The term *vascular bundle* is applied to a strand-like portion of the vascular system of a plant. Such a bundle consists fundamentally of primary tissues; with these, secondary tissues may be present. Small bundles—bundle ends and the slender bundles of leaves, fruits, etc.—are wholly primary. Larger bundles, such as those of monocotyledonous stems, may also be wholly of primary development; and those of many herbaceous dicotyledons are largely primary in nature. In these bundles the xylem and phloem are highly specialized, the conducting cells being of high type and reduced in number, and supporting cells few or lacking; protection and support are often given to the soft or weak conducting tissues, and support is given to the organ in which the strands lie, by sheaths of fibers which more or less completely surround the strand of conducting cells. Because of the frequent presence of these fibrous sheaths with the conducting cells, the fibers were at one time supposed to constitute a part of the bundle, both morphologically and physiologically, and the bundle was, therefore, called a *fibrovascular bundle*. It has long been clear, however, that the function of the bundle is primarily conduction, and that the "mechanical tissue" of an axis is not necessarily related in position to the bundle; also that the fibrous tissue adjacent to the conducting strand is morphologically not a part of the vascular tissue. (In some bundles it may in part arise from the procambium strand.) Hence, the much better term *vascular bundle* has been in recent years substituted for *fibrovascular bundle*. The new term has not, however, wholly replaced the older and less accurate term. (The vascular bundle is further discussed in Chap. XI.)

Types of Vascular Bundles.—It is characteristic of vascular tissue that phloem or xylem is rarely found alone; a bundle usually consists of both of these types of conducting tissue. When together, xylem and phloem show several types of arrangement with each other in the bundle. These fall into three general classes: (*a*) those in which the xylem and the phloem lie radially side by side; (*b*) those in which one type of tissue surrounds, or ensheaths, the other; (*c*) those in which the two types of tissue are separated from one another. In the first group are *collateral bundles* (Figs. 65A, 139A,D, 140), the phloem lying beside the xylem and external to it; and *bicollateral bundles* which are merely collateral bundles with phloem on the inside of the xylem also (Figs. 65B, 138). Bundles of the second group are known as *concentric bundles*, the type in which the phloem surrounds the xylem being *amphicribral* (Figs. 65C, 139 B), and that where the xylem surrounds the phloem, *amphivasal* (Figs. 65D; 139C). The term "concentric bundle" is often loosely used to indicate an amphicribral bundle, doubtless because the amphivasal type of concentric bundle is uncommon, and "concentric" is believed to be sufficiently specific. Concentric should be used only as a general

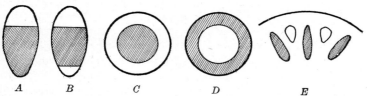

Fig. 65.—Diagrams showing types of arrangement of xylem and phloem with relation to each other. (The xylem is shaded, the phloem unshaded.) *A*, collateral bundle; *B*, bicollateral bundle; *C*, amphicribral bundle; *D*, amphivasal bundle; *E*, radial arrangement.

descriptive term covering both amphicribral and amphivasal conditions. The arrangement of the tissues in the third group is such that no definite bundles are formed. Strands of xylem and of phloem lie on different radii of an axis, separated by nonconducting tissue (Figs. 65E, 130A). These strands are commonly said to constitute *radial bundles*. Since, however, no definite bundles exist—unless each strand be considered a bundle—and the xylem strands are often not independent but are united in a central core, it is better to call such a condition in the primary xylem and phloem *radial arrangement*. The term "radial bundle" goes back to the period before the proposal of the stelar theory; the stele of a root then constituted one bundle, radial in structure. Unfortunately, the term is still occasionally applied to the primary vascular structure of roots. The terms "exarch," "endarch," and "mesarch" are sometimes also applied, descriptively, to bundles; they then merely indicate that the xylem of the bundle is of the stated type.

Occurrence of Bundle Types.—The collateral bundle is the common type of bundle and is characteristic of the stems and leaves of angiosperms and of most gymnosperms. The bicollateral bundle is uncommon, occurring in those angiosperms which possess internal phloem in their steles, as, for example, the Cucurbitaceae. In most plants that have internal phloem, this tissue is not closely associated with the xylem and external phloem in position or method of origin. On a structural as well as a morphological basis, the amphivasal bundles of these plants are not typical bundles; the internal phloem is separated from the rest of the bundle. Amphicribral bundles are common in the ferns; small bundles, such as the bundles in floral parts, ovule traces, and small leaf-trace bundles are commonly of this type. Amphivasal bundles are rather rare. They occur chiefly in the monocotyledons, and there largely in nodal regions and in rhizomes. The radial arrangement of primary vascular tissue is characteristic of roots, where it is always present, and does not occur elsewhere.

Vascular bundles are to be thought of as parts of a unit vascular system, not as fundamental structural units; in the axis they are more or less isolated segments of a stelar column or cylinder.

THE PRIMARY VASCULAR SKELETON

The primary vascular tissues of a plant form a definite vascular skeleton, which may, in a way, be compared with the skeleton of an animal. In the various organs and parts of plants, the vascular tissues differ in arrangement, position, and method of attachment from those of other organs or parts of organs, and these differences are constant and characteristic. The skeleton of a species has a definite and fixed plan and differs more or less from that of other species. The skeletons of the different larger groups of plants—as of the larger groups of animals—differ from one another in important respects; the skeletons of smaller groups differ in less important respects, but may be very varied in structure. Great diversity of vascular structure exists among plants from simple to highly complex.

The Stele.—The axis is fundamentally cylindrical, and the skeleton of this part of the plant body as a whole naturally conforms to this type. The vascular tissues of the stele in their simplest condition form a solid, rod-like column in which the phloem surrounds the xylem. A stele with its vascular tissues arranged in this way is known as a *protostele* (Figs. 66*A*, 127). The protostele is not only a very simple kind of stele, but also is clearly the primitive type from which all others have been derived in the course of evolutionary specialization. The vascular column of a protostele when seen in cross section may be circular; symmetrically angular, such as triangular; stellate, with long-projecting arms; or irregularly rounded and variously lobed. A kind of stele differing from the protostele chiefly in that a pith is present in the center is the *siphonostele*, or *solenostele* (Figs. 66*B*, 135*A*,*E*). This type of stele has been derived from the protostele and represents a stage in evolutionary advance. Like the protostele, the siphonostele shows various outlines in cross section, but is commonly rounded. Two types of siphonostele are found: *ectophloic*, where phloem occurs only on the outside of the xylem; *amphiphloic*, where phloem occurs on the inside of the xylem as well as on the outside. When the siphonostele is broken up into a network or a series of longitudinal strands, it becomes a *dissected siphonostele*, or "dictyostele" (Figs. 66C, 135B,D,F). In some plants, as in the majority of monocotyledons, the vascular bundles of the dictyostele are scattered through the pith and cortex so that the semblance of a stele or ring is lost. Such a condition has clearly developed in evolution from the siphonostele or dictyostele, and is considered to belong to this type.

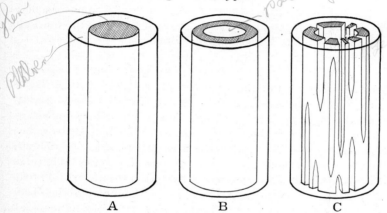

Fig. 66.—Diagrams illustrating the types of arrangement of vascular tissues in steles. *A*, protostele; *B*, siphonostele; *C*, dictyostele.

The term *monostele*, which has sometimes been used as synonymous with protostele, was applied originally to those steles in which the vascular tissues form a unit structure. In contrast with the monostele was the "polystele," a type of stele in which the vascular tissues are in the form of strands, each one of which resembles more or less the entire vascular cylinder of protostelic plants. Thus, the protostele and the unbroken siphonostele have frequently been called "monosteles," and some types of dissected siphonosteles termed "polysteles." The bundles of a dissected siphonostele—especially when the stele is amphiphloic and the bundles amphicribral—resemble small protosteles in cross section, and the stele therefore appears like a multiple protostele, hence the term "polystele" was given to such a central cylinder. It is now understood, however, that this type of stele is merely a broken-up siphonostele. Hence the term "monostele" has little use, and is commonly replaced by the more specific protostele and siphonostele, the broken siphonostele being called a "dictyostele." The term "polystele," as applied to any type of dissected siphonostele, suggests inaccurate morphology, and therefore should not be so used. True polysteles are doubtless present in some groups of fossil plants, but no living plants have this type of stele.

Two theories exist as to the method in evolutionary change by which the siphonostele has been derived from the protostele. According to the "expansion theory," the central portion of the stele does not become vascular but remains less specialized, becoming pith. Therefore, according to this theory, the pith is morphologically vascular tissue. The term "expansion" is unfortunate here since expansion of the protostele has not necessarily occurred. According to the "invasion theory," the cortex has invaded the central cylinder in the course of the phylogenetic development of vascular plants, leaf and branch gaps being the openings through which the change has occurred. Under these conditions, the pith is obviously not stelar in nature. A discussion of these theories lies outside the scope of this book. It seems, however, to have been established beyond much question that, in seed plants at least, the pith is morphologically extrastelar in nature. In most of the pteridophytes the same condition obtains; in a few the pith is possibly stelar in nature.

Occurrence of Stelar Types.—The protostele, being the primitive type, is found more commonly among the primitive plants. It occurs in many of the ancient fossil forms and in the present-day flora is characteristic of the club mosses and a few of the ferns. It is characteristic of the roots of all plants. The siphonostele or some modification thereof is found in the stem of all other living plants. The ectophloic siphonostele is characteristic of the stems of gymnosperms and of angiosperms generally, and is without question the most common type. The amphiphloic siphonostele is found in most ferns and some families in the angiosperms. The dissected siphonostele occurs in many ferns and in the angiosperms to some extent, especially in the herbaceous types. Dissection of the siphonostele is brought about in two morphologically different ways: by the overlapping of leaf and branch gaps, and by the dropping out of segments of the cylinder during reduction of the vascular cylinder in the development of some types of herbaceous stems.

Leaf Traces.—Prolongations of the stelar vascular tissues that supply the leaves (Fig. 67) constitute the *leaf* or *foliar traces*. The traces that supply one leaf constitute the *leaf supply*. The term "trace" is applied to these bundles from the point where they first become evident as foliar supply bundles to the base of the leaf. Structurally, leaf traces are strands of primary vascular tissue, the proximal part of which consists of xylem alone. The distal part of the trace is made up of both xylem and phloem, and secondary tissues may be added in late stages. Because the trace is merely an extension of the vascular system of the stem, the point of origin cannot readily be determined. As an identifiable, protoxylem-containing strand it can often be followed in the stem for some distance below the

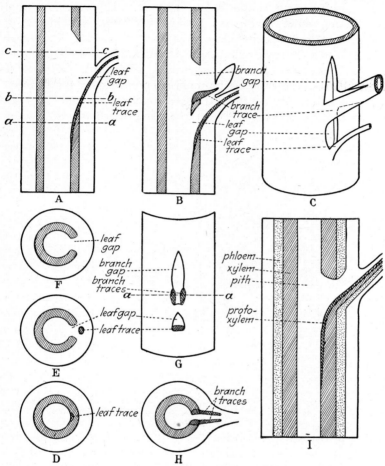

Fig. 67.—Diagrams illustrating leaf and branch traces and gaps. *A*, longitudinal section of node through leaf trace and gap; *B*, similar to *A*, but with branch trace and gap also present; *C*, view of vascular cylinder showing departure of leaf and branch traces, and the gaps associated with each; *D*, *E*, *F*, cross sections through stem illustrated in *A* at levels *a-a*, *b-b*, and *c-c*, respectively; *G*, face view of outside of cylinder shown in *C*, the leaf and branch traces cut away at the surface of the cylinder; *H*, transverse section of *G* at *a-a*. (In diagrams *A–H* the vascular tissue is not differentiated as xylem and phloem; the traces are doubly cross-hatched. In *I*, more detailed structure is shown, protoxylem, metaxylem, and phloem being indicated.)

level at which it begins to swing outward and there found to merge with other traces or with the primary xylem cylinder.

The traces supplying a leaf range in number from one to many, and the number is usually constant for a given species and often for a family and even larger group. This number is the number of bundles that leave the stele; these in their course through the cortex may fuse or branch so that the number entering the leaf is different. The vascular supply to stipules is derived from the lateral traces, usually within the cortex (Fig. 70K). Within the petiole or leaf base, where the bundles are no longer called leaf traces, forking and fusion commonly take place.

Branch Traces.—The primary vascular supply to lateral branches is also derived from the stele of the main axis, usually in the form of two bundles, less often, one bundle. These strands are termed *branch traces*, or *ramular traces* (Fig. 67B, C, G, H). Branch traces, like leaf traces, are

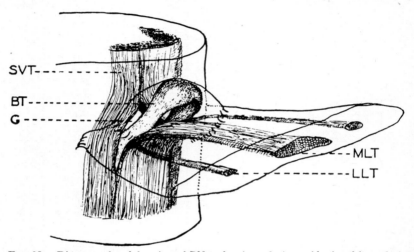

Fig. 68.—Diagram of nodal region of *Phlox* showing relations of leaf and branch traces to stem stele. The leaf trace divides at once into three bundles. The two branch traces, still in meristematic stage in the bud base, arise above and lateral to the leaf trace and show first stage of fusion to form stele of the branch. (*After Miller and Wetmore.*)

connected with the first-formed parts of the primary stelar skeleton Thus, all parts of the axis and the appendages are tied together by the primary vascular system. Where the branch supply consists of two traces, these bundles unite within a short distance, forming a complete stele (Fig. 67C); where but one trace occurs, this strand has usually the cross-sectional form of a crescent or horseshoe with the opening downward, and the cylindrical stele of the branch is formed by the closure of the opening as the trace passes out. While the branch is still in the bud stage its traces are largely in the procambium stage but their form, and relation to the stele of the mother branch is clear (Fig. 68).

Leaf and Branch Gaps.—In the majority of vascular plants the outward passage of a leaf or branch trace is associated with the formation of a break or interruption in the vascular cylinder around and above the point of departure of the trace (Figs. 67B,C,G, 69A–F). This opening, through which the cortex and pith become continuous, is known as a *gap*—a *leaf gap* accompanies a leaf trace, and a *branch gap* accompanies a

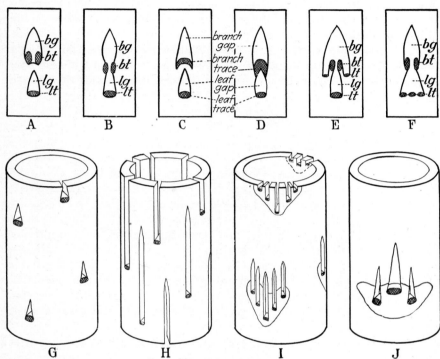

FIG. 69.—Diagrams showing variety of form of leaf and branch traces and gaps. A–F, face views of nodal region, showing traces cut away at outside of vascular cylinder, and the gaps: A–D, one leaf trace, E, three traces and three gaps, F, three traces, departing together, leaving one gap; A, the leaf gap closed below the two branch traces and the branch gap; B, the leaf gap fused with the branch gap; C, the leaf gap closed below the departure of the single branch trace; D, the single branch trace departing at the top of the leaf gap; E, three leaf gaps from as many traces, all fused with the branch gap; F, one leaf gap from three traces, fused with the branch gap. G–J, diagrams of vascular cylinders showing various amounts and types of dissection by leaf gaps.

branch trace. Leaf gaps are constant in appearance in the great group of vascular plants known as the Pteropsida. This group is made up of the ferns, the gymnosperms, and the angiosperms. Leaf gaps do not occur in the Lycopsida, a group which includes the club mosses, horsetails, and a few other similar plants. Branch gaps are present in all vascular plants which possess a pith. In protosteles, gaps, of course, do not occur, since no pith is present. Gaps are not associated with root traces.

Leaf gaps vary much in width and in longitudinal extent. Their size is not directly related to the size, type, or persistence of the leaf. Commonly, in the gymnosperms and angiosperms, leaf gaps are small, extending but a short distance above the point where the trace leaves the vascular cylinder. In the ferns, leaf gaps are generally larger and may extend for considerable distances, even through several internodes. Branch gaps are commonly larger than leaf gaps and extend for greater distances in the axis.

The Breaking of the Vascular Cylinder by Gaps.—The vascular cylinder is broken up in different degrees by the presence of leaf gaps. Where the gaps are small and of limited longitudinal extent, the cylinder is but slightly broken by them (Fig. 69G); but where they are large and elongate, extending through one or more internodes, the siphonostele is dissected (Fig. 69H). The degree to which a siphonostele is dissected depends upon the number and extent of the gaps and upon the closeness of the nodes. Where the gaps overlap, due to their great length or to the shortness of the internodes, the cylinder consists of a network of bundles which in cross section appear as a circle of discrete strands (Fig. 69H,I). By the presence of leaf gaps alone, the cylinder may thus be broken into distinctly separated strands. Branch gaps still further complicate the structure. These leaf and branch gaps together make the primary vascular cylinder complex in the arrangement of its vascular tissues, especially in nodal regions where the leaf and branch traces are departing. It is further complicated in many herbs and vines by the presence of the breaks formed in the cylinder by reduction of vascular tissue that accompanied the evolutionary development of the special stelar structure of these plants.

Branch traces are separated abruptly and depart as soon as freed from the cylinder (Fig. 67C); leaf traces, on the other hand, may be freed from the cylinder and yet maintain their position in the circle for some distance before passing out into the cortex (Fig. 70A). Below the point where the trace is separated from the cylinder, it is frequently evident for some distance as a distinct—though not isolated—strand, chiefly of protoxylem (Fig. 70A–E). This distinctness of the strand in the xylem cylinder is due to the type and size of its cells that are different from those of the adjacent xylem. Externally, the trace is not definitely limited, merging into the xylem of the cylinder (Fig. 70E). Such a downward extension of the trace may be very short or lacking, or may be several internodes in length.

A leaf-trace bundle is commonly freed from the stelar cylinder on both sides simultaneously, but one side may remain attached for a time, even while the trace swings out into the cortex (Fig. 70F,G,H). The trace then appears to depart from the side of the gap rather than from its base. Where a trace departs from the side of a gap or other opening in the cylinder already present, the gap theoretically associated with the new trace is not evident, being merged with the larger gap.

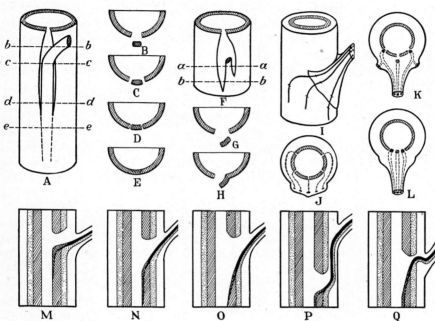

Fig. 70.—Diagrams showing variations in the method of departure of leaf traces. *A–E*, the trace freed some distance below its departure from the cylinder: *A*, face view; *B–E*, cross sections at levels *b-b* to *e-e*, respectively. *F–H*, the trace freed on one side before the other: *F*, face view; *G* and *H*, cross sections at levels *a-a* and *b-b*, respectively. *I* and *J*, traces departing from the cylinder in regions not underlying the leaf attachment—"girdling traces": *I*, course of traces seen from the outside, the points of departure from the cylinder and of entrance to the petiole indicated; *J*, the course of the traces projected from their points of departure to the petiole. *K*, projected course of three traces arising separately; stipule supply arising from lateral traces. *L*, projected course of three traces arising from one gap. *M–Q*, longitudinal sections of nodes showing various courses of the departing trace.

The Number of Leaf Traces in the Plant Groups.—The number of traces supplying the leaf ranges, as above stated, from one to many. In the pteridophytes there is usually one, though sometimes there are two or many; in gymnosperms, one or two; and in angiosperms, one, three, five, or many. Three is perhaps the primitive number for the last group; where there is one this is either the result of fusion of the original three in evolutionary development, or of reduction of the three to one by the loss of the lateral bundles. Three and one are the most common leaf-trace numbers in the flowering plants, three being characteristic of nearly all the Amentiferae, and of such families as the

Rosaceae, Aceraceae, and Compositae; and one of the Lauraceae, Ericaceae, and Labiatae. The number of traces is largely independent of the size, type, and duration of the leaf, and of the nature of its attachment. Many plants with large leaves, such as *Fraxinus*, have one trace; others, such as *Juglans*, three; and still others, such as *Aralia*, many; plants with small leaves likewise have one, few, or many traces. The floral bracts of *Salix*, which are minute, ephemeral leaves, have three traces. Leaves with clasping bases may have many, as in the Umbelliferae, or one, as in the Caryophyllaceae.

The Departure of the Leaf Trace from the Vascular Cylinder.—The traces of a leaf normally depart individually, often at considerable distances from one another laterally about the stem, as well as vertically on the axis (Fig. 70*I*,*J*,*K*). Frequently, however, the traces depart side by side (Figs. 69*F*, 70*L*). Only one gap is then formed by the group. Traces most frequently depart from that segment of the stele which lies directly beneath the attachment of the leaf. Where there is more than one trace, the median trace appears opposite the center of the leaf, the laterals arising successively higher and higher in the stem and farther and farther around the cylinder from the median trace (Fig. 70*I*,*J*). Traces may depart even from the side of the stele opposite to that on which the leaf stands. Such traces enter the base of the petiole directly if the latter clasps the stem extensively, or may "girdle" the twig in the cortex, swinging around the stem as they pass upward through the cortex to the petiole base. When traces thus pass around the stem for some distance in their course from vascular cylinder to petiole, they have been called "girdling traces" (Fig. 70*I*, *J*). The leaf traces of the cycads are girdling traces of an extreme type. Of the bundles of a leaf supply, the median trace is commonly the largest; the laterals form a series progressively smaller toward the margins of the leaf base. Lateral traces may, however, be stronger than the median trace, as in the potato plant (Fig. 73).

The angle at which the trace leaves the central cylinder varies greatly, being usually very small, the strand departing gradually from the pith and passing obliquely, sometimes almost vertically, through the cortex (Fig. 70*N*,*O*,*P*). Less commonly, the trace passes outward at nearly right angles to the stele (Fig. 70*M*), entering the leaf base after a very short course through the cortex. Branch traces also commonly pass out at nearly right angles.

The Dissection of the Vascular Cylinder by Reduction.—The primary vascular cylinder may be broken into isolated strands by leaf gaps alone, the number and width of the strands varying with the pattern of leaf-trace number and phyllotaxy and with the width and length of the gaps

(Fig. 69*H*,*I*), but in many plants the primary cylinder is still further dissected. This additional breakup, which took place in evolutionary development, has been brought about by a progressive reduction in the amount of the primary xylem: at first radially, so that the cylinder became thin (Fig. 71*A*); secondly, tangentially, so that some longitudinal strips of the cylinder—those shown thin in Fig. 71*A*—lost all primary vascular tissue. The nonvascular areas are sometimes called *inter-*

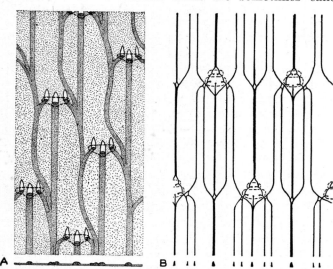

FIG. 71.—Diagrams of primary vascular systems of stems, the cylinder split longitudinally and shown in one plane, and cut transversely at the level of the base of portion shown in face view. *A*, *Populus canadensis*, the cylinder broken only by leaf gaps, and thin between the protoxylem strips that constitute the basic skeleton; *B*, *Pilea pumila*, the cylinder consisting of the protoxylem-containing bundles, the areas between lost in reduction. Leaf- and branch-trace origin shown in *B*, leaf trace only in *A*.

fascicular rays (Chap. XI). Tangential reduction is found in most herbaceous and some woody angiosperms. In the majority of these plants secondary growth soon closes the interfascicular strips and buries the primary vascular cylinder (Chap. VI). The primary-xylem skeleton is easily seen only in early stages of stem development. In a small proportion of herbs—those often erroneously considered typical of the group anatomically, such as *Ranunculus, Curcurbita, Impatiens, Pilea*— the nonvascular areas are very wide and are not closed by secondary growth (Fig. 71*B*). In these the primary skeleton is easily seen even in the mature plant. (Further discussion of stem structure is found in Chap. XI.)

The primary vascular cylinder, dissected by leaf and branch gaps or by both these and interfascicular rays, is a network of anastomosing strands

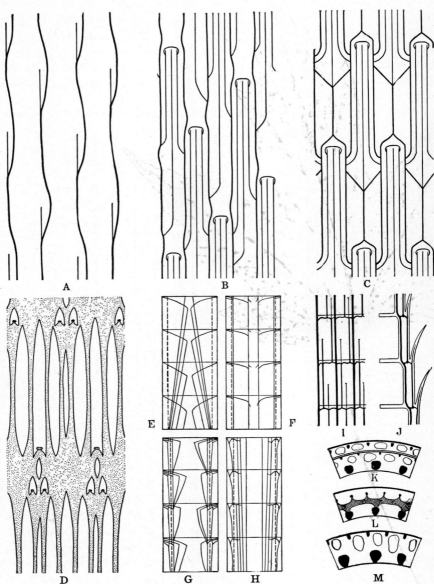

Fig. 72.—Diagrams of the primary vascular system (or of a part of the system) of stems. *A–D*, the cylinder split open and shown in one plane: *A, Thuja occidentalis* (the lines represent broad bands, separated by narrow breaks); *B, Chenopodium glaucum; C, Iresine paniculata; D, Ephedra distachya. E–H,* types of bundle course in the monocots—the horizontal lines indicate nodes, the dotted vertical lines imaginary cortical limits: *E,* a palm type: *F, Tradescantia virginica; G,* rhizome of *Acorus Calamus; H, Scirpus cyperinus. I–M, Dulichium arundinaceum; I,* bundle system seen in one plane; *J,* longitudinal section; *K–M,* cross sections: *K,* above the node; *L,* through the node; *M,* below the node. (*B, C, after Wilson; D, after Thompson; E–M, after Plowman, E, from Von Mohl, F and G, from Guillaud, H, modified.*)

THE PRIMARY BODY 151

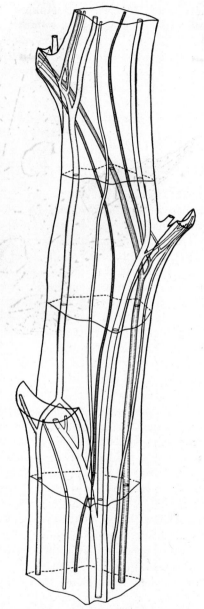

FIG. 73.—The primary vascular system of the stem of *Solanum tuberosum*. The complex nodeal structure is repeated at each node, so that, except for the position of the bundles in the stem, all nodes and all internodes are alike. The large bundles are cauline bundles; the small ones, common bundles. All these bundles are later tied together by thin flanges of late-developing primary vascular tissue (not shown), and, still later, the entire primary xylem skeleton is buried by secondary tissue. (*After Artschwager.*)

that is characteristic of each species. Anastomosis takes place chiefly in nodal regions, and long straight bundles may extend through the internodes (Figs. 71*B*, 72*B,C*). In many monocotyledons and in some dicotyledons the bundles do not lie in a cylinder but are arranged somewhat in the form of a loose sheaf, so that they appear in cross section as though scattered. In some ferns, such as *Pteridium*, and in a considerable number of dicotyledons, for example, *Dianthera*, the bundles form an open cylinder, but a few of the strands run through the cortex or the pith. In all stems the nodal regions are more complex than internodal. In monocotyledons the many bundles form extremely complex nodal structure (Fig. 72*E–M*). There is little clear information as to the basic plan in these nodes. In some there are resemblances to the nodal structure of herbaceous dicotyledons.

The bundles of a dissected primary cylinder have been interpreted as *leaf-trace bundles, cauline bundles*, and *common bundles*. The use of these terms is not consistent, and the distinction of the types is difficult and has doubtful value. A bundle that distally connects only with the vascular system of a leaf, even if its point of origin is far below the leaf, is a leaf-trace bundle. (The term "leaf trace" is restricted in definition to that part of a bundle between the point of its departure from the stele to the base of the leaf.) The term "cauline bundle" is usually applied to bundles that form the major vascular system of a stem; to bundles that branch, uniting with other similar bundles, and may give rise to leaf-trace bundles but do not terminate as leaf traces. The term "common bundle" is often applied to bundles that pass through the stem unbranched for some distance and terminate as leaf traces; they are bundles common to both stem and leaf. The term common bundles is merely another name for the part of a leaf-trace bundle that lies in the vascular cylinder. The bundles of a stem can be simply termed "stem bundles" and "leaf-trace bundles." A logical use of terms for these bundles depends upon an interpretation of their fundamental morphology—whether the vascular cylinder of the stem is a basic unit or whether it consists of fused leaf traces. A discussion of this question is beyond the scope of this book.

General Structure of the Primary Vascular Cylinder.—In most woody and some herbaceous angiosperms, the primary vascular cylinder of the stem is not so extensively broken up by interfascicular parenchymatous bands as in the forms just discussed. It is perforated only by comparatively small and more or less remote leaf and branch gaps. In such cylindrical sheets of tissue, there are commonly more or less prominent ridges projecting into the pith (Figs. 71*A*, 74*F–K*). These ridges correspond morphologically to the bundles of the more dissected stele. In the ontogeny of the vascular cylinder these ridges mature first, the larger and most deeply projecting ones earliest. The innermost parts of these early formed strands consist of protoxylem. Later, less prominent ridges develop, and these have little or no protoxylem. When the stem is very young and the connecting flanges of tissue are still undifferentiated,

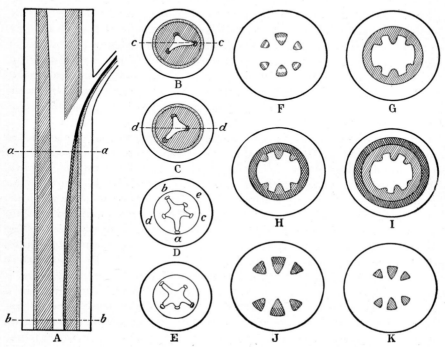

FIG. 74.—*A–E*, diagrams showing the effect upon pith form of the gradual departure of the trace. *A–C*, based on *Alnus*; *D, E*, on *Quercus*. *A*, longitudinal section (in plane *c-c* of *B*); *B* and *C*, cross sections of stem shown in *A* at levels *a-a* and *b-b*, respectively (the traces depart according to one-third phyllotaxy); *D, E*, cross sections of a similar stem with two-fifths phyllotaxy, *D* being one internode lower than *E*; the traces are labeled, in order of their departure, *a, b, c, d, e*. *F–K*, cross-section diagrams showing the form of the primary vascular cylinder, the inwardly projecting ridges (here exaggerated) representing cauline bundles and leaf-trace bundles as seen in an internode (*F–I* based on *Populus*): *F*, young stem in which only the first primary xylem, forming the ridges, is mature; *G*, the ridges connected by later-formed primary xylem; *H*, the ridges connected only by secondary xylem; *I*, like *G*, but with secondary xylem also; *J*, the first-formed primary bundles not later connected, and secondary xylem formed on the outside of the bundles; *K*, like *J*, but without secondary xylem. (Phloem is not shown in *F–K*.)

the ridges constitute discrete bundles (Fig. 74*F*). When the primary vascular tissue has all matured, uniting the bundles into a cylinder (Fig. 74*G*), these first-formed strands are still prominent, owing to their position nearer the center of the pith and the presence of protoxylem. In those plants where, during primary growth, the early-formed bundles are not later connected laterally by primary vascular tissue, a cylinder of isolated primary strands is formed, but the bundles may be united by secondary tissues (Fig. 74*H*). A complete cylinder may thus be formed, the first-formed primary bundles united either by primary or by secondary growth (Chap. VI). Where these connecting bands, either primary or secondary, do not develop, a stele of the so-called "herbaceous type," with isolated strands, is formed (Fig. 74*J,K*). This condition is structurally comparable with the early stage of a complete vascular cylinder when only the first-formed bundles have appeared. In some dicotyledonous herbs, for example, *Asclepias*, *Hypericum*, *Digitalis*, there are no prominent, early-formed protoxylem strands except leaf traces close to the point of departure; protoxylem arises nearly simultaneously and uniformly around the cylinder.

As seen in a cross section of the mature woody cylinder of a stem, the most prominent of the ridges projecting into the pith are usually leaf-trace bundles, the largest being median traces of the leaves next above. As the node is approached these largest traces become less and less pronounced as wedges projecting into the pith because they swing outward, either gradually (Fig. 74*A*) or abruptly, and come to lie in indentations of the vascular cylinder rather than on projecting points (Fig. 74*B–E*). Ridges of the pith then project into the vascular cylinder. As the traces pass farther and farther out, the pith ridges increase in extent outward and break the vascular cylinder above the trace as it passes out, forming a leaf gap. The shape of the pith as seen in cross section may indicate clearly the phyllotaxy of the plant in question. Thus a one-third phyllotaxy may be evident in a three-lobed pith, as in *Alnus* (Fig. 74*A–C*); a two-fifths in a five-lobed pith, as in *Quercus* and *Populus* (Fig. 74*D,E*); a one-half by an oval or elliptical pith, as in *Ulmus*. In long internodes projecting ridges of the pith are prominent only where the traces gradually pass obliquely outward; the more gradual the departure of the trace, the longer, and often the deeper, the lobe. Where the number of traces per leaf is more than one, the lobing may be obscured, but lateral traces commonly are smaller and pass out much more abruptly than do median traces, and have little effect upon pith shape.

The number of first-formed strands, whether ultimately still free or fused into a cylinder, may be few or many. These become more or less obscure where an unbroken cylinder is formed, whether this is of primary nature only, or whether secondary growth is also concerned in the completion of the cylinder. In monocotyledons the large number of bundles and the complexity of the system render the scheme of arrangement usually difficult of interpretation.

THE PITH

The pith is a roughly cylindrical body of tissue in the center of the axis, enclosed by the vascular tissues (Fig. 135). Its outer surface is furrowed more or less deeply by the inwardly projecting strands of protoxylem and ridged in some plants by ray-like extensions between bundles and by projections where its tissues extend out through leaf and branch gaps. The number, prominence, and arrangement of these furrows and ridges depend upon the skeletal plan of the plant in question and upon other features. (The pith was long known as the *medulla*, but this Latin name has become obsolete. As a descriptive word, the name still persists, as in "medullary bundle.")

Structure of the Pith.—Histologically, the pith is made up of rather uniform tissue, chiefly parenchymatous, in which the cells are arranged rather loosely, often with pronounced intercellular spaces, and in some forms tending to be in longitudinal rows (Fig. 43A,C,D). In shape, the cells of the pith vary greatly, but they are mostly isodiametric or cylindrical, with thin, cellulose walls. Not uncommonly, thick-walled, lignified parenchyma cells and sclereids are also present, often arranged in groups forming disks or diaphragms of firm tissue. There are two types of "diaphragmed pith": that in which the smaller, thinner walled cells between the disks persist indefinitely (*Liriodendron, Nyssa*), and that in which the interdisk cells collapse toward the end of the growing season in which they were formed (*Juglans*). Fibers occur only rarely and then in the peripheral portions where they are associated morphologically with primary vascular tissue, especially with internal phloem. When the pith is developing, all the cells are active and in leafy shoots may contain chlorophyll, but when the region is mature, the cells become less active, and all may have lost their protoplasts, or living and dead cells may be present in any proportion. The proportions vary in different parts of a plant, such as the node and the internode, and in different species. Usually, the smaller cells and those nearer the vascular tissue remain alive. The small living cells may form a fairly definite pattern with the nonliving cells. In woody plants the living cells of the pith serve as storage cells in resting periods, becoming filled with starch and fatty substances.

In kinds of cells, intercellular spaces, secretory tissues, and cell contents, the pith is usually closely similar to the cortex of the same plant, except that protective, supportive, and photosynthetic cells are scarce or lacking in the pith.

In the ontogeny of the stem, the pith cells in many plants mature very early. The surrounding tissues are meristematic and continue to elongate so that the pith may be torn apart longitudinally to a greater or less extent. If marked radial increase is taking place at the same time, a prominent "hollow pith" is formed, with the broken cell walls lining the cavity (Figs. 135D, 136C). This condition is common among herbs but rare in woody plants. Where the destruction is less extensive, cavities or canals of various extent and shape are formed. Nodal diaphragms are due to the presence of thicker walled cells at the node, or to the very slight elongation of the nodal regions as contrasted with the extensive growth of the internode (the expansion of a bud being due in large degree to the development of internodes).

The Medullary Sheath.—The peripheral layers of the pith-like central core of many stems consist of smaller cells, usually thicker walled, more closely packed, and more richly protoplasmic. Although this region merges into the central part, it is often distinguished as the *medullary sheath*, or *perimedullary zone*. If an inner endodermis is present, the medullary sheath is separated from the pith proper by this layer. The outermost parts represent parenchyma of the primary xylem, since the groups of protoxylem tracheids and vessels project into the sheath; when internal phloem is present, the sieve-tube clusters lie near the middle of the sheath. This layer is not, therefore, morphologically, a part of the pith.

The cells of the medullary sheath may be parenchymatous, as in many Chenopodiaceae, Boraginaceae, and Euphorbiaceae, or sclerenchymatous, as in members of the Umbelliferae and Compositae—or both parenchymatous and sclerenchymatous cells may be present. Fibers occur only rarely, and then chiefly in connection with internal phloem.

The Pith of Roots.—Roots characteristically lack a pith. Where such a core is present, it is similar in structure to that of the stem of the plant but tends to be more homogeneous and does not break down. It is more nearly cylindrical because the primary xylem points do not project into it, and there are no gaps in the vascular cylinder.

Duration of the Pith.—The pith persists indefinitely in nearly all plants. In woody stems the changes taking place in heartwood formation of the first annual rings affect the pith cells in a similar way. But until this stage is reached, some of the pith cells in most woody plants remain alive. In other forms all the pith cells die very early. The pith is not crushed by the crowding in upon it of the vascular bundles as secondary growth takes place. Only in a few woody vines of anomalous stem structure, such as *Aristolochia*, is there found this peculiar condition of crushing during secondary growth. Furthermore, no growth or structural change occurs in the pith after primary growth of the axis is complete. Therefore, the pith is present in tree trunks and other old stems in size, shape, and structure as it was in the young twig when secondary growth began. Only in lack of protoplasts and in chemical nature is it different.

THE PERICYCLE

The pericycle is a thin cylinder of tissue, at most only a few rows of cells wide, sheathing the vascular tissues. It is limited internally by the primary phloem and externally by the endodermis. Where the latter is lacking, the pericycle merges with the cortex. Typically this layer consists of parenchyma, as in most roots and in the stems of the pteridophytes. The abundant "pericyclic fibers" of some plants have in recent years been shown to be a part of the primary phloem—cells surrounding protophloem elements that disappeared early in the ontogeny of the tissue. To what extent the so-called "pericyclic fibers" generally are phloem fibers is at present unknown. Certainly many of them—such as those of *Cannabis* (hemp) and *Linum* (flax)—belong to the phloem. Whether those of plants in which the fibrous strands alternate with the protophloem groups also belong to the phloem is apparently unknown. (Study of the ontogeny of the region is necessary for the determination of the tissue to which so-called "pericyclic fibers" belong.)

It has been claimed that no pericycle is present in the stems of many angiosperms because the fibers which were thought to make up much of this layer belong to the phloem. A narrow well-marked pericycle is present in the vascular cryptogams, in both root and stem, and in seed plants, in the roots. Only in the stems of some seed plants does it appear to have been lost. In the stems of many angiosperm seedlings and in the bases of the stems of many herbs, a true or a vestigial endodermis is present and a narrow band of parenchyma separates the endodermal sheath from the phloem. This band represents either the outermost layer of the primary phloem or a pericycle closely like that of the root with which the stem is continuous. In other plants, especially woody herbs, the protophloem fibers lie against the endodermis, and no pericycle is present. The pericyclic region is in need of critical detailed and especially comparative study.

In roots, the pericycle is normally parenchymatous. This is the region of origin of the meristems which form lateral roots and commonly of the first phellogen layers of roots, as well as of secondary cambium in anomalous steles. It is doubtless for this reason that the pericycle of roots was formerly known as the *pericambium*. Adventitious roots and stems also arise commonly in the pericycle. In old roots the cells of the pericycle may become lignified or even suberized.

The parenchyma cells of the pericycle share the function of storage with similar cells in other regions. Secretory cells and canals, laticiferous cells, and other specialized cell types may occur in the pericycle.

THE ENDODERMIS

The endodermis is a uniseriate sheet of cells separating the stele from the cortex, without intercellular spaces and with structural features unlike those of other cells. The cells constituting this layer are typically elongate, with the long axis parallel with the course of the vascular tissue and with the end walls mostly transverse. In cross section they are usually more or less oval or elliptical with the longer axis in a tangential direction. The protoplasts are those of typical parenchyma cells. Starch, tannin, and mucilage are frequently present; sometimes, as in *Apios*, crystals are abundant.

The important peculiarity of endodermal cells is the presence in restricted parts of the wall of a waxy substance similar to cutin and suberin, which makes those parts impervious to water. The "waterproofing" substance takes lignin stains to some extent and has recently been called lignosuberin. Endodermal cells are of two types, thin-walled and thick-walled. In the former, the modified wall areas are in the form of bands known as *Casparian strips* that run completely around the cell on the radial and end walls (Fig. 75). These bands range in width from minute threads to broad bands that occupy the entire radial wall. They may be merely chemically different from the rest of the wall but are usually slightly thicker. In cross section (Fig. 75A–C) the strips are often called *Casparian dots*, or *radial dots*. This thin-walled type of endodermal cell, which is the more common type, is sometimes called *primary*. It seems to represent a first stage—after the meristematic—in development, a stage that persists as the mature condition in pteridophytes and most dicotyledons.

In the thick-walled type of endodermis, the radial and inner tangential walls—sometimes all the walls—are thickened (Fig. 76) by suberin lamellae laid down over the earlier formed wall with its Casparian strips. For this reason this type is called *secondary*—it is not a secondary tissue. The thickening of the wall may be so great as nearly to close the lumen (Fig. 76A). The thick wall, like the Casparian strips, is strongly "suberized." In late stages in this thickening, the lamellae may be largely or wholly of cellulose. Frequently in this type of endodermis there occur occasional isolated, thin-walled cells, known as *passage cells* or *transfusion cells*, which have no suberized areas. In roots these cells are situated opposite the protoxylem cells. Both radial and tangential walls may be heavily pitted, pits occurring in the Casparian strips when the strips are wide; the end walls have few or perhaps no pits.

The primary type occurs in the pteridophytes and in most of the dicotyledons; the secondary type is characteristic of the monocotyledons.

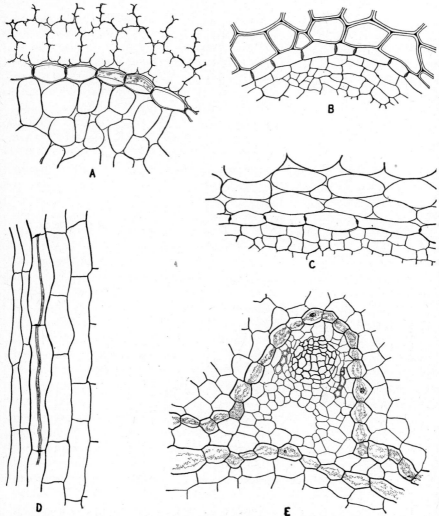

Fig. 75.—The endodermis. *A*, cross section in leaf of *Pinus*, showing the Casparian strips in cross section, and in face view on the end walls; *B*, cross section in rhizome of *Polypodium;* *C, D*, cross and longitudinal sections in stem of *Lobelia*, showing the radial dots, and the Casparian strips in face view on the radial walls; *E*, cross section of outer and inner endodermis in stem of *Equisetum*, showing the protoplasts plasmolized, but retaining connection with the Casparian strips.

The latter has been called a *phloëoterma*, but this term has been applied in various ways and is now rarely used.

Occurrence and Position of the Endodermis.—In position in the plant body the endodermis is always in rather close association with the vascular system. Typically, it lies external to the vascular cells, either abutting directly upon them or lying external to the pericycle. It, therefore, separates the stele from the cortex. In some plants an *inner*

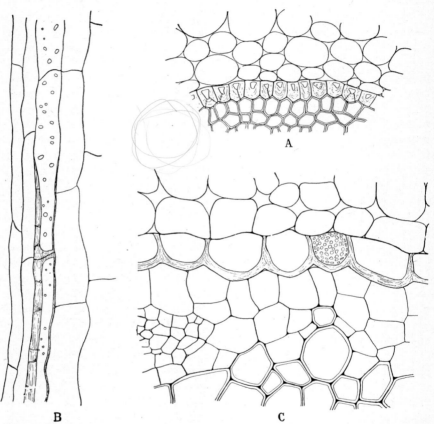

FIG. 76.—The endodermis, thick-walled types. *A, B*, cross and longitudinal sections from root of *Smilax rotundifolia; C*, cross section from root of *Musa sapientum*.

endodermis is also present in the stem. This endodermis which, morphologically, is doubtless only a part of the external endodermis, separates the stele from the pith.

The endodermis has been considered both the innermost layer of the cortex and the outermost layer of the stele. It has been claimed to be cortical on the basis of histogenesis because in roots it is derived commonly from the same mother cells as are the innermost cortical cells.

However, it has been shown that it may be formed in the stem from procambium mother cells which also form pericyclic cells and protophloem, and that in roots, it may belong histogenetically to either the stele or the cortex. Functionally it is obviously a layer separating these two regions. For convenience in description, it is here considered the limiting layer of the stele.

An endodermis that limits vascular tissue on the inside and shuts it off from the pith is called an *inner endodermis* (Fig. 75E). Such a layer is not structurally distinct from the outer endodermis since the two are continuous through leaf and branch gaps. The endodermis not only limits the vascular tissue of the axis but may also surround the vascular bundles in leaves, as in *Plantago,* and in perhaps modified form, in grasses, as the "mestome sheath," but it is not a constant feature in vascular plants. It is present in all roots and throughout the plant body in most pteridophytes. In many gymnosperms, it is characteristic of the leaves but is mostly absent in stems. In angiosperms, it occurs in some parts of the stems of a majority of herbaceous species. It is present in aquatic plants and species of moist habitats; in creeping plants and in many seedlings; in rhizomes and some leaf bases; in the bases of some stems that lack it in the upper parts. Woody stems and the leaves of angiosperms generally have no endodermis. In herbaceous plants the endodermis occurs more commonly in the advanced families, especially in the gamopetalous and apetalous dicotyledons. In the monocotyledons, where it is more common than in the dicotyledons, the secondary type is usually present. In stems of many plants where a typical endodermis is absent, a layer of cells closely similar to an endodermis, but with simple cellulose walls, takes its place. Such a layer is probably a vestigial endodermis.

The occurrence of the endodermis in a given genus or even in an individual plant may be highly variable. In the genus *Piper,* for example, it may form a continuous cylinder of typical cells or be represented by a vestigial layer. Again, the endodermis may be discontinuous: present in some internodes and lacking in others; present opposite vascular bundles and lacking between them.

Function of the Endodermis as Related to Structure.—The function of the endodermis has been a matter of much discussion since this layer was first studied, and is still largely in question. Many functions have been ascribed to it. These are based largely on its obvious relation to water and to vascular tissue. The endodermis appears to be a "watertight layer" between vascular cells and surrounding tissues, especially where the organs in question are in a moist or wet situation, but it occurs also in many plants of dry soil and in the leaves of xerophytic gymnosperms.

Its absence in woody twigs is possibly the result of loss in evolution. Further evidence that the endodermis is a sort of water dam is (*a*) it always lacks intercellular spaces, and (*b*) the walls are more or less cutinized or suberized so that in the thin-walled type, water can pass through the endodermis only through the tangential walls and through the protoplast, that is, through a semipermeable membrane. For this reason it has been looked upon as a diffusion layer separating regions of different osmotic pressure and as a diffusion layer that prevents loss of mineral nutrients, or of food, from vascular tissue.

Among the many and diverse functions that have been assigned to the endodermis at various times are the following: (*a*) It is a mechanical protective layer, a sort of inner, accessory epidermis. (And, as such, the secondary type in monocotyledonous roots may serve to some extent when the cortex has sloughed away, but the majority of endodermal layers are of delicate nature.) (*b*) It is connected with the maintenance of root pressure. (*c*) It is an air dam, preventing the water-conducting cells from becoming clogged with air.

The persistence of starch in endodermal cells has suggested theories—now long abandoned—that the endodermis is a "starch sheath," a carbohydrate-conducting or storage layer, or a layer limiting storage to the inside or to the outside. The presence, type, and behavior of starch grains in the endodermis of certain plants have also led to the theory that this layer is an orienting "organ," the starch grains being statoliths which, by alteration of position in the cytoplasm, cause sensory stimuli leading to change in orientation of the organ. Such a function can be characteristic of the endodermis only of certain plants, and of parts of organs, since starch grains are not always present in endodermal cells, and in many types of such cells cannot "fall" from one position to another as the orientation of the organ is changed because of the presence of gums and crystals in those cells.

The function of the endodermis is thus in question. That it frequently has to do with the relations of water and xylem seems to be without doubt. Its specific functions in different plants probably vary greatly since types of endodermis characterize families to some extent. Secondary functions doubtless also exist, and the primary function may have been lost in some plants.

Many of the facts regarding the distribution of the endodermis in plant groups and its structural variations suggest the possibility that the endodermis is an ancient structure of much physiological and perhaps morphological value, which has become modified in the course of phylogenetic development. In part, it may still retain its original relations and functions; in part, it may be more or less vestigial in nature. In some plants it may have become specialized or modified in adaptation to new functions, as vestigial structures frequently are. It has disappeared in woody twigs where continuing secondary growth occurs.

The cells of the endodermis, except those of the thick-walled type, may become meristematic at any time. In them the cork cambium of the root develops frequently—that of the stem, occasionally. The initials of lateral roots and of adventitious buds commonly arise in this layer. Since the endodermis matures early in the development of the primary tissues, its continuity as a sheath must be maintained during primary increase in length and diameter of the region. If it is to persist long during increase in diameter resulting from secondary-tissue development, continued new-cell formation must take place. Radial divisions in the mature cells maintain the layer for a time, but in woody stems and roots secondary growth soon crushes this layer, and all evidence of its former presence may be lost.

THE CORTEX

The portion of an axis that surrounds the central cylinder and is separated from the cylinder by the endodermis is the *cortex*. It is limited on the outside by the external uniseriate layer of the axis, the epidermis. In good usage the term cortex applies only to the definite region thus morphologically distinct, and only to the *primary* cells and tissues developed therein. Distinction is sometimes made between "primary cortex" and "secondary cortex"—the latter term being applied to secondary tissues developed within the primary cortex from cortical cells, chiefly to periderm. Such a use is, however, misleading, since similar layers, such as those formed in the secondary phloem, are also loosely termed "secondary cortex." This usage is, of course, due to confusion resulting from the use of the term "cortex" in the loose sense of an outer part or covering, and in the physiological sense of any protective outer layer. The term cortex should be restricted to the definite primary region.

In thickness, the cortex varies from a few to a great many rows of cells (Fig. 77). It is esentially parenchymatous in nature, but may contain many kinds of cells arranged in many ways. Collenchyma often constitutes a large proportion of the tissues. Ridges—and in angular stems, the corners—may consist largely of this temporary supporting tissue. Fibers and sclereids, as well as secretory and storage cells of various types, occur freely. Fibers commonly are in sheets or in large strands; they are often associated with the epidermis, forming below it an accessory outer protective layer, a *hypodermis*. The term "hypodermis" is applied to a layer of supporting or protecting cells of any type which lie immediately under the epidermis and reinforce this layer in some way. The cortical parenchyma cells usually contain chloroplasts, and definite specialized photosynthetic tissue may be formed (Fig. 181). The cells are arranged loosely or closely, and commonly no definite plan of arrangement occurs. Tangential rows may be conspicuous, and in specially modified roots and stems, such as those of aquatic plants, symmetrical radiating sheets of cells may occur. The cortex of roots is more homogeneous than that of stems and usually consists of parenchyma only (Figs. 127B, 133).

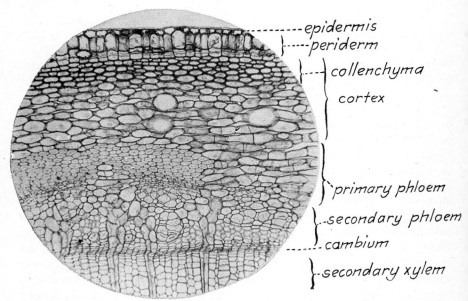

Fig. 77.—The cortex in a woody stem, *Magnolia acuminata*. The cortex consists of collenchyma and parenchyma. The cells of the latter are in tangential rows formed in part by continued radial division as an accommodation to diameter increase in the stem. Secretory cells are present. The thick band of fibers belongs largely or wholly to the primary phloem. Beneath this, lie other primary phloem groups, more or less crushed, and, beneath these, secondary phloem. The cambium and secondary xylem are also shown. Below the epidermis a periderm layer is beginning to develop.

The tissues of the cortex are strictly primary and, as a whole, mature with the primary tissues of the stele, but there is considerable overlapping of development with secondary-tissue formation within the stele. Collenchyma develops early, but sclerenchymatous cells are usually late in reaching maturity. Parenchyma cells may continue to multiply, even into the second year in many woody plants. Additional tissue is thus developed as the volume of the cortex is rapidly increased with increase in diameter of the axis. Most of the divisions of the parenchyma cells are radial, forming tangential rows or sheets of cells. The later divisions of this type do not result in rounded daughter cells, and the new walls appear to divide the elongated mother cells (Fig. 77). The cortex of an axis in which marked secondary growth has occurred has tissues crowded and even more or less crushed radially. This is especially true in herbaceous stems that have strong xylem cylinders, as in *Linum* and *Solidago*.

The various cell types in the cortex serve various functions, but the cortex is primarily a protective layer. Such functions as support, photosynthesis, and storage are secondary.

THE EPIDERMIS

The epidermis constitutes a layer over the entire outer surface of the plant body, continuous except for stomatal and lenticellular openings. In meristematic regions it is, of course, undifferentiated, and in older stems and roots it may have been destroyed by secondary growth. Typically it consists of a single layer of cells; in a few plants it is biseriate or multiseriate. The epidermis of many plants has been described as multiseriate. In most of these the interpretations are incorrect: in some the outer region of the cortex resembles the epidermis in structure and function and is therefore assumed to be a part of the epidermis; in others, a periderm layer has just been initiated; in others, the epidermis itself is thin, weak, and ephemeral, and the abutting cortical layer functions as an epidermis; in still others, especially aquatics and plants of moist habitats, the epidermis closely resembles the outer cortex. Interpretation of a series of outermost cell layers as a biseriate or multiseriate epidermis is usually an interpretation based on function and not on morphology. In plants where the epidermis seems to be more than one layer thick, a study of development is necessary for correct interpretation.

Epidermal cells have a large central vacuole and thin peripheral cytoplasm. Minute leucoplasts are present, but chloroplasts are absent except in the guard cells of stomata and in plants of aquatic or moist and deeply shaded habitats. Mucilage, tannin, and crystals occur occasionally. The cells vary much in size and outline but are essentially tabular. They are closely fitted together and are often lobed, toothed, or flanged in various ways, the projections dovetailing into those of other cells and strongly interlocking the cells (Figs. 79, 170). The cells of leaves, and especially those of petals, are more complex in this respect than are those of other organs. Epidermal cells often have unevenly thickened walls, the outer and the radial walls being much thicker than the inner wall (Fig. 179). This additional thickness and the cutinization of the walls is of much importance from the standpoint of mechanical protection and of prevention of loss of water. The cuticle (Chap. II) adds greatly to the efficiency of the epidermis in the latter respect.

Ontogeny and Duration of the Epidermis.—The cells from which the epidermis develops are set off very early in the apical meristem as a superficial layer. In stems this is commonly the first zone to become distinct. In terms of the tunica-corpus theory, the epidermis is formed by anticlinal divisions of the outer layer of the tunica when a distinct tunica is present. Where tunica and corpus are not distinct, the epidermis is formed by anticlinal and periclinal divisions of the dermatogen. Later increase in the number of cells in the epidermis is the result of anticlinal division. The early established uniseriate condition is maintained throughout the life of the epidermis. Where the stem increases in diameter during secondary growth, continued slow division may go on in the epidermis, which in this way is accommodated to the increasing surface. Changes in epidermal cells, other than those involved in the development of phellogen layers, are rare. In roots, the epidermal cells, and also those of the outer cortex, generally become lifeless and lignified or suberized after the root hairs cease to function. In perennial stems epidermal cells live until the development of a periderm layer cuts off the water and food supply. In leaves, flowers, and most fruits, epidermal cells normally live as long as the organ of which they are a part.

Function of the Epidermis.—In function, the epidermis is primarily a covering layer which protects against rapid loss of water and mechanical injury. It may also serve secondarily in photosynthesis and secretion. Parts of the epidermis are structurally modified to serve some important physiological function, for example, the secretory tissue of nectaries, the stomata of leaves and stems, and the absorbing hairs of roots.

Root Hairs.—In roots, many or all of the epidermal cells become *root hairs*. In the ontogeny of such cells the outer wall expands, forming a long, tube-like process, in shape and structure a typical hair (Fig. 78). The walls of epidermal cells of roots in general, and especially those of the projecting hairs, are thin and delicate. They are commonly of cellulose and permit ready diffusion of water and dissolved substances. It is by absorption through root hairs that the chief supply of water and mineral nutrients is obtained. These structures are ephemeral, persisting usually but a few days or weeks, after which they collapse and the remains of the cells and of adjacent cells become suberized or lignified. Root hairs with thick lignified walls are reported in certain Compositae where they are said to persist into the second growing season.

The Stoma.—The openings in the epidermis through which gaseous interchange takes place between the intercellular spaces of the subepidermal cells and the atmosphere are known as *stomata*. These openings are spaces between two specialized epidermal cells which are known as *guard cells*, because changes in size and shape of these cells determine the opening and closing of the stoma (Fig. 79). Ordinarily, these two cells alone of the epidermal cells adjacent to the stoma differ from typical epidermal cells. Where other cells also are modified because of the presence of the stoma, and contribute in some way to the functional activity of the guard cells, such cells are known as *accessory cells*. This type of cell is restricted almost wholly to xerophytic plants (Chap. XIV). The term stoma is also, and perhaps preferably, applied to the

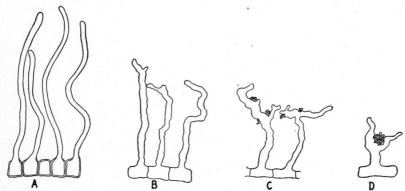

Fig. 78.—Root hairs. *A*, grown in water or moist air; *B*, grown in moist soil; *C*, *D*, grown in dry soil. (Particles of soil adhere in *C* and *D*.) (*After Schwarz*.)

opening in the epidermis plus the surrounding guard and accessory cells. In this usage the orifice is known as the *stomatal opening*, or *stomatal aperture*.

Structure and Action of the Guard Cells.—In all stomata the opening and closing is dependent upon changes in the turgor of the guard cells, increased turgor causing the stoma to open. The way in which the guard cells operate varies in different species with the shape of the guard cells and the thickening of parts of the wall. In a common type the walls of the guard cells are uniformly thickened and the guard cells are elliptical as seen in transverse section (Fig. 79*F*). With an increase in turgor these cells tend to become round in cross section, thus separating the closing walls of the stoma. In other types that open in a similar manner, the wall is unevenly thickened (Fig. 79*E*). A modification of this same principle of change in the shape of a cell that is asymmetrical in cross section is found also in other types of stomata in which the walls are unevenly thickened. In these types the cell wall may be very thick, except at two points or regions known as "hinges." The lumen of such cells is elliptical or oval as seen in cross section, and with an increase of turgor tends to become round, thus causing a stretching or flattening of the thin parts of the wall with consequent opening of the stoma (Fig. 79*B*,*G*, the hinges being the median point on the side of

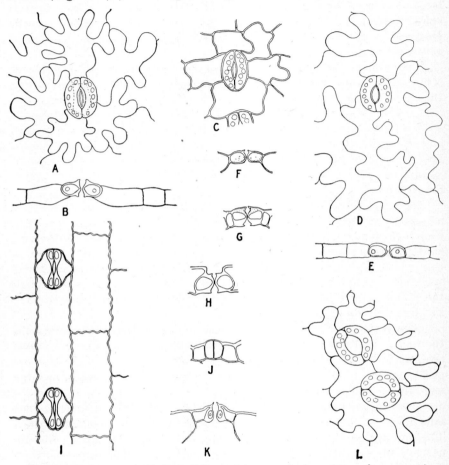

Fig. 79.—Stomata. *A*, *B*, from *Solanum tuberosum*, in face view and cross section; *C*, from *Malus pumila*; *D*, *E*, from *Lactuca sativa*; *F*, from *Medeola virginica*; *G*, from *Aplectrum hyemale*; *H*, from *Polygonatum biflorum*; *I*, *J*, *K*, from *Zea mays*: *I*, face view, *J*, cross section through end of stoma, *K*, median cross section; *L*, from *Cucumis sativus*. (*F*, *G*, *H*, and *J*, after Copeland.)

the aperture and the entire opposite wall). In another common type the guard-cell walls next to the aperture are somewhat thicker than those on the opposite side (Fig. 79*H*). Increase in turgor in guard cells of this type causes the thinner walls to stretch, and this, in turn, causes the entire cell to become more sharply curved and the opening enlarged. In the grasses and some other plants the ends of the guard cells are thin-walled and enlarged, whereas the central parts next the aperture are thick-walled and rigid (Fig. 79*I,J,K*). With increase in turgor, these bulbous ends of the guard cells become distended and press against each other, forcing the rigid centers to separate and the stoma to open. Decrease in the turgor of the guard cells always results in the closure of the stoma.

The above statements are descriptive only of general stomatal structure. Other more or less distinct types exist, with variations in the shape of guard cells and position of the thickened areas of their walls. The details of structure about the opening vary greatly with different species, but are constant for some large groups, for example, the grasses and the gymnosperms. Differences in behavior and fundamental form are not controlled by habitat, but marked modifications are found in xerophytes (Chap. XIV). The essential structural features are the same for all stomata.

The guard cells differ from other cells of the epidermis of which they are a part in that their protoplasts are more richly cytoplasmic, with a prominent nucleus, and chloroplasts and starch grains are usually present.

Functionally, stomata are of the greatest importance since it is through these openings that gaseous interchange between the intercellular space systems and the outer air takes place. Upon this diffusion through the stomata, the functions of respiration, transpiration, and photosynthesis largely depend.

Occurrence of Stomata.—Stomata occur on all parts of the plant except the root. On floral organs and in aquatic plants they may be few, abortive, or lacking. They are most numerous on leaves except in those plants in which the leaves are reduced, and photosynthesis is in large measure restricted to the stems.

In leaves stomata may occur in both the upper and the lower epidermis; in woody plants they are commonly absent from the upper epidermis. The number per unit area tends to be greater in woody plants then in herbs, and greater in plants of dry and exposed, than of moist and sheltered habitats. It increases toward the apex and margin of the leaf where the cells decrease in size, the proportion of stomata to epidermal cells remaining the same. In woodland habitats the number per unit area increases with height above the ground. The number is apparently related to the humidity of the environment; variations in amount of light seem to have no effect.

Ontogeny of the Stoma.—During the development of the epidermis a guard mother cell is formed by the division of one of the young epidermal cells. The guard mother cell divides to form the pair of guard cells.

170   AN INTRODUCTION TO PLANT ANATOMY

Accessory cells are derived from the same cell as the guard mother cell or from several adjacent cells.

Hairs.—Appendages of the axis and leaves which consist only of epidermal cells are known as *hairs*, or *trichomes*. Such appendages are unicellular or multicellular and occur in very many forms (Fig. 80). All

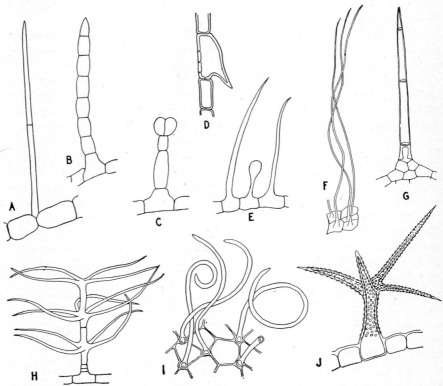

Fig. 80.—Hairs. *A*, from corolla of *Epigaea*; *B*, from leaf of *Coreopsis*; *C*, from corolla of *Phryma*; *D*, from leaf of *Avena*; *E*, from calyx of *Heliotropium*; *F*, from stem of *Onopordum*; *G*, from leaf of *Cucumis*; *H*, from young leaf of *Platanus*; *I*, from fruit of *Rubus strigosus*; *J*, from stem of *Aubrietia*.

transitional stages occur between the typical epidermal cell and one in which the outer wall is prolonged into a tube sufficiently pronounced to be considered a hair. Such intermediate cells, called *papillose cells*, are common upon petals (Fig. 169) and many leaves. So-called "unicellular" hairs are formed in two ways: by the prolongation of the outer wall of the epidermal mother cell, as in root hairs (Fig. 78); by the transverse division of the elongating mother cell—the outer daughter cell forming the hair, the inner a basal cell in the epidermis. Unicellular hairs assume numerous shapes, including elaborately forked and branched types (Fig. 80). Multicellular hairs likewise are found in innumerable types, and range from simple linear hairs of a few cells to complex, intricately branched, or massive structures, involving considerable areas of the epidermis. Stinging hairs (Fig. 56*A*), scales, and many glandular hairs (Fig. 56*B–E*) are complex multicellular structures. Root hairs are morphologically typical hairs.

The cells of hairs may be dead or living. If alive, the protoplasts contain little cytoplasm, unless the hair is associated with secretion of some type when the cytoplasm is abundant and richly granular. The thickness of the wall and its chemical nature vary greatly. Such hairs as those on the fruit of the peach and raspberry (Fig. 80*I*), the flower of the willows, and the bud of the grape are very thick-walled. Cutinized and lignified hairs are frequent. Cotton "fibers" are cellulose hairs.

Hairs are connected with many functions, major and minor, but it is in the reduction of transpiration through the additional coating they provide that they are probably most important.

References

(See also references in Chap. III; and for vascular tissue, in Chaps. IV, VII, and VIII)

ARTSCHWAGER, E. F.: Anatomy of the potato plant, with special reference to the ontogeny of the vascular system, *Jour. Agr. Res.*, **14**, 221–252, 1918.

BAILEY, I. W., and W. W. TUPPER: Size variation in tracheary cells: I. A comparison between the secondary xylems of vascular cryptogams, gymnosperms, and angiosperms, *Proc. Amer. Acad. Arts and Sci.*, **54**, 149–204, 1908.

BARKLEY, G.: Differentiation of vascular bundle of *Trichosanthes anguina*, *Bot. Gaz.*, **83**, 173–184, 1927.

BOND, G.: The occurrence of cell division in the endodermis, *Proc. Roy. Soc. Edinburgh*, **50**, 38–50, 1930.

———: The stem endodermis in the genus *Piper*. *Trans. Roy. Soc. Edinburgh*, **56**, 695–724, 1931.

BOND, T. E. T.: Studies in the vegetative growth and anatomy of the tea plant (*Camellia thea* Link) with special reference to the phloem, *Ann. Bot.*, **6**, 607–630, 1942.

BUGNON, P.: Origine, évolution et valeur des concepts de protoxylème et de metaxylème, *Bull. Soc. Linn. de Normandie*, 7 sér., **7**, 123–151, (1924) 1925.

CHANG, C. Y.: Differentiation of protophloem in the angiosperm shoot apex, *New Phyt.*, **34**, 21–29, 1935.

CHEADLE, V. I.: Specialization of vessels within the xylem of each organ in the Monocotyledoneae, *Amer. Jour. Bot.*, **31**, 81–92, 1944.

——— and N. B. WHITFORD: Observations on the phloem in the Monocotyledoneae. I. The occurrence and phylogenetic specialization in structure of the sieve tubes in the metaphloem, *Amer. Jour. Bot.*, **28**, 623–627, 1941.

COL, A.: Recherches sur la disposition des faisceaux dans la tige et les feuilles de quelques dicotylédones, *Ann. Sci. Nat. Bot.*, 8 sér., **20**, 1–288, 1904.

COPELAND, E. B.: The mechanism of stomata, *Ann. Bot.*, **16**, 327–364, 1902.

CORMACK, R. G. H.: Investigations on the development of root hairs, *New Phyt.*, **34**, 30–54, 1935.

———: The effect of environmental factors on the development of root hairs in *Phleum pratense* and *Sporobolus cryptandrus*, *Amer. Jour. Bot.*, **31**, 443–440, 1944.

CRAFTS, A. S.: Vascular differentiation in the shoot apices of ten coniferous species, *Amer. Jour. Bot.*, **30**, 382–393, 1943.

DAMM, O.: Ueber den Bau, die Entwicklungsgeschichte und die mechanischen Eigenschaften mehrjähriger Epidermen bei den Dicotyledonen, *Beih. Bot. Centralbl.*, **11**, 219–260, 1901.

DANGEARD, P. A.: Essai sur l'anatomie comparée au liber interne dans quelques familles de dicotylédones, *Le Botaniste*, **17**, 225–364, 1926.

DORMER, K. J.: Shoot structure in angiosperms with special reference to Leguminosae, *Ann. Bot.*, **9**, 141–143, 1945.

ESAU, K.: Ontogeny and structure of the phloem of tobacco, *Hilgardia*, **11**, 343–424, 1938.

———: Development and structure of the phloem tissue, *Bot. Rev.*, **5**, 373–432, 1939.

———: Vascular differentiation in the vegetative shoot of *Linum*. I. The procambium, *Amer. Jour. Bot.*, **29**, 738–747, 1942. II. The first phloem and xylem, *Amer. Jour. Bot.*, **30**, 248–255, 1943   III. The origin of the bast fibers, *Amer. Jour. Bot*, **30**, 579–586, 1942.

———: Ontogeny of the vascular bundle in *Zea mays*, *Hilgardia*, **15**, 327–368, 1943.

———: Origin and development of primary tissues in seed plants, *Bot. Rev.*, **9**, 125–206, 1943.

———: Vascularization of the vegetative shoots of *Helianthus* and *Sambucus*, *Amer. Jour. Bot.*, **32**, 18–29, 1945.

FLOT, L.: Recherches sur la zone périmédullaire de la tige, *Ann. Sci. Nat. Bot.*, 7 sér., **18**, 37–112, 1893.

FRANZ, H.: Beiträge zur Kenntnis des Dickenwachstums der Membranen, (Untersuchungen an den Haaren von *Humulus Lupulus*), *Flora*, **129**, 287–308, 1935.

GRIS, A.: Sur la moelle des plantes ligneuses, *Ann. Sci. Nat. Bot.*, 5 sér., **14**, 34–79, 1872.

GROB, A.: Beiträge zur Anatomie der Epidermis der Gramineenblätter, *Bibl. Bot.*, **36**, 1–123, 1896.

GUILLAUD, A.: Recherches sur l'anatomie comparée et le développement des tissues de la tige dans les monocotylédones, *Ann. Sci. Nat. Bot.*, 6 sér., **5**, 1–176, 1877.

GUTTENBERG, H. VON: Die Aufgaben der Endodermis, *Biol. Centralbl.*, **63**, 236–251, 1943.

HÉRAIL, J.: Recherches sur l'anatomie comparée de la tige des dicotylédones, *Ann. Sci. Nat. Bot.*, 7 sér., **2**, 203–314, 1885.

HIRSCH, W.: Untersuchungen über die Entwicklung der Haare bei den Pflanzen, *Beitr. Wiss. Bot.*, **4**, 1–36, 1900.

JEFFREY, E. C.: The morphology of the central cylinder in the angiosperms, *Trans. Canad. Inst.*, **6**, 599–636, 1899.

———: The structure and development of the stem in the pteridophyta and gymnosperms, *Phil. Trans. Roy. Soc. London*, **195B**, 119–146, 1903.

———: "The Anatomy of Woody Plants," Chicago, 1917.

JONES, W. R.: The development of the vascular structure of *Dianthera americana*, *Bot. Gaz.*, **54**, 1–30, 1912.

Kaplan, R.: Uber die Bildung der Stele aus dem Urmeristem von Pteridophyten und Spermatophyten, *Planta*, **27**, 224–268, 1937.

Kaufman, K.: Anatomie und Physiologie der Spaltoffnungsapparate mit verholzten Schiesszellmembranen, *Planta*, **3**, 26–59, 1927.

Kerl, H. W.: Beitrag zur Kenntnis der Spaltöffnungsbewegung, *Planta*, **9**, 407–463, 1929.

Kroemer, K.: Wurzelhaut, Hypodermis und Endodermis der Angiospermenwurzel, *Bibl. Bot.*, **59**, 1–151, 1903.

Kundu, B. C.: The anatomy of two Indian fibre plants, *Cannabis* and *Corchorus*, with special reference to fibre distribution and development, *Jour. Indian Bot. Soc.*, **21**, 93–128, 1942.

Lestiboudois, T.: Phyllotaxie anatomique, ou recherches sur les causes organiques des diverses distributions des feuilles, *Ann. Sci. Nat. Bot.*, 3 sér., **10**, 15–105, 136–189, 1848.

Linsbauer, K.: Die Epidermis, *In* "Handbuch der Pflanzenanatomie," IV, 1930.

Louis, J.: L'ontogénèse du système conducteur dans la pousse feuillée des dicotylées et des gymnospermes, *La Cellule*, **44**, 87–172, 1935.

Meyer, F. J.: Bau und Ontogenie der Wasserleitungsbahnen und der an diese angeschlossenen Siebteile in den vegetativen Achsen der Pteridophyten, Gymnospermen und Angiospermen, *Prog. Rei Bot.*, **5**, 521–588, 1917.

Miller, H. A., and R. H. Wetmore: Studies in the developmental anatomy of *Phlox Drummondii* Hook., II. The apices of the mature plant, *Amer. Jour. Bot.*, **33**, 1–10, 1946.

Morot, L.: Recherches sur le péricycle ou couche périphérique du cylindre central chez les phanérogames, *Ann. Sci. Nat. Bot.*, 6 sér., **20**, 217–309, 1885.

Mylius, G.: Das Polyderm. Eine vergleichende Untersuchung über die physiologischen Scheiden, Polyderm, Periderm und Endodermis, *Bibl. Bot.*, **79**, 1–119, 1913.

Netolitzke, F.: Die Pflanzenhaare, *In* Linsbauer, K.: "Handbuch der Pflanzenanatomie," IV, 1932.

Olivier, L.: Recherches sur l'appareil tégumentaire des racines, *Ann. Sci. Nat. Bot.*, 6 sér., **11**, 5–133, 1881.

Pfitzer, E.: Beiträge zur Kenntniss der Hautgewebe der Pflanzen, III. Ueber die mehrschichtige Epidermis und das Hypoderma, *Jahrb. Wiss. Bot.*, **8**, 16–74, 1871.

Plowman, A. B.: The comparative anatomy and phylogeny of the Cyperaceae, *Ann. Bot.*, **20**, 1–33, 1906.

Priestley, J. H.: The mechanism of root pressure. *New Phyt.*, **19**, 189–200, 1900. **21**, 41–47, 1922.

―――― and E. E. North: Physiological studies in plant anatomy, III. The structure of the endodermis in relation to its function, *New Phyt.*, **21**, 113–139, 1922.

Salisbury, E. J.: On the causes and ecological significance of stomatal frequency with special reference to the woodland flora, *Phil. Trans. Roy. Soc. London*, **216B**, 1–65, 1927.

Schwarz, F.: Die Wurzelhaare der Pflanzen, *Untersuch. Bot. Inst. Tubingen*, **1**, 135–188, 1883.

Schwendener, S.: Die Spaltöffnungen der Gramineen und Cyperaceen, *Sitzungsb. Konig.-Preuss. Akad. Wiss. Berlin*, **1889-1**, 65–79, 1889.

Scott, L. I., and J. H. Priestley: The root as an absorbing organ, I. A reconsideration of the entry of water and salts in the absorbing region, *New Phyt.*, **27**, 125–140, 1928.

SHARMAN, B. C.: Developmental anatomy of the shoot of *Zea mays* L., *Ann. Bot.*, N.S. **6**, 245–282, 1942.

SINNOTT, E. W.: The anatomy of the node as an aid in the classification of angiosperms, *Amer. Jour. Bot.*, **1**, 303–322, 1914.

——— and I. W. BAILEY: Investigations on the phylogeny of the angiosperms. 3. Nodal anatomy and the morphology of stipules, *Amer. Jour. Bot.*, **1**, 441–453, 1914.

SKUTCH, A. F.: Origin of endodermis in ferns, *Bot. Gaz.*, **86**, 113–114, 1928.

———: Anatomy of the axis of the banana, *Bot. Gaz.*, **93**, 233–258, 1932.

SNOW, L. M.: The development of root hairs, *Bot. Gaz.*, **40**, 12–48, 1905.

SOAR, I.: The structure and function of the endodermis in the leaves of the Abietineae, *New Phyt.*, **21**, 269–292, 1922.

STAUDERMANN, W.: Die Haare der Monocotylen, *Bot. Arch.*, **8**, 105–184, 1924.

STERLING, C.: Growth and vascular development in the shoot apex of *Sequoia sempervirens* (Lamb.) Endl., III. Cytological aspects of vascularization, *Amer. Jour. Bot.*, **33**, 35–45, 1946.

STRASBURGER, E.: Ein Beitrag zur Entwicklungsgeschichte der Spaltöffnungen, *Jahrb. Wiss. Bot.*, **5**, 297–342, 1866.

———: "Ueber den Bau und die Verrichtungen der Leitungsbahnen in der Pflanzen. Histologische Beiträge," III, Jena, 1891.

THOMPSON, W. P.: The anatomy and relationships of the Gnetales, I. The genus Ephedra, *Ann. Bot.*, **26**, 1077–1104, 1912.

TRAPP, G.: A study of the foliar endodermis in the Plantaginaceae, *Trans. Roy. Soc. Edinburgh*, **57**, 523–546, 1933.

VAN FLEET, D. S.: The development and distribution of the endodermis and associated oxidase system in monocotyledonous plants, *Amer. Jour. Bot.*, **29**, 1–15, 1942.

VAN TIEGHEM, P., and H. DOULIOT: Sur la polystélie, *Ann. Sci. Nat. Bot.*, 7 sér., **3**, 275–322, 1886.

VESQUE, J.: Mémoire sur l'anatomie comparée de l'écorce, *Ann. Sci. Nat. Bot.*, 6 sér., **2**, 82–198, 1875.

WILSON, C. L.: Medullary bundle in relation to primary vascular system in the Chenopodiaceae and Amaranthaceae, *Bot. Gaz.*, **78**, 175–199, 1924.

WISSELUNGH, C. VAN: Beitrag zur Kenntniss der inneren Endodermis, *Planta*, **2**, 27–43, 1926.

ZIEGENSPECK, H.: Ueber die Rolle des Casparyschen Streifens der Endodermis und analoger Bildungen, *Ber. Deut. Bot. Ges.*, **39**, 302–310, 1921.

CHAPTER VI

THE ORIGIN AND DEVELOPMENT OF THE SECONDARY BODY AND ITS RELATION TO THE PRIMARY BODY. THE CAMBIUM

That the primary plant body is in itself structurally and functionally complete is evident from the fact that in many forms it alone constitutes the plant, for example, the majority of monocotyledons and the pteridophytes. In the gymnosperms and in most dicotyledons, primary growth is soon followed by secondary growth, which usually becomes, both structurally and functionally, the more important. Secondary growth is effected by definite layers of initials—in vascular tissues, by the cambium; in other tissues, by similar meristems. These growing layers provide additional and constantly renewed conducting, supporting, and protecting tissues. Primary growth chiefly increases the length of the axis and adds the appendages; secondary growth increases the diameter of the axis (after the initial increase), and is responsible in most plants for the greater part of the mature plant body, supplying the requisite support, protection, and conduction for the large body of woody plants. Only in the tree ferns and a few of the monocotyledons is a large body present which is wholly primary in nature. The majority of the larger monocotyledons, including some of the palms, the woody yuccas, and other lilies, possess secondary growth of a special type.

Secondary tissues fall into two groups: the vascular tissues, those formed by the true cambium; and those, such as cork, formed by other similar but secondary meristems.

The Origin of the Cambium from Procambium.—As explained in the preceding chapter, the primary vascular skeleton is built up by the maturing of the cells of the procambium strands or cylinder to form xylem and phloem. In plants that have no secondary growth, all cells of the procambium strands mature to form vascular tissue, and there is no further increase in the amount of this tissue, except by unusual methods of development. In plants in which secondary growth later appears, a part of the procambium strand remains meristematic and gives rise to the cambium proper. The cambium, then, commonly represents a persistent portion of the apical meristem (Fig. 58), a section which remains meristematic and which becomes transformed into a growing layer of a different type.

Since the maturation of procambial cells usually proceeds progressively toward the center of the procambium strand, the cells in the central region are the last to mature. Where cambium develops, these last central cells are not transformed into permanent xylem or phloem cells but retain their meristematic activity indefinitely as cambium cells. In the early stages of cambium development there is a more or less irregular zone of meristematic tissue between the areas of primary xylem and phloem. At this time the protoxylem elements are usually mature and the metaxylem cells are maturing. The beginnings of cambial activity and the development of the last of the primary xylem therefore occur simultaneously.

As commonly understood, the cambium does not exist as such until there is established a definite tangential row of initials that divide repeatedly in the tangential plane. Just before the formation of such a sheet of tissue, there is a transitional period during which cell division is taking place in various planes in the central procambial zone but tending to occur in the tangential plane only in the later stages of development. As these tangentially dividing cells mature, those retaining their meristematic activity become arranged in a definite tangential row, becoming the true cambium, which then forms secondary tissues. In some plants, both woody and herbaceous, early tangential divisions in the procambium bring about radial arrangement of nearly all the primary vascular cells, especially those of the protoxylem (Figs. 62B, 140A). It may be impossible, therefore, to distinguish sharply between primary and secondary vascular tissues on the basis of position and arrangement, but with the change in method of development, there is an abrupt change in cell length—the secondary elements are much shorter than the primary.

In roots, the formation of the cambium differs from that in stems because of the radial arrangement of the alternating xylem and phloem strands (Chap. X). Here the cambium arises as discrete strips of tissue in the procambial strands inside the groups of primary phloem (Fig. 130A). Later, by lateral extension the strips of cambium are joined in the pericycle opposite the rays of primary xylem. Because of the place of origin of the cambium in roots, this meristem does not, in its early stage of development, form a symmetrical cylinder, but rather, as seen in cross section, a band of tissue that curves outward around the ends of the xylem rays and inward inside the strands of phloem. Commonly, secondary-tissue formation is most rapid beneath the groups of phloem so that the cambium, as seen in the transverse section of older roots, soon forms a circle. In some roots development of secondary tissue by the cambium that lies inside the phloem groups goes on to a considerable extent before lateral extension around the xylem is complete, so that the early undulate form of the cambium cylinder is omitted.

ORIGIN AND DEVELOPMENT OF THE SECONDARY BODY

Fascicular and Interfascicular Cambium.—In stems the first procambium to develop from promeristem is usually in the form of more or less isolated strands. In some plants these first-formed strands soon become, in part or wholly, united laterally by additional similar strands formed between them and by the lateral extension of the first-formed ones. A continuous or broken cylinder of procambium is formed. In the course

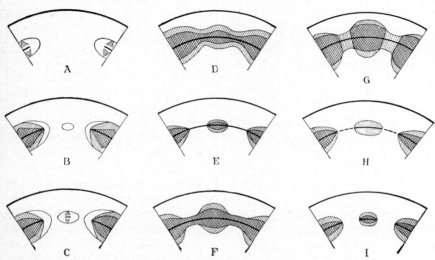

Fig. 81.—Diagrams to show the formation of the cambium cylinder in stems. (The primary vascular tissue is singly cross-hatched, the secondary, doubly.) *A–D*, successive stages in type where complete primary cylinder is formed: *A*, the first primary tissues are mature in the procambium strands and the cambium has appeared in the center of the bundle; *B, C*, the bundles have increased in size, secondary tissues are formed, and new small bundles are arising between the first; *D*, complete primary, secondary, and cambial cylinders are formed by the union of the bundles. *E–F*, type with complete cambium cylinder, but incomplete primary cylinder: *E*, the fascicular cambium has built up secondary tissues in the bundles; the interfascicular cambium has just arisen; *F*, secondary tissues complete the vascular cylinder. *G*, similar to *E–F*, but the interfascicular cambium forming only parenchyma or sclerenchyma; *H*, similar to *E*, but with a vestigial interfascicular cambium, which forms few or no vascular cells; *I*, primary cylinder of discrete strands, each with cambium and secondary tissues; no union by secondary growth.

of development this procambium cylinder gives rise to a cylinder of primary vascular tissue and cambium, similarly complete or broken. Ultimately, a cylinder of secondary vascular tissue may be formed, arising in strands as does the primary cylinder (Fig. 81*B–D*). In some herbaceous forms, such as *Ranunculus* and *Impatiens*, the procambium strands, and hence the primary vascular tissues, do not fuse laterally but remain as discrete strands. In these plants the cambium likewise is in the form of longitudinal strips, since it does not extend laterally beyond the limits of the primary xylem and phloem between which it arises (Fig. 81*I*). Such **strips of cambium then constitute the entire cambial meristem, but more** often in herbaceous stems the cambium extends laterally across the intervening spaces until a complete cylinder is formed (Fig. 81*E*, *F*). Where such extension occurs, the cambium arises from interfascicular meristematic cells derived from the apical meristem. To the strips of cambium that arise within collateral bundles the term *fascicular cambium* is applied, and the strips between the bundles are known as *interfascicular cambium* (Fig. 140*A*). The latter term is sometimes restricted to those strips of cambium lying between primary bundles which do not in their activity form xylem and phloem, but merely parenchyma (Fig. 135*D*), as in *Clematis*.

Between the condition in some plants, where a complete cambium cylinder is apparently formed from a complete procambium cylinder, and the condition frequent in herbs, where the cambium, even in the mature plant, is in the form of discrete strips, all intergrading conditions are found relative to the formation and activity of the interfascicular cambium. In herbaceous plants with woody cylinders the interfascicular cambium may be identical with the fascicular cambium in origin and function, but merely delayed in development from normal procambium, which in those regions may form little primary vascular tissue. Such cambium forms normal secondary vascular tissues in the same way as the fascicular cambium, although often the amounts formed are not so great (Fig. 81*F*). The same structure may form from interfascicular cambium developed from more or less permanent parenchymatous tissues; but more frequently no real vascular tissue is formed, the cambium giving rise to parenchyma only, as in *Clematis*. Under these conditions, the discrete vascular bundles are separated by secondary parenchyma. This is apparently a specialized condition found for the most part in woody vines and some herbaceous forms. It has undoubtedly been derived in phylogeny from the woody condition and is not, as is frequently stated, a stage in the development of a woody cylinder by the fusion of vascular bundles. In some herbaceous stems, as in species of *Geum* and *Agrimonia*, for example, an incomplete, vestigial interfascicular cambium is found; its disconnected cells divide but once or twice (Fig. 81*H*).

Time of Cambium Development in Stems.—In the stems of plants with well-developed secondary growth, the cambium begins to differentiate from the procambium in a given region just before that region ceases to elongate. The cambial derivatives are not, of course, established as such at this time, but some divisions may take place so that as soon as elongation ceases there may be simultaneous maturation of cambial derivatives and of primary metaxylem, although a considerable proportion of the latter usually matures first. In the majority of plants the formation of cambium is going on in the new parts of axes as long as the axes are increasing in length. In plants which have both intercalary meristems and secondary growth, as in some of the mints, cambium may be in an early stage of development in regions other than those near the tips of the axis. In general, it can be said that in plants which have secondary growth, stem elongation, whenever it may occur, is either accompanied by or immediately followed by cambium development. In some of the reduced herbs, which have very little secondary growth, cambial activity may be delayed for some time.

Time of Cambium Development in Roots.—In roots, cambial development frequently does not take place so quickly after elongation ceases as in the stems of the same plant. In many of the smaller feeding roots, a cambium may not be formed at all, even though secondary growth occurs abundantly in the stems and larger roots. Lack of secondary growth is apparently correlated with the function of the root as an absorbing organ, since wherever extensive secondary growth arises the root is no longer capable of absorption in that region because of the destruction of the root hairs, endodermis, and cortex, and the usual immediate formation of periderm. Roots of some herbaceous plants may be without secondary thickening even though such growth occurs in the stems; and in some woody species a considerable proportion of fibrous rootlets contain only primary growth. In the main root system of both woody and herbaceous plants the cambium arises soon after elongation has ceased.

Extent of the Cambium.—In the normal woody plant and also in many herbs the cambium forms a layer over the entire inner part of the body, except at the growing tips of the axis where the cambium has not yet been differentiated. The cambium of a part of a stem or root has the form of a hollow cylinder, and that of the entire plant, a branching tubular structure. The whole layer is frequently spoken of as the *cambium cylinder*. Strand-like extensions of the cambium often occur in leaf traces, and at leaf and branch gaps above such traces there are breaks in the continuity of the cambium while the axis is young. Usually within a few weeks after the initiation of cambial growth—or at most within a few months, the length of time depending upon the size of the gap and upon other factors—the cambium extends across the gaps, gradually closing them from the edges. The cylinder is henceforth unbroken except for the occurrence of wound areas.

In plants in which the stele is dissected, as in some herbaceous forms, the extent of the cambium is, as already stated (page 177), no greater than the width of the bundles of which the stele is composed. In these plants the cambium consists of a broken cylinder of strips of tissue. The strips follow various courses, according to the pattern of the primary vascular system (Chap. V). The width of these strips or plates of cambium necessarily varies as greatly as does the width of the vascular bundles of which they are a part. In some reduced herbaceous forms, the total width of the strips is only a small fraction of the circumference of the stele. The cambium may extend in strips into the petiole and larger vascular bundles of leaves.

Duration of the Cambium.—The extent of the functional life of the cambium varies greatly in different species and also in different parts of the same plant. In a perennial woody plant the cambium of the main axis lives from the time of its formation until the death of the plant. It is only by the continued activity of the cambium in producing new xylem and phloem that such plants can maintain their existence, since the functioning life of a given portion of these tissues is comparatively short. In leaves, inflorescences, and other deciduous parts the functional life of the cambium is short—in many leaves possibly a few days only, and in peduncles at the most a few weeks. Here all the cambium cells mature as vascular tissue. The secondary xylem then abuts directly upon the secondary phloem in the vascular bundle. In the annual stems of perennial plants and in the stems of annuals generally, the cambium is of this type; it is functionally active for a short time only, and all its cells mature into vascular tissues. In some of the specialized dicotyledonous herbs which have small, isolated vascular bundles or very thin vascular cylinders, very little secondary growth takes place. Some such forms lack secondary growth altogether or show so little that it is difficult to determine whether or not a true cambium is ever formed, since the last-formed metaxylem cells may be arranged more or less in even, radial rows and simulate secondary xylem.

Effect of Cambial Activity upon the Primary Body.—Since the cambium arises between the primary xylem and primary phloem, a part of the primary body is enclosed by the newly formed tissues. This inner part, the pith and primary xylem, is completely shut off from the outer parts. It persists within the cloak of secondary tissues unchanged except for the ultimate disappearance of cell contents and for certain chemical changes accompanying the death of pith and wood parenchyma cells. The primary-xylem skeleton is not distorted as the stem becomes older. In many herbs the pith is destroyed close to the growing point during elongation of the axis and the rapid expansion of the tissues outside as the pith is maturing, but this is not the result of secondary growth. The original primary body lying within the position first occupied by the cambium is to be found in the oldest stems and roots, structurally as in the axis before secondary growth began; the xylem skeleton, pith, gaps, and inner parts of leaf traces of the seedling tree are still present in the base of the old tree trunk.

ORIGIN AND DEVELOPMENT OF THE SECONDARY BODY

The primary tissues lying outside the cambium—primary phloem, pericycle, endodermis, cortex, and epidermis—are, on the other hand, pushed outward by the development of secondary tissues. Since the increase in circumference to which these tissues must be accommodated quickly surpasses the extent to which plasticity, or the slow primary

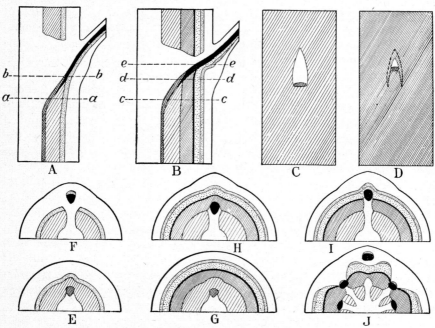

Fig. 82.—Diagrams showing the burial of leaf-trace bases by secondary growth. *A, B*, longitudinal sections: *A*, before the beginning of secondary growth; *B*, after considerable secondary growth; the gap is partly closed and the phloem has been pushed away from the base of the trace, only the xylem being buried. *C, D*, face views of the surface of the vascular cylinder, the trace cut away at the surface level: *C*, stage shown in *A*; *D*, stage shown in *B*, the gap partly closed and the base of the trace buried. *E, F*, cross sections at levels *a-a* and *b-b* in *A*; *G, H, I*, cross sections at levels *c-c*, *d-d*, and *e-e* in *B*; *J*, cross section below a node, showing departure of five traces, the bases buried in varying degrees. (Primary xylem is lightly cross-hatched; secondary, heavily. Primary phloem is finely stippled; secondary, coarsely. The xylem of the leaf trace is not differentiated in kind; its downward continuation is doubly cross-hatched.)

growth which may still be going on, enables accommodation, these tissues are commonly either ruptured or crushed. The primary phloem in most plants is quickly flattened and crushed and appears as disorganized strips of crushed tissue. Often the crushed cells are absorbed early and no traces of the tissue remain. The endodermis may increase somewhat in extent but likewise is destroyed soon after secondary growth begins. The pericycle and cortex, owing to their somewhat firmer structure, and to their partial accommodation by slow primary growth to the diameter increase caused by secondary growth, often persist for a longer time. But in most plants with well-developed perennial secondary growth these outer parts are sooner or later crushed or broken open and killed by exposure to drying and other types of injury—especially by the shutting off of food and water supplies by cork layers which develop within them (Chap. IX). The dead parts are soon sloughed off by decay or by abscission, and, after a variable interval, ranging from a few weeks to several or many years in different species, the outer primary body disappears. The development of secondary tissues preserves the inner part of the primary body intact but is responsible for the ultimate destruction of the outer portion. Exceptions to the loss of the entire outer part of the primary body occur in a few woody plants where the cortex persists for many years through the capacity of its parenchyma for slow primary growth. Such slow and long-continued growth of primary tissues obtains in the epidermis of some woody twigs; in the cortex and pericycle of the same and of similar plants; in a few plants more or less throughout the stem, as in the trunks of some palms that lack secondary growth.

Outer primary tissues that persist unchanged after marked secondary growth has taken place are most conspicuous in herbaceous plants. In these plants the overlapping, in time of formation and in position in the axis, of primary and secondary growth is apparently greater than in woody forms. The outer tissues are accommodated to the increase in diameter due to secondary growth with less distortion than in typical woody plants. But often as the stem becomes old, the softer cells of the cortex, pericycle, and phloem become much compressed radially, as in *Aster, Linum, Cannabis*, and many similar herbs with thick, woody cylinders.

The Relation of Secondary Growth to Leaf Traces.—At nodes the projection of leaf traces makes the stem structure complex. With the increase in thickness of secondary xylem the bases of the leaf traces (those parts within the cambium cylinder) are buried, and since the cambium lies always between the xylem and the phloem (Fig. 82A), the formation of new xylem causes the outward movement of all phloem as well as of the cambium itself, not only on the axis, but also on the trace (Fig. 82B). Because of the place of origin of the cambium, secondary growth buries the proximal parts of the leaf traces—the innermost part, lying in the primary xylem, is without phloem and is buried without change; the phloem is stripped away from the part of the trace adjacent to this and pushed outward, leaving only the xylem of the trace embedded in the secondary xylem of the stem (Fig. 82B). The length of the buried part depends largely upon the angle at which the trace departs.

ORIGIN AND DEVELOPMENT OF THE SECONDARY BODY 183

Only short pieces of the trace lying outside the original position of the cambium are buried, since continued secondary growth, in forcing outward the cortex and phloem in which the outer parts lie embedded, breaks

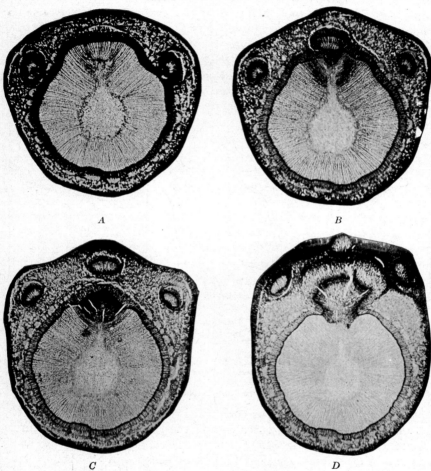

FIG. 83.—Cross sections at successive levels upward of the nodal region of a one-year-old twig of *Malus pumila*. *A*, the lateral leaf traces have passed into the cortex and their gaps are closed; *B*, the median trace has passed into the cortex, leaving the gap open, the dark-staining, meristematic branch traces (supplying the bud) are arising from the sides of the gap; *C*, the three leaf traces well out in the cortex, the branch traces have united, the gap is evident; *D*, the leaf traces entering the base of the petiole, the branch traces have formed a nearly complete vascular cylinder.

the trace in two, the distal part of the strand being torn away and carried outward with the tissues in which it lies. Rupture is due to the outward thrust of secondary growth laterally upon the trace (Fig. 84*A–D*). This rupture does not occur until sometime after the leaf has fallen, commonly the first or second growing season thereafter. The secondary xylem of the first year embeds the inner part of the trace without injuring it, the phloem being slowly pushed outward but not broken. The time of rupture of the trace—its elements dead and nonfunctioning after the fall of the leaf—depends upon a number of factors: the rate of secondary growth, the size and cross-sectional shape of the trace, and especially the angle at which the trace departs (Fig. 84B,E,F). The more nearly the course of the trace approaches a right angle with the stem (Fig. 84E) the longer the period before rupture, since the surrounding tissues are stripped away and the xylem strand buried. Where the trace passes

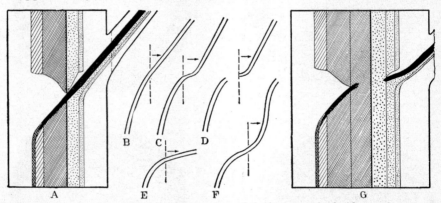

Fig. 84.—Diagrams illustrating the rupture of the leaf trace in deciduous plants. *A*, longitudinal section of node at end of first season's growth, the gap has been closed and the base of the trace buried in secondary xylem; *B, C, D*, stages in the condition of the trace (other parts omitted) during the following season: *B*, the trace unchanged, *C*, the trace stretched and bent, *D*, the trace broken in two, the outer part carried outward. The dotted lines represent the position of the cambium. *E*, form of trace which ruptures very slowly; *F*, form which ruptures quickly; *G*, longitudinal section of node at end of second season, the trace ends separated by secondary xylem and phloem. (Shading as in Fig. 82.)

upward vertically through the cortex, it is quickly broken, owing to the lateral exposure of a long outer part of the trace to the outward thrust of secondary growth (Fig. 84F). Large traces are broken later than small ones, and those crescent-shaped or horseshoe-shaped in cross section later than strap-shaped ones. The inner and outer parts of leaf traces are thus separated (Fig. 84G); the outer part is ultimately lost with the destruction of the cortex, whereas the inner is preserved indefinitely, embedded in the xylem.

In evergreen leaves the traces are extended by a type of secondary growth that increases them in length by additions of new tissue in the middle. The primary xylem of the trace is ruptured gradually and in an oblique direction and new cells to replace those destroyed are added meanwhile by the cambium of the "armpit" region (Fig. 85). As long as the leaf persists, the upper older xylem cells of the trace are continuously broken and new cells are added below. The gap is often not quite closed but remains open until the trace is broken. Thus, where the leaves are long-persistent, as in *Araucaria*, the trace may become very long and be evident in secondary wood in annual rings far from the center of the tree. After an evergreen leaf dies, there is complete rupture of the trace, as with deciduous leaves. Normally, the ruptured end of leaf traces, both in deciduous and in evergreen plants, is quickly covered by

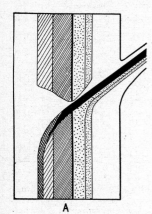 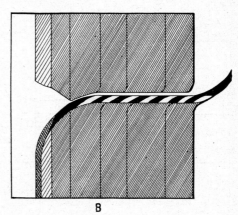

FIG. 85.—Diagrams showing the extension of the trace in evergreen leaves. *A*, the condition at the end of the first season; *B*, at the end of the fifth season (the parts outside the cambium omitted). In *B*, the primary parts of the trace (in solid black) are separated by a central, secondary part (in alternate black and white); this part is ruptured, like the primary, but the breaking is continuous, the trace being built up anew, obliquely, by an "armpit" section of the cambium, an extension along the under side of the trace at the point of its contact with the secondary xylem. (Shading as in Fig. 82. See text for further description.)

typical cambium cells. There is soon no evidence in the later formed secondary xylem of the position of the trace but in *Agathis* and *Araucaria* the trace cambium continues to form trace-like cells after the leaf has fallen, and secondary wood, even of old tree trunks, always shows buried leaf traces.

Branch traces are buried in the same way as are leaf traces but are, of course, not ruptured. The embedding of branch bases is further discussed later in the chapter.

Relation of Secondary Growth to Leaf and Branch Gaps.—Leaf gaps are closed by the gradual lateral extension of the cambium; the new meristematic cells arise apparently out of the parenchyma cells of the gap. The size and the shape of the gap determine in part the length of time before the gap is closed, wide gaps being closed more slowly than long narrow ones. In most angiosperms, leaf gaps are closed in the first season (Figs. 83, 84*A*, *G*). Branch gaps, which are often large, are closed more slowly than leaf gaps; some branch gaps remain open until the second to fourth years.

Function of the Cambium.—Meristems that form secondary tissues are commonly looked upon as uniseriate sheets of initials that form new cells usually on both sides. The cambium, a meristem of this type, forms xylem internally and phloem externally. The tangential division of the cambium cell forms two apparently identical daughter cells (Fig. 86*B*):

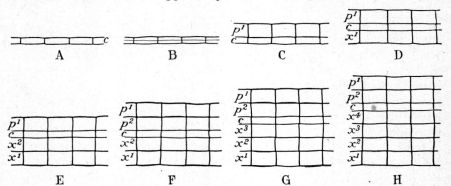

Fig. 86.—Diagrams illustrating the formation of xylem and phloem by the cambium, and the changes in position of the phloem and the cambium brought about by this activity. *A*, the cambium; *B*, the cambium cells divided, each forming two daughter cells; *C*, one daughter cell enlarged and matured as a phloem cell (p^1), the other enlarged to cambium-cell size, remaining a meristem cell (*c*); no change in the position of the cambium occurs. *D*, the cambium cells have divided again, the inner daughter cell in this case having matured as a xylem cell (x^1), the outer becoming the cambium cell (*c*); the cambium and the phloem have been moved outward the width of the xylem cell. *E–H*, further divisions occur, resulting in the formation of one more phloem cell and three more xylem cells. (The xylem and phloem cells are here represented as maturing before the formation of the next cell, whereas a number of cells are normally present in an immature condition.)

one remains a meristematic cell, the persistent cambial cell; the other becomes a *xylem mother cell* or a *phloem mother cell* (Fig. 86*C*), depending upon its position internal or external to the initial. The cambium cell continues to divide in a similar way: one daughter cell always remaining a cambium cell; the other becoming either a xylem or a phloem mother cell. The sequence of xylem and phloem formation during this process, if there be any uniform sequence, is not known. It is probable that there is no definite alternation and that for brief periods, only one kind of tissue is formed. Evidence for lack of alternation is that often—in both woody and herbaceous plants—several times as many xylem cells as phloem cells are formed. Adjacent cambium cells apparently divide at nearly the same time, and the daughter cells belong to the same tissue. In this way the tangential continuity of the cambium is maintained.

In the formation of xylem, the enlargement of the developing cells causes the outward movement of the cambium and of all cells lying outside of this layer. This increases the diameter of the cambium cylinder. The maturing of phloem cells causes the outward movement of the phloem cells and cells external to these only; the position of the cambium is unchanged by phloem formation. The activity of the cambium causes it to move outward each season to the extent of the thickness of the mature xylem formed in that season.

Structure of the Cambium.—There are two general conceptions of the cambium as an initiating layer: one, that it consists of a uniseriate layer of permanent initiating cells with derivative cells that—though they may divide a few times—soon become transformed into permanent tissue; the other, that there are several rows of initiating cells, forming a cambium "zone," some of the individual rows of which persist as cell-forming layers at least for some little time. Because cells mature continuously during growing periods on both sides of the cambium, it is obvious that only a single layer of cells can have permanent existence as a cambium. Other layers, if present, function only temporarily and become completely transformed sooner or later into permanent cells. The question involved is apparently how many times a xylem or a phloem mother cell and its derivatives may divide; this is difficult to ascertain and is probably not constant. Divisions appear to take place more freely in phloem initials than in xylem initials. (Transverse divisions, such as those resulting in the formation of wood and phloem parenchyma are not considered here.) Phloem and xylem mother cells may not divide or may divide a few times. During rapid growth, there may be more divisions than at other times. Loose application of the term "cambium" to the entire differentiating region between the mature xylem and the mature phloem (Figs. 87A, 90), often leads to the conception that the cambium (in the stricter sense) is a multiseriate layer. Though the term was first applied to a wide layer of differentiating substance, believed to be at least in part without cellular structure, best usage now applies it only to the initiating layer.

Cellular Structure of the Cambium.—Cambium cells, in general, are of two fundamentally different types: the *ray initials*, which are more or less isodiametric and give rise to the vascular rays; and the *fusiform initials*, the elongate, tapering cells that divide to form all cells of the vertical system (Figs. 87, 90). The ray initials show little variation in cell shape; the number concerned with the formation of a ray is few or many, dependent upon the size of the ray, which varies greatly in different species and frequently in the same species. The fusiform initials are uniform in shape as seen in cross section, but show great differences in

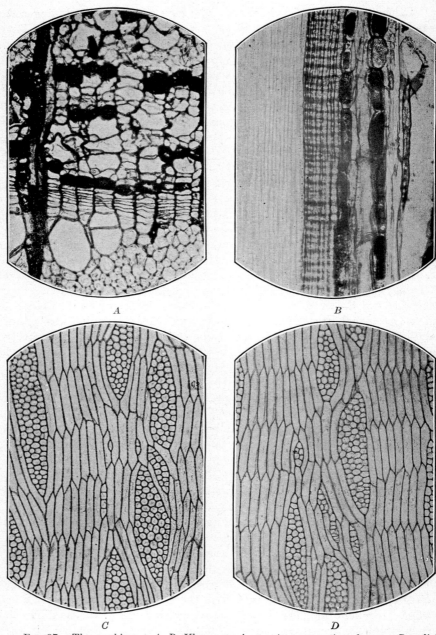

Fig. 87.—The cambium. *A, B, Ulmus americana: A,* cross section, *dormant; B,* radial section, *dormant;* the cambium is in the middle of the figure; the primary pit-fields of the radial walls of the cambium and immature phloem cells are seen in face view. (For tangential section of *U. americana,* see Fig. 88*B.*) *C, D, Robinia Pseudo-Acacia: C,* tangential section, *dormant,* the primary pit-fields are seen in section giving a beaded appearance to the radial walls; *D,* tangential section, like *C,* but *growing,* the radial walls are much thinner and the pit-fields obscure or lacking. × 100.

ORIGIN AND DEVELOPMENT OF THE SECONDARY BODY

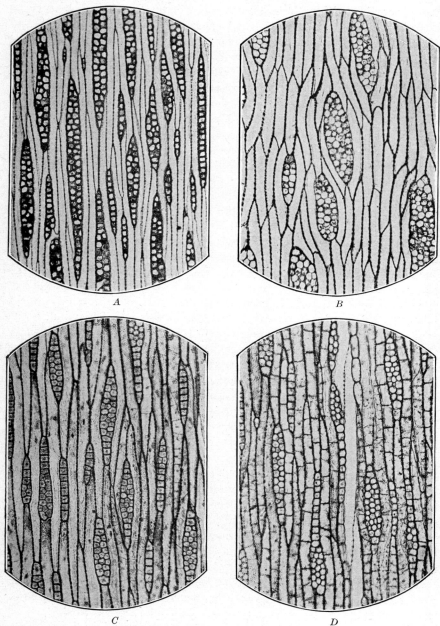

FIG. 88.—The cambium in tangential section. *A, Pyrus communis; B, Ulmus americana.* In *A* and *B* the radial walls are thick and the pit-fields prominent. *C* and *D, Juglans cinerea.* *C* and *D* together form a series, from the left of *C* to the right of *D*, showing the ontogeny of the phloem. At the left of *C* is typical cambium; at the right of *C* the transverse divisions resulting in phloem-parenchyma formation are taking place; at the left of *D* the parenchyma cells are maturing; at the center and right of *D* the sieve tubes and parenchyma are nearly mature. All × 100.

the proportion of length to tangential width. In woody plants with relatively short cambium cells, for example, *Robinia* (Fig. 87C,D) and *Ulmus* (Fig. 88B), the length is from five to ten times the tangential width. In other plants, such as *Pyrus* (Fig. 88A), *Juglans* (Fig. 88C), and other types in which the vascular tissues are relatively unspecialized, the ratio of length to tangential width is twenty-five (or more) to one. Gymnosperms show an extreme condition in which the ratio may be anywhere from fifty to one to one hundred (or more) to one, dependent upon the species and other factors. In herbaceous plants, as a class, the common type is that with the shorter initials. Cambium with short initials is clearly the most specialized type, phylogenetically the most recent.

Size of Cambial Cells.—The size of the individual elements of the cambium varies through wide limits. In the specialized woody dicotyledons, such as *Robinia*, the elongate cambium cells of the mature plant are about 175 micra in length by 20 micra in tangential width by 7 micra in radial width. In *Juglans* and *Liriodendron*, which have longer initials, the size is about 600 by 25 by 8 micra. The gymnosperms show the extremes in large size. In *Pinus Strobus*, for example, the dimensions of 4000 by 42 by 12 micra have been given; a maximum length of 5000 micra is reported in *Larix*. The size of the ray initials is fairly uniform, the tangential diameter generally being about the same as, or a little less than, that of the adjacent fusiform initials. It has recently been shown that, in some groups of plants at least, the length of the fusiform initials increases with the age of the plant. Thus, in the gymnosperms the length may increase from 1 to nearly 4 mm during the first 60 years, after which it remains constant. In the dicotyledons this increase is much less; in unspecialized woody plants, such as *Juglans*, only from 0.8 to 1.2 mm during the first 30 years. In highly specialized types, such as *Robinia*, the increase may be only from about 0.145 to 0.175 mm before the maximum is reached. The size of cambium cells varies to some extent in the same plant, dependent on position relative to branches, buds, or wound tissue, and also on different ecological factors. In crotch angles extreme variation in size and distortion in shape occur. Curly grain in wood is the result of abnormalities in the arrangement of the cambium cells or in the form of the cambium cylinder. Spiral grain also is related to the structure of the cambium.

In some woody plants, such as *Robinia* (Fig. 87C,D) and *Diospyros*, the fusiform cambium cells, as seen in tangential section, are in more or less definite transverse rows. Such a cambium is called *storied*. This stratified arrangement is correlated with short-length initials and with the formation of highly specialized vessels. It is responsible for a similar stratified condition of the cells of the xylem and phloem. Some genera, for example, *Fraxinus* and *Ulmus* (Fig. 88*B*), have short cambial initials that are not storied. This condition is apparently intermediate between the storied type and the extreme nonstoried type with long narrow initials, such as that of *Juglans* (Fig. 88*C*), *Salix*, *Populus*, *Pyrus* (Fig. 88*A*). The gymnosperms have cambium of the nonstoried type. Herbaceous plants with well-developed secondary growth, for example, *Solanum* (potato), have short initials that are nonstoried. The prevalent opinion that cambial initials are brick-shaped is based upon the study of transverse and radial sections alone.

Structure of Cambial Cells.—The protoplast of the cambium cell is strongly vacuolate, usually with one large vacuole and thin peripheral cytoplasm. Except during dormant periods, the cytoplasm streams actively. The nucleus is large and in the fusiform cells is much elongated. The apparently multinucleate condition seen in tangential sections of the resting cambium is due to the fact that the tangential walls of the radially extremely narrow cells are very thin and transparent, so that the contents of several cells may be seen more or less clearly at the same focal level. The tangential walls are at all times without definite thin areas. The radial walls of the cambium cells, on the other hand, are much thicker and, while the cambium is dormant, show abundant thin areas, the primary pit-fields (Figs. 87*B*, 88*A*,*B*).

Cell Division in the Cambium.—Division of the cambial initials must provide not only for the formation of new xylem and phloem cells radially on each side, but also in large part for the increase in the circumference of the cambium cylinder itself. The former is accomplished by the tangential division of the cambium initials, and by the subsequent division of the xylem and phloem mother cells; the latter, by typical radial division of the cambial initials in some plants (Fig. 89*E*); in others, by transverse division or oblique radial division followed by increase in size and gliding growth (Fig. 89*A*); and in small part, by the increase in the tangential dimension of the initials as the plant grows older. Increase in the number of vascular rays is also an important factor contributing to the increase of the cambium cylinder in circumference. New ray initials are formed from fusiform initials by transverse division of an entire cambium cell or sometimes of part of the cell only. Longitudinal divisions also occur in the formation of a new biseriate or multiseriate ray. In the formation of broad and very high rays more than one fusiform initial takes part. A combination of these methods brings about increase in girth of the cambium. Radial division of the cambial initials (Fig. 89*E*) is characteristic of those plants that have short, storied cambium cells and in which the derived vascular tissues are of the highly specialized type, for example, some of the Leguminosae. In plants with longer, nonstoried initials, there exist all transitional conditions in respect to the position of the newly formed wall or cell plate, from the transverse plane to the radial-longitudinal plane (Fig. 89*A,B,C*). Apparently in evolutionary specialization, the new cell wall tends to approach the radial position as the initials become shortened and approach the storied

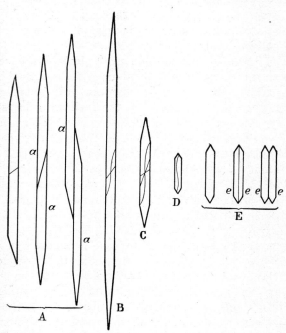

Fig. 89.—Diagrams of cambium cells, tangential view, illustrating the method of increase in girth of the cambium. *A*, fusiform initial dividing nearly transversely, *a, a,* products of this division which elongate and slide by one another as they mature (two stages); *B, C, D*, three forms of initials showing position of new wall in "radial" (anticlinal) division: *B, C*, showing two and three positions respectively, *D*, the only position; *E*, short initial, *e, e*, products of the radial division of *E*, which enlarge tangentially but not longitudinally. (*After Bailey.*)

condition. Division in the long cambium cells and their daughter cells shows in extreme form the slow progressive building of the new wall by kinoplasmosomes (Figs. 14, 15) from the dividing nucleus to the ends of the cell. When division is tangential, the daughter cells that persist as cambium initials increase in radial diameter only. New cambium initials formed by transverse or oblique divisions increase greatly in length (Fig. 89*A,B,C*); those formed by radial divisions do not increase in length (Fig. 89*E*).

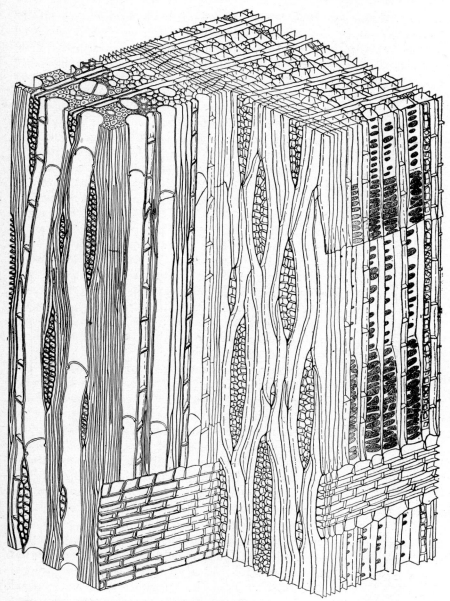

FIG. 90.—The cambium and developing and mature xylem and phloem in *Malus pumila*. (The region of maturing xylem is here represented greatly shortened. The contents and the pits of all cells have been omitted.)

Gliding and Intrusive Growth of Cambial Cells and Cambial Derivatives.—Strong gliding or intrusive growth occurs when new cambium initials increase in size to that of the mother cell; when the cambium derivatives mature as xylem and phloem cells; when adjustment takes place among unequally enlarging cambium derivatives. When a fusiform initial divides transversely or nearly so (Fig. 89A) in the formation of new initials, the daughter cells enlarge to about twice their original length, extending past each other and between adjacent initials. In the maturation of the elongate cells of xylem and phloem, there is great increase in length—even to four or five times the length of the initial (Fig. 91)—and contact with new cells far above and below. Developmental changes in size and shape in the zone of plastic cells near the cambium are doubtless in some measure those of symplastic growth (page 12), but rapid and great enlargement of the vessel elements brings a pushing aside and tearing apart of the adjacent small cells with consequent formation of new cell contacts by many cells. This displacement of cells is especially great in ring-porous woods with simply perforate vessels, such as *Quercus* and *Robinia*.

Ontogeny of Secondary Vascular Tissues.—The xylem mother cells cut off by the cambium may develop into permanent xylem elements without further division or, as frequently happens, may divide once or several times before mature cells are formed. In simple gymnosperm wood all xylem derivatives from fusiform cambial cells become tracheidal cells and are essentially alike except for the differences between early and late wood. The tracheids are formed directly from the xylem mother cells by increase in radial diameter and in length, thickening of the wall, and the loss of the protoplast. In gymnosperms with wood parenchyma and in all plants containing vessels the xylem mother cells differentiate into two or more cell types—tracheids, wood parenchyma, vessels, and wood fibers. Wood parenchyma cells are formed by the transverse division of the mother cell into a number of segments (Fig. 90), and by the subsequent radial enlargement and the thickening of the walls of these segments. The transverse divisions occur in a vertical row of

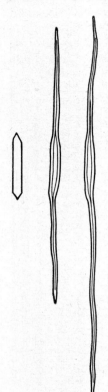

FIG. 91.—Cambium cell and wood fibers derived therefrom in *Robinia Pseudo-Acacia*. (Drawn to scale.) Elongation of the developing fiber involves penetration of the tips between cells above and below; see also Fig. 92A.

mother cells, so that the resulting parenchyma cells form a vertical series extending for some distance in the axis. Commonly the vertical series of derivatives of a single mother cell retain the prosenchymatous shape of that cell so that the series from a single initial can be readily recognized in the mature tissue. The arrangement of wood parenchyma and its relation to other cells are discussed in Chap. VII. The ontogeny of the vessel is discussed in Chap. IV. The initials of vascular rays divide

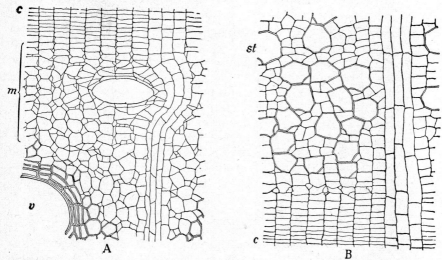

FIG. 92.—The ontogeny of secondary vascular tissue in *Robinia*. *A*, early stages in xylem formation. *c*, the cambium. The median portion (*m*) of the figure shows the region of major cell elongation and of adjustment to vessel development which involve tangential and radial enlargement. The tips of elongating cells appear from above and below between the radial rows of cells, increase in size, and crowd the cells from symmetrical arrangement. The great enlargement of vessel elements crowds the surrounding cells from position, flattening and otherwise distorting them; the rays often are turned at right angles to their course. *v*, a vessel nearly mature, as are surrounding cells. *B*, stages in phloem development to mature tissue. The companion cells appear early; they and the sieve tubes later enlarge greatly. *c*, the cambium, *st*, sieve tube.

tangentially, and the daughter cells increase greatly in radial diameter but little or not at all in other diameters.

Time of Cambial Activity.—In perennial stems and roots of plants that have dormant periods, cambial activity commonly ceases well before the beginning of this period. Division in cambium cells begins again at the end of the dormant period—in early spring, April and May in the northeastern United States—before, during, or after the breaking of the buds. In evergreen trees it may occur somewhat earlier. There is no uniformity in the position of first growth in the cambium; this is often in the central part of the tree, extending then to all parts, but cambial growth may also start at the base of the tree or at the tips of the twigs. Likewise, of two trees of the same species growing side by side, one may have cambial activity while the other is still dormant. On bright, sunny days in late winter, cambial activity may be started on the southwest side of exposed trunks as a result of absorption of heat from the sun by the dark-colored bark. If the sunny days are followed by a rapid temperature fall, the active cambium cells are likely to be killed, and the type of winter injury known as "sun scald" results.

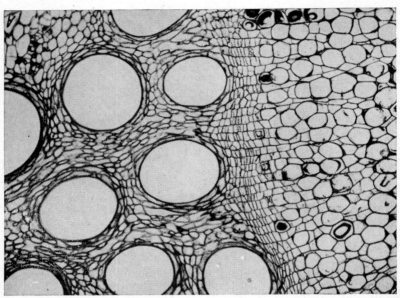

Fig. 93.—Transverse section of part of the cambium region of a vascular bundle of *Cucurbita*, phloem at the right, xylem at the left. Developmental stages of both tissues shown. Enormous enlargement of vessel elements has destroyed radial arrangement, changed position, and distorted form of surrounding cells. × 90. (*After Esau and Hewitt.*)

The first vascular cells to mature in the spring are phloem mother cells. These are cells that were cut off from the cambial initials during the previous growing period and have remained over winter in an immature condition. The new sieve tubes are formed at a time when the translocation of stored food materials is most actively going on. From histological evidence, it is probable that in many woody plants all sieve tubes functioning in a given season are matured during that season, and all sieve tubes matured the preceding season have already ceased to function, though they may not lose their protoplasts or become crushed until rapid growth begins in the spring. The period of most rapid phloem formation may come several weeks after the differentiation of the first phloem elements. This period coincides, in general, with that of the most active growth in the xylem. The duration of seasonal cambium activity varies with different ages of the plant and of the plant parts, with different species, and with environmental conditions. Generally, after the "flush" of growth in the early summer, there is a gradual slowing down in the formation of new cells, and in the trunk and main branches of some species there is cessation of cambial growth by midsummer. Cambial activity continues longest in the small rapidly growing twigs or shoots that are late in completing their apical growth. In nursery trees, growth in diameter may continue until late in the fall; in such trees, terminal growth also usually continues late. Abundant nitrogen and water in the soil are apparently important causal factors in such late growth.

Burial of Branch Bases.—As successive annual layers of xylem are laid down by the cambium, all tissues within the cambium cylinder are buried more and more deeply. In this way the bases of branches become embedded in the wood of the tree trunk. In a living branch, the buried portion has the shape of an inverted cone because, as new layers of xylem are laid down over the branch, the cambium is moved by the increase in the diameter of the trunk (Fig. 94A) farther and farther away from the point of insertion of the branch in the trunk. The buried portion of the branch cannot increase further in diameter; hence the inner portions are progressively of less diameter as the attachment to the primary cylinder is approached. At the apex of the cone-shaped mass thus formed is the pith of the branch at its point of union with the pith of the main axis (Fig. 94A). When a branch dies, there is of course, no further increase in diameter, and its base is buried as a cone of dead tissue (Fig. 95). Knots found in lumber are sections of embedded branch bases, loosely or tightly held in the board, depending on whether the branch was dead or living at the time it was embedded.

As the base of a branch is buried by the formation of new xylem on the main axis, the phloem about its insertion is forced outward—that in the angle of the crotch more rapidly than that below the crotch—and the base of the branch is stripped of its phloem. In small branches in which the increase in diameter is relatively small as compared with that of the main axis, as in the fruit spurs on the larger limbs of apple trees, this stripping is most marked. In this process the phloem is thrown up into folds which often appear as concentric rings about the base of the partly buried branch. With larger branches and more rapid growth, the older phloem is mechanically broken and crowded out of the crotch angle, and the younger, more plastic tissue is thrown up into irregular folds. Great distortion in the shape and arrangement of the cambium cells results, and some of the cells apparently are crowded out and destroyed. Adjustment to the necessary reduction in area of the cam-

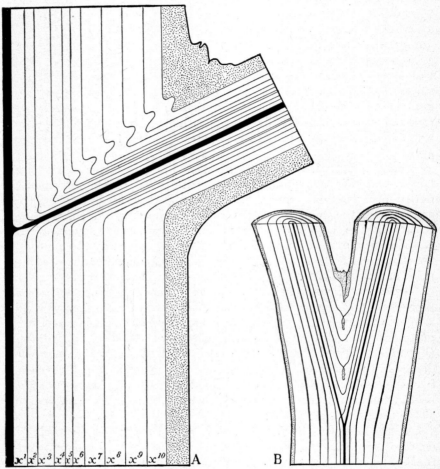

Fig. 94.—*A*, diagram illustrating the burial of branch bases by secondary xylem. The phloem is pushed away from the buried portion and thrown up in the crotch angle into folds. The cambium layer is also distorted in position. The pith is shown in solid black; the phloem and cortex stippled. x^1 to x^{10} successive annual rings. *B*, diagram of crotch with narrow angle, showing union of xylem of trunk and branch, the pushing out of the phloem in the angle, and the occasional enclosure of "pockets" of phloem within the xylem. The enclosure of phloem, together with the formation of abundant wood parenchyma in the xylem of the crotch union, renders the crotch weak. (*B, after MacDaniels.*)

bium in the crotch angle probably takes place during the growing season while the surrounding cells are in a plastic condition. At this time some

ORIGIN AND DEVELOPMENT OF THE SECONDARY BODY 199

of the cambium cells are apparently forced laterally out of the crotch angle to the sides of the crotch where they help to compensate for the increase in the girth of the cambium which takes place in this region.

In the crotch angle there is usually a well-defined region where the tissues of the trunk meet those of the branch. Here the conducting tissues of branch and trunk may remain more or less distinct, and, between the two, masses of thick-walled wood-parenchyma may be formed. Thus the branch is not "tied to" the trunk on the upper side

Fig. 95.—Transverse section of decaying trunk of *Pinus Strobus*, showing the form of buried branch bases. The inner xylem and pith of the trunk have been destroyed but the branch bases, resin-filled and resistant, have persisted. $\times \frac{1}{2}$.

and as a result, there is in the angle of many crotches a zone of weakness which may cause the branch to break away when under stress (Fig. 96). Where the angle of the crotch is very narrow and growth is rapid, the bark on the two sides is forced together before cambium growth in the angle pushes it out beyond the region of xylem union. "Pockets" of dead phloem are then enclosed in the angle (Fig. 94B), and are another cause of weak crotch union. The nature of the crotch tissue, the angle at which the branch meets the trunk, and the relative size of trunk and branch all have a practical bearing in the practice of pruning fruit trees.

200 AN INTRODUCTION TO PLANT ANATOMY

Cambium Growth about Wounds.—Among the important functions of the cambium is the formation of *callus*, or *wound tissue*, and the healing of wounds. When wounds occur in roots or stems, masses of soft parenchymatous tissue quickly form on or below the injured surface; this tissue is known as callus. Callus may be formed by the division of parenchyma cells in the phloem and the cortex, but its most frequent

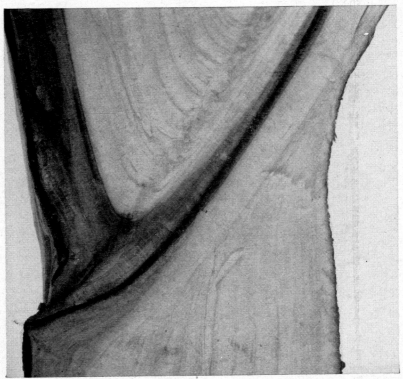

FIG. 96.—Radial section of part of trunk of *Liriodendron*, showing buried branch base. Pith of branch and trunk (black) united at "shoulder" on lower left; xylem of branch separated from that of trunk on upper side by the contorted tissue of the crotch angle. Heartwood and pith, dark; sapwood and bark, light. $\times \frac{1}{2}$.

source is the cambium. In the formation of callus in the healing of a wound, there is at first abundant proliferation of the cambium cells, with the production of masses of parenchyma. The outer cells of this tissue either become suberized themselves, or periderm (Chap. IX) develops within them, so that a protecting bark is formed beneath which the cambium is active in forming new vascular tissue in the normal way. In wounds due to pruning, callus is formed about the edges early in the growing season. As a new annual ring is formed in the uninjured

ORIGIN AND DEVELOPMENT OF THE SECONDARY BODY

surrounding tissues the cambium layer assumes a position at an angle to the face of the wound at the point of intersection with it. In this position new tissue formed in the normal way will extend the growing layer over the cut surface until the two opposite sides meet. The cambium layers then unite and the wound is completely covered. Subsequently formed growth-rings bury the wound more and more deeply.

The Cambium in Budding and Grafting.—The important practices of budding and grafting have as their basis the ability of the cambium of both stock and scion to develop callus and unite, forming over the union of stock and scion a continuous cambium layer which will give rise to normal conducting tissue. There is apparently an actual union of the cambium of the two plants. Where there is so-called "incompatibility" of stock and scion, such as occurs between some varieties of apple and certain dwarf-apple stocks, the cambia of the stock and scion fail to unite to form a normal growing layer that will, in turn, produce normal xylem and phloem but produce, instead, masses of parenchyma which make the union weak and conduction slow. Cambium activity and structure in relation to graft union is not well understood.

Where the cambium is injured during the growing season—for example, when branches are ringed—it may be regenerated from the immature xylem cells beneath, provided the tissues are protected from desiccation soon after the injury. In ringing experiments, it is sometimes difficult to prevent the formation of new cambium even by scraping the surface of the wound with a knife because callus tissue is formed by the living, immature cells of the xylem, and in this callus a new cambium is differentiated. At first this new meristematic tissue is not normal cambium in shape and size of cells, but eventually the normal condition is attained. The formation of galls and some types of burls may be due to local stimulation of the cambium to abnormal activity by insects or disease.

The Cambium in Monocotyledons.—The monocotyledons as a class are without secondary thickening; the plant body consists of primary tissue only. Vestiges of typical cambial activity occur, however, in some genera, especially in the vascular bundles of nodal regions and of leaf bases. A special type of secondary thickening occurs in a few forms (some woody and a few herbaceous Liliaceae—*Dracaena*, *Aloë*, *Yucca*, *Veratrum* and some other genera), whereby the stem is increased in diameter by the formation of a cylinder of new bundles embedded in a tissue of less specialized nature (Fig. 97). Here a cambium layer is formed from the meristematic parenchyma of the pericycle or of the innermost cortical cells. In roots, at least in some plants, a cambium of this type forms in the endodermis. The initials of this cambium vary greatly

202   AN INTRODUCTION TO PLANT ANATOMY

in shape in different species, ranging in form through polygonal and rectangular to fusiform, and are commonly variable even in a small region in a single plant. They stand in tiers forming a storied cambium as in the normal cambium of some dicotyledons.

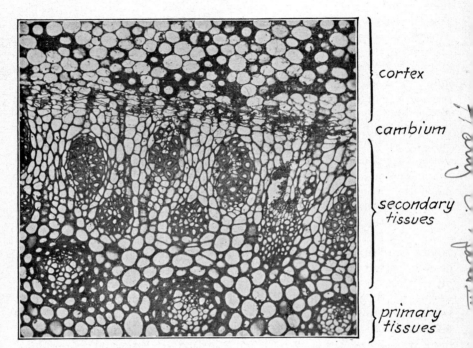

Fig. 97.—Special type of cambial growth of "woody" monocotyledons; stem of *Dracaena fragrans*.

The development by cambial activity in steles with anomalous structure is discussed in Chap. XI.

References

(See also References for Chaps. III, IV, V, VII, and VIII)

Artschwager, E. F.: Anatomy of the potato plant, with special reference to the ontogeny of the vascular system, *Jour. Agr. Res.*, **14**, 221–252, 1918.

Bailey, I. W.: The cambium and its derivative tissues, II. Size variations of cambial initials in gymnosperms and angiosperms, *Amer. Jour. Bot.*, **7**, 355–367, 1920.

———: The cambium and its derivative tissues, III. A reconnaissance of cytological phenomena in the cambium, *Amer. Jour. Bot.*, **7**, 417–434, 1920.

———: The cambium and its derivative tissues, IV. The increase in girth of the cambium, *Amer. Jour. Bot.*, **10**, 499–509, 1923.

———: The significance of the cambium in the study of certain physiological problems, *Jour. Gen. Physiol.*, **2**, 519–533, 1920.

BROWN, H. P.: Growth studies in forest trees, I. *Pinus rigida* Mill., *Bot. Gaz.*, **54**, 386–403, 1912. II. *Pinus Strobus* L., *Bot. Gaz.*, **59**, 197–241, 1915.

CHEADLE, V. I.: Secondary growth by means of a thickening ring in certain monocotyledons, *Bot. Gaz.*, **98**, 535–554, 1937.

ESAU, K.: Vessel development in celery, *Hilgardia*, **10**, 479–488, 1936.

——— and W. B. HEWITT: Structure of end walls in differentiating vessels, *Hilgardia*, **13**, 229–244, 1940.

HILL, A. W.: The histology of the sieve tubes of angiosperms, *Ann. Bot.*, **22**, 245–290, 1908.

JACOB DE COURDEMOY, H.: "Recherches sur les monocotylédones à accroissement sécondaire," 108 p., Lille, 1894.

KLINKEN, J.: Uber das gleitende Wachstum der Initialen im Kambium der Koniferen und den Markstrahlverlauf in ihrer sekundären Rinde, *Bibl. Bot.*, **84**, 1–41, 1914.

KNUDSON, L.: Observations on the inception, season, and duration of cambium development in the American larch [*Larix laricina* (Du Roi) Koch], *Bull. Torr. Bot. Club*, **40**, 271–293, 1913.

KOSTYTSCHEW, S.: Der Bau und das Dickenwachstum der Dikotylenstämme, *Beih. Bot. Centralbl.*, **40**, 295–350, 1924.

MACDANIELS, L. H.: The apple-tree crotch. Histological studies and practical considerations, *Cornell Univ. Agr. Exp. Sta. Bull.*, 419, 1923.

———: The histology of the phloem in certain woody angiosperms, *Amer. Jour. Bot.*, **5**, 347–378, 1918.

MISCHKE, K.: Beobachtungen über das Dickenwachsthum der Coniferen, *Bot. Centralbl.*, **44**, 39–43, 65–71, 97–102, 137–142, 169–175, 1890.

NÄGELI, C.: "Dickenwachsthum des Stengels und Anordnung der Gefässstränge bei den Sapindaceen," München, 1864.

NEEFF, F.: Über die Umlagerung der Kambiumzellen beim Dickenwachstum der Dikotylen, *Zeitschr. Bot.*, **12**, 225-252, 1920.

SCHMIDT, E. W.: "Bau und Funktion der Siebröhre der Angiospermen," 108 p., Jena, 1917.

SCHOUTE, J. C.: Uber Zellteilungsvorgänge im Cambium, *Verhandel. Akad. Wetenschappen Amsterdam*, 2s. **9**, 1–60, 1902.

SCOTT, D. H., and G. BREBNER: On the secondary tissues in certain monocotyledons, *Ann. Bot.*, **7**, 21–62, 1894.

TISON, A.: Les traces foliaires des conifères dans leur rapport avec l'épaississement de la tige, *Mém. Soc. Linn. de Normandie*, **21**, 59–82, 1903.

———: Sur la mode d'accroissement de la tige en face des faisceaux foliaires après la chute des feuilles chez les dicotylédones, *Mém. Soc. Linn. de Normandie*, **21**, 1–17, 1902.

CHAPTER VII
SECONDARY XYLEM

Secondary xylem often makes up the bulk of the vascular tissue of a plant, and, in most woody plants, constitutes the bulk of the entire plant body. In such plants it is a tissue of great importance, because through its functions a huge plant body may be maintained in the atmosphere with its various parts favorably situated as to light and air. Mechanically, the xylem supports and anchors the body; physiologically, it conducts water, and perhaps other inorganic materials absorbed by the roots, to all living parts of the plant. Further, it provides in its living cells storage space for considerable quantities of food. Of the total amount of xylem in the body of most trees, however, a considerable part (heartwood) is not functioning except in the way of mechanical support. The secondary xylem of tree trunks is of great economic importance, since it constitutes the timber and wood of commerce. It also has many other uses of less, though often of much importance.

GROSS STRUCTURE OF SECONDARY XYLEM

Secondary xylem consists of a closely compacted mass of thick-walled cells so arranged as to form two systems: a longitudinal, that is, vertical, and a transverse, radiating system.[1] The longitudinal system consists of elongate, overlapping, and interlocked cells—tracheids, fibers, and vessel elements—and of longitudinal rows of parenchyma cells. All these cells have their long axes parallel with the long axis of the organ of which they are a part. The cell types of this system are discussed under Xylem in Chap. IV. In secondary xylem, these cells are in general similar to those of the primary xylem of the same species. They are considerably shorter and may differ otherwise in form and in structure—for example, in vessel-element perforation. Annular and spiral tracheids and vessels are of course absent. The radial system consists chiefly of parenchyma cells with their long axes at right angles to the long axis of the central cylinder. These horizontally elongate cells constitute the xylem rays. Rays are, of course, absent in primary xylem.

[1] The distinction of two such systems is made merely for convenience in description. As distinct tissues, the systems, of course, do not exist. The basis of distinction is that of arrangement of cells and of direction of conduction.

Xylem Rays.—*Xylem rays*, or *wood rays*, are sheets of tissue, more or less strap-like or ribbon-like, extending radially in the xylem (Figs. 98, 99). Xylem rays are a part of a system of conducting tissue which extends as a continuous band through the cambium into the secondary phloem. These radiating strips of cells have been generally known as *medullary rays*, or *pith rays*, since in position and parenchymatous nature they to some extent suggest radiating portions of the pith. The use of this term is based upon a supposed homology of these structures with actual projections of the pith ("medulla"), such as occur in herbaceous stems of the *Ranunculus* type (Fig. 135*F*). The term "medullary ray" for these radiating bands of xylem and phloem is, however, clearly a misnomer since very few of these rays of tissue have any connection with the pith and, further, are not morphologically homologous with structures, such as the radiating pith 'arms' of stems of the *Ranunculus* type, which may perhaps logically be called medullary rays. These radial bands are best called *vascular rays* since the bands are rays of vascular tissue, partly of xylem, partly of phloem. The terms *wood ray*, or *xylem ray*, and *phloem ray* for those parts of the ray confined to xylem and phloem respectively, are already in use and are particularly appropriate.

Vascular rays lie at right angles to the long axis of the stem or root and are always continuous through the cambium and into the phloem. They are straight except when crowded aside by unequal growth of the surrounding cells or tissues. All vascular rays are initiated by the cambium and, once formed, are increased in length indefinitely by the cambium. New ray tissue is thus added in a region near the middle of the ray, the older parts being separated by the formation of new cells in the cambium between them. The length of a ray therefore depends upon the length of time since it was initiated and upon the rate of growth of the secondary tissues. At the beginning of secondary growth a certain number of vascular rays are initiated. As the circumference of the vascular cylinder increases with the addition of new secondary tissues, the distal parts of these rays become more and more widely separated. Soon the width of the segments of xylem and phloem so increases that the cells formed in the center of the segments are remote from the existing rays. New rays are then formed at frequent intervals tangentially, so that all xylem and phloem cells are fairly close to rays. The scattered vertical and tangential distribution of the rays brings into contact with every tracheid and similar elongate cell at least one ray (Figs. 98*C*, 99*C*). In gymnosperm woods where no wood parenchyma is present, every tracheid is in direct contact with at least one ray. Vessels, also, in their longitudinal extent, come into contact with many rays. In

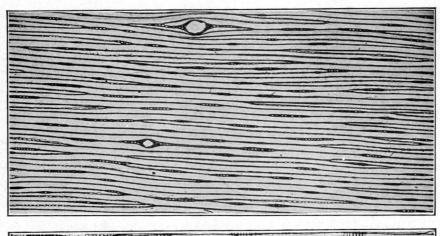

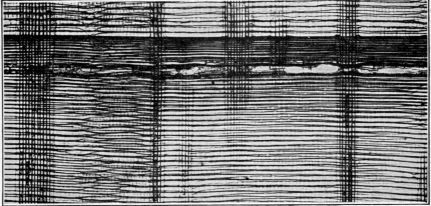

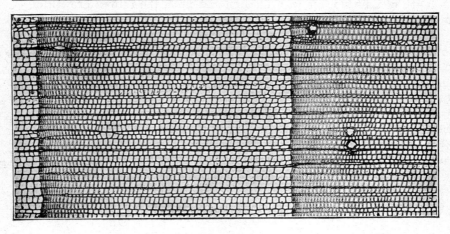

Fig. 98.—Secondary xylem of *Picea rubra*. *A*, transverse; *B*, radial; *C*, tangential section. ×50. (*Courtesy of the U.S. Forest Products*

SECONDARY XYLEM

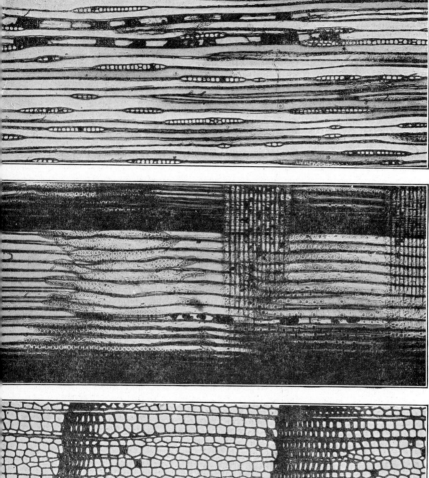

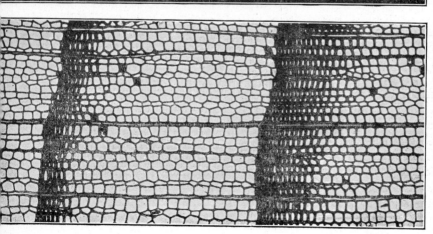

Fig. 99.—Secondary xylem of *Sequoia sempervirens*. *A*, transverse; *B*, radial; *C*, tangential section. ×50. (*Courtesy of the U.S. Forest Products Laboratory.*)

herbaceous stems, such as those of *Ranunculus*, where vascular bundles are separated by projecting parenchymatous wedges (which may perhaps be called true "medullary rays"), and in vines, such as *Clematis*, where the bundles are separated by bands of secondary parenchyma, vascular rays are lacking. In such plants the bundles are so small that no conducting cells are far removed from radial sheets of parenchyma. The absence of vascular rays in these plants is doubtless the result of loss during evolutionary specialization. The relation of contact or of closeness in position of living cells (ray cells or wood parenchyma) to conducting cells (tracheids and vessels) is apparently of the greatest importance to the functioning ability of these nonliving cells.

Annual Rings; Growth Rings.—The secondary xylem in perennial axes commonly consists of concentric layers (Fig. 3), each one of which represents a seasonal increment. When seen in cross section of the axis, these layers appear as rings, and the terms *annual ring* and *growth ring* or *growth layer* are applied to each layer. These layers are usually called annual rings because in the woody plants of temperate regions and in those of tropical regions where there is an annual alternation of growing and dormant periods, each layer represents the growth of one year. The term "growth ring" is best for general use because under some climatic conditions the growing periods may not be one per year.

An annual ring or growth ring of xylem is a layer of secondary xylem formed in one growing season over the entire plant and is, therefore, an extensive tubular structure having the general form of the axis of the plant. It is open at the ends where meristems occur.

The width of growth rings varies greatly and depends upon the rate of growth of the tree, which is, of course, controlled by many factors. Wide rings are formed in young trees and under favorable growth conditions generally. Unfavorable growing seasons produce narrow rings, and favorable seasons wide ones. Such injury as defoliation during the growing period also causes narrow-ring formation. Abrupt changes in the width of rings successively formed are produced by sudden changes in the growth conditions of the tree; serious accident to the tree, severe pruning, changes in soil drainage, fertilization, and removal of shading trees leave evidence of the changed conditions in the width of the ring. The growth rings of the tree provide a real record of some phases of the history of the tree.

A growth ring varies in width in different parts of the plant and in different parts of the circumference of the axis at a given level. The thickness around the axis is most uniform over parts that are free from branches, as the smooth trunks of trees. Below the insertion and along the underside of branches, in some types of crotches, above large roots, about wounds and other abnormalities, there may be a very decided thickening of the ring. Local increased food or water supply is apparently responsible for this condition. In trees, such as the apple tree with prominent large branches low on the trunk, the trunk may be built up in definite ridges or segments, each composed of the conducting tissues leading to the limb directly above. Where there are several large limbs near the same level, the entire trunk below may be obscurely divided into radial segments of tissue leading mainly to those limbs. The upper central part of the tree may thus be robbed of sufficient water and nutrients. Of course, the segments are not distinct, except as flutings on the stem, and merge into one another laterally. Experiments have shown that, even in smooth trunks, conduction is more or less restricted to vertical segments of the trunk; that is, roots on one side of a tree to a large extent supply the branches on the same side. Conduction laterally is slow in the trunk, the 'main current' passing directly upward. Tangential conduction takes place where no roots or branches occur on one side, as well as where wounds cut off direct communication. Buttressed tree trunks, such as are seen in elm trees and many others, especially tropical forms, are due to strong local thickening by the cambium above the main roots.

Growth rings are characteristic of woody plants of temperate climates; they are weakly developed or lacking in tropical forms except where there are marked climatic changes such as distinct wet and dry seasons. Annual plants and herbaceous stems of perennials show, naturally, but one layer.

Early Wood and Late Wood.—The presence of growth rings is due to seasonal variations in growing conditions. Tissue formed in the early part of the growing period, *early wood,* differs in cell size, type, and arrangement, and in proportion of kinds of cells from that formed later in the period, the *late wood.* The terms *spring wood* and *summer wood* are in common use for early wood and late wood, respectively, but are not strictly applicable because in some plants the two parts of the ring are not so restricted seasonally, even in temperate regions. The terms "annual ring," "spring wood," and "summer wood" have been long in use and are generally apt for temperate climate plants; they will doubtless continue in use. (The terms *summer wood* and *fall wood,* formerly used for early and late wood respectively, have ceased to be used, since in most trees no wood is formed in the autumn, and some wood commonly develops in the spring.) An annual ring, therefore, consists of two parts: an inner layer, early wood, and an outer layer, late wood. No line exists

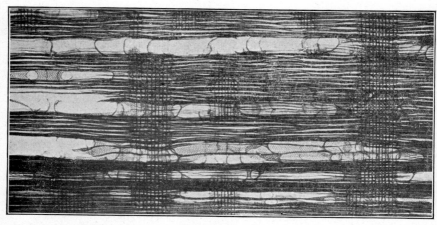
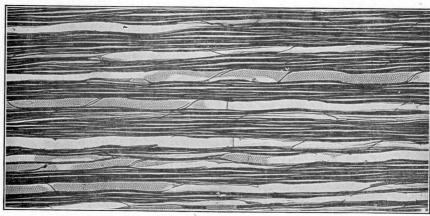
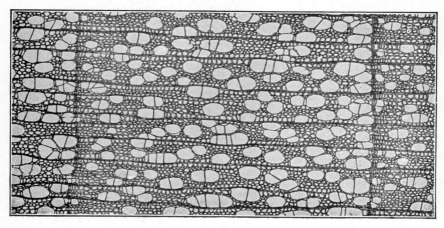

Fig. 100.—Secondary xylem of *Populus tremuloides*. *A*, transverse; *B*, tangential; *C*, radial section. ×50. (*Courtesy of the U.S. Forest Products Laboratory.*)

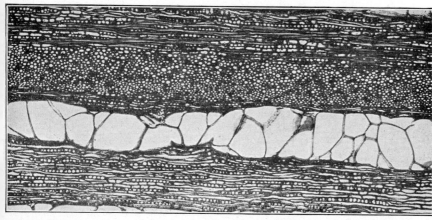

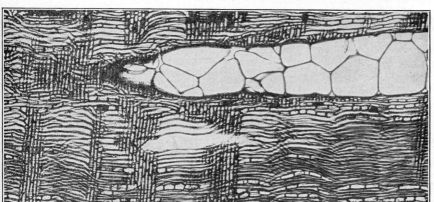

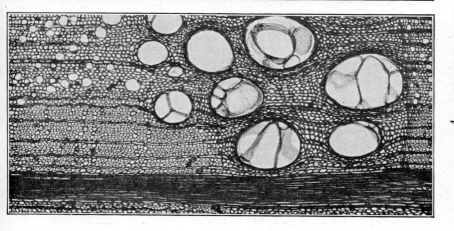

Fig. 101.—Secondary xylem of *Quercus minor*. *A*, transverse; *B*, radial; *C*, tangential section. ×50. (*Courtesy of the U.S. Forest Products Laboratory.*)

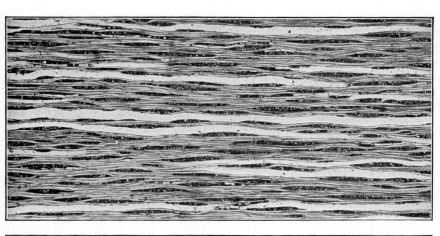

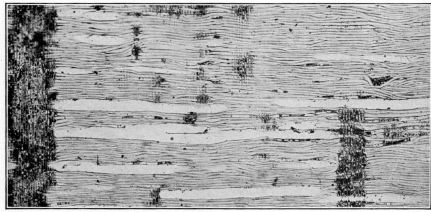

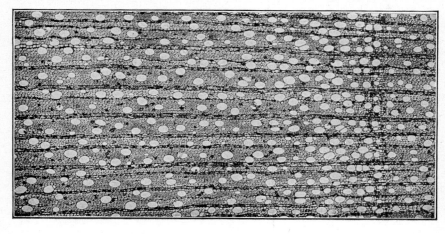

Fig. 102.—Secondary xylem of *Malus pumila*. *A* transverse; *B* radial; *C*, tangential section. ×50. (*Courtesy of the U.S. Forest Products Laboratory.*)

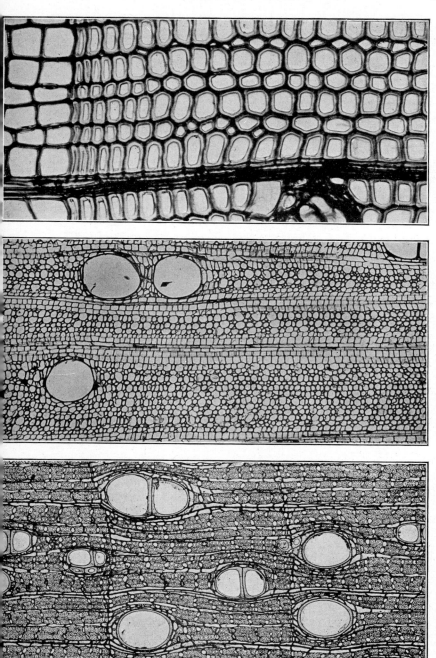

Fig. 103.—Secondary xylem in transverse section. *A*, a heavy wood, *Diospyros*, × 250. *B*, a light wood, *Ochroma*, × 50. *C*, *Picea sitkensis*, × 50. (*Courtesy of the U.S. Forest Products Laboratory.*)

between these two parts of the ring because early wood merges into late wood. However, the line between the late wood of one year and the early wood of the next year is sharp; this line renders the annual ring distinct.

False Annual Rings.—The formation of *false annual rings* frequently occurs as the result of a check in normal development of xylem followed by a resumption of growth in the same period. Defoliation, drought, and other disturbances in development bring about premature formation of late wood. In trees with determinate growth, such as the oak, the winter buds, especially the terminal buds, may begin growth prematurely in late summer; the consequent growth activity is accompanied by the formation of a false ring. False rings can be readily detected by the less sharp delimitation at the outer edge of the late wood. An annual ring that is made up of two or more false rings is sometimes called a *double* (or *multiple*) *annual ring.*

Ring-porous and Diffuse-porous Wood.—Annual rings are often rendered conspicuous by the restriction of vessels to the early wood, or by the formation there of very large vessels, as in *Quercus* (Fig. 101*A*), *Catalpa*, *Ulmus* (frontispiece), or of more numerous and somewhat larger vessels there than in the late wood. When, for these reasons, xylem shows the early wood conspicuously distinct from the late wood, it is said to be *ring porous*, as in *Fraxinus* and *Quercus* (Fig. 101*A*); when the vessels are fairly uniformly scattered through the ring, as in *Betula*, *Acer*, and *Populus* (Fig. 100*A*), and where the transition in size or abundance is gradual from the early-formed to the late-formed cells, as in *Juglans* and *Malus* (Fig. 102*A*), the wood is called *diffuse porous*. Between these two types no line can be drawn and many woods are of distinctly intermediate type in this respect.

GENERAL HISTOLOGICAL STRUCTURE OF SECONDARY XYLEM

Cell Types and Cell Arrangement in Secondary Wood.—In both the longitudinal and radial systems, living and nonliving cells occur, the proportions and the arrangement of the two kinds varying greatly with the species, and to some extent also with the time of year when formed, with the organ in question, and with the individual plant. The radial system, that is, the wood rays, consists in most plants wholly of living cells; the longitudinal system, on the other hand, commonly possesses a rather small proportion of living cells. These, in the form of wood parenchyma, make up longitudinal, uniseriate strands of cells placed end to end, extending indefinitely in the wood (Figs. 99*C*, 101*B*). The water-conducting and the supporting cells of the different types occur in various proportions and arrangements. Usually, in a given wood only a few kinds of cells are present, but some species possess several cell types.

In *Abies*, for example, the wood (with the exception of the wood rays) consists wholly of tracheids, or, in some species, of tracheids and wood parenchyma; in *Picea*, of tracheids and fiber-tracheids, and in some species of wood parenchyma also; in *Larix*, of tracheids, fibers, and wood parenchyma; in *Liriodendron*, of vessels, fiber-tracheids, and wood parenchyma; in *Acer*, of vessels, fibers, and wood parenchyma; in some species of *Quercus*, of tracheids, fiber-tracheids, typical fibers, libriform fibers, gelatinous fibers, vessels, and wood parenchyma.

The different cell types are arranged in various ways. (The radial arrangement of all cells is, of course, the result of the method of development of secondary tissues.) The cell types may be fairly uniformly distributed through the wood (Figs. 100*A*, 103*A*), or there may be formed

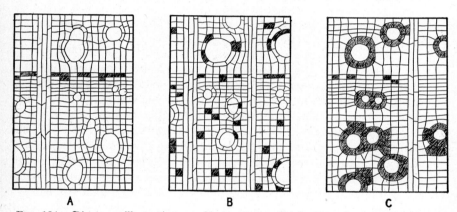

Fig. 104.—Diagrams illustrating wood-parenchyma distribution, the parenchyma cells shaded. *A*, terminal; *B*, diffuse; *C*, vasicentric.

a somewhat definite pattern of rows and masses of the different types (frontispiece). More or less definite tangential rows constitute a frequent type of arrangement; for example, in *Carya* and *Diospyros* (Fig. 103*A*), the wood parenchyma cells are so placed (in the figure the parenchyma cells are the small cells with large lumina). Vessels are often found in clusters, as in *Betula*, *Ulmus* (frontispiece), and *Robinia*.

Wood-parenchyma Distribution.—Wood parenchyma is distributed in three ways. These constitute types that are constant in genera and larger groups. In some gymnosperm woods, wood parenchyma is absent; in others, such as that of *Larix* and *Pseudotsuga*, and in some angiosperm woods such as *Magnolia* and *Salix*, wood parenchyma cells occur only in the last-formed tissue of the annual ring,—in other words "on the face of the late wood." Such woods have *terminal wood parenchyma* (Fig. 104*A*). Where parenchyma occurs not only in this location, but is also scattered throughout the annual ring, some of the cells lying among the tracheids and fiber-tracheids, the plant has *diffuse*, or *metatracheal, wood parenchyma* (Fig. 104B). *Malus* (Fig. 102A), *Quercus* (Fig. 101A), and *Diospyros* (Fig. 103A) have diffuse wood parenchyma. Where parenchyma occurs at the edge of the annual ring and elsewhere only about vessels, that is, in direct contact with the latter, or in contact with other parenchyma cells which are directly or indirectly in contact with vessels, and does not occur isolated among tracheids and fibers, the plant has *vasicentric*, or *paratracheal, wood parenchyma* (Fig. 104C). *Acer*, *Fraxinus*, and *Catalpa* have parenchyma of this type.

Structure of Gymnosperm Wood.—Certain general types of secondary xylem characterize the larger plant groups. That of the gymnosperms is simple and homogeneous, consisting of very few cell types, often of tracheids only—with the exception, of course, of the rays—as in species of *Abies* and *Agathis*. In such xylem the late wood differs hardly at all from the early wood; in other genera, as in *Larix*, *Sequoia* (Fig. 99A), and in the hard pines, the late wood is conspicuously different from the early wood. Some genera show intermediate conditions. In all genera the nonliving cells of the late wood, fiber-tracheids or fibers, differ from typical tracheids only in thickness of wall, width of lumen, and size and number of pits. Typical fibers are rare in gymnosperm woods. Wood parenchyma—often called *resin cells* in gymnosperm wood—is absent in a few genera, for example, *Araucaria*, *Taxus*, species of *Picea*, and *Pinus* (except for that about resin canals). It is scarce in many other genera, such as *Larix* and *Tsuga*, where it is terminal. More abundant parenchyma, arranged diffusely, characterizes such genera as *Juniperus*, *Thuja*, *Sequoia*, and *Podocarpus*. Vessels occur only in the Gnetales. Gymnosperm wood is simple in structure.

Structure of Angiosperm Wood.—The wood of the angiosperms is characterized by the presence of vessels, and, in general, by a complexity of structure far greater than that of the gymnosperms. This complexity is due to the presence of several kinds of cells—tracheids, vessels, fiber-tracheids, fibers of various types, wood parenchyma cells (or some of these); to variety in ray form; and to the arrangement and interrelations of the different kinds of cells with one another. In most genera, vessels are abundant, and in some woods they constitute a considerable percentage of the wood, as in *Tilia* and *Populus* (Fig. 100). They are also abundant in vines and many herbs. In herbs with complete woody cylinders they are frequently less abundant, and are small. In roots, vessels usually make up a large proportion of the wood. The wood of only a few groups of angiosperms—the Trochodendraceae and a few other primitive members of the Ranales; some xerophytes and hydrophytes;

secondary wood of monocotyledons—lacks vessels. Wood parenchyma is probably present in nearly all woody angiosperms, being abundant in some, as in *Carya* and *Platanus*, and scarce in others, as in *Acer* and *Liriodendron*. Typical wood parenchyma is absent in herbs with isolated vascular bundles and in some vines with bundles separated by wide parenchymatous rays. Fibers of several kinds may occur, even in a single species. Complexity and variety of structure are prominent features of angiosperm wood structure; in this respect no tissue, unless it be angiosperm phloem, surpasses secondary xylem.

Xylem Rays.—The xylem ray partakes in the general function of the vascular ray, of which it constitutes the part inside the cambium. The vascular ray, from its structure and position, seems to serve for transverse intercommunication in the living parts of vascular tissue, and perhaps also, through accompanying, radially extended intercellular spaces, to some extent makes possible an interchange of gases with the outside atmosphere. By the agency of the vascular ray, water may be readily transferred from the xylem to the cambium and the phloem, and food supplies moved from the phloem to the cambium and to the wood parenchyma. To the parenchyma and the inner ray cells, food for storage is readily passed inward from the phloem by the means of the rays. In the older, longer rays the terminal parts are not functional, the inner being included in heartwood, and the outer cut off, together with the surrounding cells, by periderm layers (Chap. IX).

The cells constituting a xylem ray are for the most part elongated in the direction of the long axis of the ray. Where all the cells are of this type the ray is *homogeneous*; where the cells are of different morphological types, some being vertically elongated or subcubical, the ray is *heterogeneous*. Ray cells are typically prismatic, often distinctly rectangular, though the corners are often rounded, and in very broad rays, as in *Quercus*, the cells may be nearly round in cross section. Xylem rays vary in width, height, and longitudinal extent. They may be one cell wide, as in *Picea* (Fig. 98C) and *Populus* (Fig. 100), when they are known as *uniseriate;* or two to many, *multiseriate*. Both uniseriate and multiseriate rays may be either homogeneous or heterogeneous. In angiosperms heterogeneous rays are apparently more primitive than homogeneous ones, and the condition where all rays are uniseriate and homogeneous is the most advanced. Groups of closely placed, narrow rays, such as are found in *Alnus* and *Carpinus*, are called *aggregate rays;* and multiseriate rays of certain types are sometimes known as *compound rays*. The latter term is a poor one because for some such rays it is a misnomer. The cells of a xylem ray lie in definite horizontal rows, and the cells of contiguous rows are so placed that the ends of a cell rarely coincide with the ends of the cells above or below. Neither do the end walls of the ray cells bear any relation to the walls of the cells of the vertical system. Structurally, the ray is more or less like a brick wall, the individual cells representing the bricks; the uniseriate ray is like a wall one layer thick; the biseriate, two layers thick, etc. In height, the rays range from one cell to very many, from a fraction of a millimeter to 8 or 10 centimeters. Certain so-called "medullary rays" of vines and herbs may much exceed the maximum height given but these are of different morphological nature. In a given species the rays may be of various widths and heights (Figs. 99C, 102C), or all may be of approximately the same width and the same height. For example, in nearly all gymnosperms the rays are uniseriate, but vary much in height (Fig. 98C); in many species of *Quercus*, uniseriate and very large multiseriate rays occur (Fig. 101A,C), no intermediate types being found, and the height of the rays varies. In *Betula* and *Acer*, the rays range from two to ten cells in width and are fairly constant in height; in species of *Diospyros*, the rays are alike in width and in height. The shape of the cross section of the ray as seen in tangential sections of the wood varies greatly in different plants, being linear, oblong, fusiform, or round-fusiform. The type is usually constant for a species.

The length (radial extent) of a xylem ray is dependent upon the position of the point of origin of the ray in the xylem (for origin of rays, see Chap. VI). From that point the ray is continuous to the cambium; xylem rays are discontinuous between the point of origin and the cambium only in case of injury to the cambium. (False beginnings and endings of rays appear in sections because the ray passes out of the plane of section and seems to end.)

In most woods the rays are arranged without definite plan, except that of a fairly uniform distribution through the vascular tissue (Figs. 98C, 102C), such that two rays never come into contact, and no considerable areas exist between them. In certain woods, especially those of tropical genera, such as *Diospyros*, the rays occur more or less definitely in tiers. Aside from their position as radiating bands, rays bear in their arrangement no relation to general stem structure; but the distribution of certain of the first-formed and of the very large rays may be dependent upon phyllotaxy, these rays having a definite relation to the position of leaf traces.

Ray Tracheids.—The cells of the wood rays are usually all living, and of a type that differs from typical parenchyma chiefly in the presence of a thick, more or less lignified wall. However, in a few genera of the gymnosperms, for example, *Pinus*, the rays consist of both living and nonliving cells. The latter are termed *ray tracheids* because they are tracheid-like in the lack of protoplasts and in the pitting and chemical nature of their walls, but resemble the living ray cells in their general shape and position in the ray (Fig. 105). In shape they are less uniform than the living cells, tending to be considerably longer, narrower, and

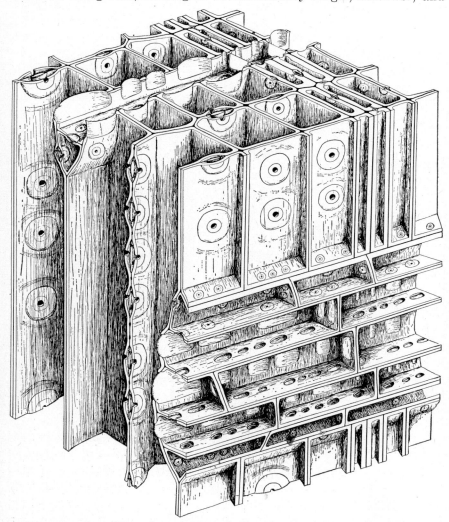

Fig. 105.—Secondary xylem of *Pinus Strobus*, showing details of structure. (The proportionate size of the ray is increased; only small parts of the tracheids are shown; the contents of the living ray cells are omitted.)

lower. Ray tracheids occur normally at the upper and lower margins of the rays; hence they are often spoken of as *marginal ray tracheids*. The marginal rows of ray tracheids consist entirely of this type of cell;

only very rarely here or elsewhere in the ray are two kinds of cells formed in one row. The number of rows of cells which may be of ray tracheids varies from one to several, one to three being most common. Rays that consist of only one, two, or three rows of cells may be entirely of ray tracheids; and where the rays are very high, there are often, near the middle of the ray, rows of *interspersed ray tracheids* in addition to the normal marginal rows. Ray tracheids, from their structure and position, apparently serve to conduct water radially.

Marginal Ray Cells of Angiosperms.—Ray tracheids do not occur in the angiosperms but the rays of this group are not always homogeneous. The wood of many genera, such as *Salix* and *Nyssa*, possesses rows of *marginal cells* different in size, shape, and contents from the other cells, and obviously different in function. These cells are always living. Their longest diameter is vertical, or the cells are shorter than the typical ray cells. The pits of the lateral walls, especially those leading to vessels, are larger and more numerous than the corresponding pits of other ray cells. Marginal cells may form continuous rows, when they are said to be *conterminous,* or occur scattered along the marginal rows, when they are *interspersed.* The function of marginal cells in the angiosperms is not understood. In many woods these cells contain special secretions, such as essential oils, as in *Sassafras.*

Tyloses.—Balloon-like enlargements of parts of cell walls, projecting into adjacent cell lumina through pit cavities, are known as *tyloses.* These structures are found in both primary and secondary xylem but are largely features of secondary xylem and are, therefore, discussed here. Tyloses are formed by the enlargement of the pit membranes of the half-bordered pit-pairs between wood parenchyma or wood-ray cells, and vessels or tracheids. The delicate membrane is expanded and grows, apparently by intussusception, pushing out of the pit cavity and protruding far into the lumen of the nonliving cell (Fig. 106). Into these bladder-like extensions of the parenchyma cells pass part of the cytoplasm and even the nucleus. After the tyloses are full grown, starch, crystals, resin, and gum may be formed within them, but the presence of these substances in quantity is uncommon. The tylosis may remain small or become very large; its size and shape depend in part upon the size of the lumen of the tracheid or vessel into which it extends, and in part upon the number of other tyloses present. The wall of the tylosis may remain thin and delicate, becoming wrinkled and partially collapsed in heartwood, or may become thick and even lignified; pits may be formed in it when it comes in contact with other tyloses. The tyloses in a given cell may be few, as in *Populus* (Fig. 100C), or many, as in the white oaks (Fig. 101), when they fill the lumen and become angular by compression. Tyloses are said to undergo division in some plants and form "multicellular tissue," which fills the lumen compactly, as in *Robinia* and *Maclura*. That such division is responsible for the multicellular appearance is doubtful. The appearance of separate cells may be brought about by the presence of multitudes of tyloses, each slender, and all mutually compressed, as are soap bubbles in a crowded mass. Tyloses may develop from occasional pit-pairs only, as in *Juglans, Liriodendron*, and *Sassafras*, or from great numbers of pit-pairs, perhaps from every pit

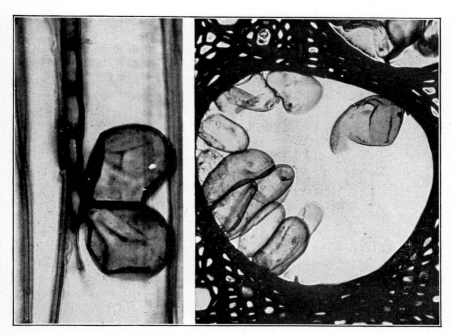

A *B*

Fig. 106.—Tyloses. *A*, two mature tyloses, derived from pits in the same ray cell, in *Aesculus octandra*, × 680; *B*, young tyloses in vessel of *Quercus lobata*. (*After Gerry.*)

(leading to a living cell) in a vessel, as in *Robinia, Rhus*, and *Catalpa*. More than one tylosis may develop from one parenchyma cell. Tyloses sometimes develop in the lumina of gymnosperm tracheids, as in the soft pines.

In the wood of conifers there is also found to some extent a closing of the cavity of resin canals by the enlargement of the epithelial cells (Fig. 103C). These enlarged cells are often called tyloses, but are better termed *tylosoids*. Similar tylosis-like growths occur in protoxylem, where entire parenchyma cells or parts of cells protrude into the weakened or ruptured tracheid or vessel between the rings or turns of the spiral.

Tyloses are of common occurrence in angiosperm wood. They are characteristic of certain species, and always absent in others. In many woods tyloses develop about the time of transformation of sapwood into heartwood, but they may be present to some extent even in the outermost rings of sapwood. They are most abundant in heartwood, but may occur also in the sapwood whenever normally present in the heartwood. Tyloses occur in the vessels of herbs also. They have been described or reported in such genera as *Cucurbita, Coleus, Canna, Portulaca, Rumex, Asarum,* and *Convolvulus.* Not only is tylosis development normal, but it may be induced in many plants by wounding. For example, they may develop in the region near wounds on the surface of a tree trunk or in the region where a branch has been removed; after the felling of a tree, tyloses may form in the sapwood throughout the log as it lies upon the ground. They may be present in the inner part of leaf traces after the leaf has fallen. Such tyloses, however, are usually sporadic in occurrence, and irregular in shape and size.

The development of tyloses either normally or as a result of wounding is said to be due to a difference in pressure in the cells on each side of a pit membrane; to the reduction of pressure; or to the cessation of conduction in the vessel, permitting the membrane to expand into the cell.

The distribution of tyloses in wood is not determined by type of wood, by rate of growth, by age of the plant, or by habitat. Where wood parenchyma is scarce, tyloses likewise are few.

Tyloses are of considerable economic importance in the use of woods. They are a factor, though a minor one, in durability; durable species, with few exceptions, possess abundant tyloses—for example, *Maclura, Robinia, Juglans nigra, Morus, Catalpa,* and the white oaks. By blocking the vessel lumina they prevent rapid entrance of water, air, and fungal filaments. (Durability, however, is largely dependent upon the chemical nature of the wall.) The presence of tyloses prevents rapid penetration of artificial preservatives. For example, under treatment with creosote, red oak wood, which is without tyloses, is penetrated for long distances through the vessels, whereas white oak wood, with tyloses, is hardly penetrated at all. Similarly, white oak wood can be used in "tight" cooperage, whereas red oak with its open vessels is of little value for this purpose.

Sapwood and Heartwood.—Although little is known definitely about the conductive activity of xylem of different ages, it is probable that cells when first mature are most active, and that there is a gradual slowing down of conduction until functional activity ceases. So long as xylem contains living cells and is apparently conducting, at least to some extent, it is known as *sapwood;* after all activity ceases, it becomes *heartwood.* The terms *alburnum* and *duramen*, formerly much used for sapwood and heartwood respectively, are passing out of use. The sapwood of a tree serves for conduction, support, and food storage; the heartwood only for support. In the transformation of sapwood into heartwood a number of important changes occur: all living cells lose their protoplasts; the cell sap is withdrawn, and commonly the water content of the cell walls greatly reduced; any food materials present in the living cells are removed; tyloses, if characteristic of the wood, are formed; the partly lignified walls of parenchyma cells may become more strongly lignified; there are formed within or brought into the changing cells certain substances new to the tissue, such as oils, gums, resins, tanniferous compounds, various aromatic and coloring substances; flexible pit membranes become fixed, usually in the closing position. In short, the xylem becomes physiologically functionless, being as heartwood merely a solid supporting column.

The extent to which the water content of the xylem is reduced with change to heartwood varies greatly: in a few plants—for example, *Ulmus* and *Malus pumila*—the heartwood remains "wet," that is, saturated with water (but such water is probably not in process of conduction); in other plants the heartwood may become very dry, as in *Fraxinus* (where in some species it is said even to become checked in the living tree). Such substances as oils, resins, and coloring materials may infiltrate the walls, and gums and resins may fill or partly fill the lumina of the cells. In *Diospyros* (ebony) and *Swietenia* (mahogany), the cell cavities are to a greater or less extent filled with a dark-colored, gummy substance. The color of heartwood, in general, is the result of the presence of these substances within the cell walls and, in some woods, the lumina. The fact that the heartwood is darker in color than the sapwood is incidental to the formation of these substances. Some genera, such as *Betula, Populus, Picea, Agathis,* have heartwood hardly if at all darker in color than the sapwood.

Heartwood, as timber, is more durable than sapwood, because the reduction of food materials available for fungi and bacteria by the absence of protoplasm and starch, the formation of such substances as resins, tannins, and oils, and the blocking of the vessel cavities by tyloses and gums, render the wood less pervious to water and less subject to attack by the organisms of decay. For this reason, as well as for many others— such as the presence of desirable color or odor, or of substances of commercial value which may be removed, such as haematoxylin—heartwood is commonly of more value commercially than is sapwood. The latter, however, is preferred in some woods, as in some grades of quarter-sawed oak, and in hickory and ash for tool handles and spokes. Owing to lack of resin, gums, and coloring substances, sapwood is also preferred for pulp wood, and for wood to be impregnated with preservatives.

Apparently, there is no fixed length of functional life in xylem. The functional period is controlled by the physiological activity of the tree or of the organ or segment of the organ in question. Young, vigorous plants or vigorously growing parts of older plants have little or no heartwood, whereas slow-growing, weak, and most old trees have a very large proportion of heartwood in stem and root. In old trees the xylem remains sapwood but a very few years. In case vigorous growth is renewed, a large amount of sapwood is built up by the retention of all sapwood as such for many years. All parts of an annual ring are not changed into heartwood simultaneously; that is, the outer limit of heartwood bears no relation, necessarily, to the annual rings. Where strongly developed roots or branches are present on one side of a trunk, the xylem of the segment on that side remains alive long after other parts of the same age have died.

Relation of Microscopic Structure to Properties and Uses of Wood.—The wood of different species varies greatly in its properties and in its value for different purposes. Characteristic qualities, and hence specific economic uses, depend largely upon the histological structure and upon the chemical nature of the tissue. Variations in histological structure which affect the properties of wood are chiefly those of the kinds, proportion, and arrangement of cells—for example, the presence or the absence of fibers and of large vessels and their restricted or widespread distribution; the diameter and the thickness of the wall of fibers; the length of fibers and the extent to which they overlap other fibers; the straightness of fibers; the abundance and the width of the wood rays; the presence of tyloses. The amount and the distribution of wood parenchyma seem to bear little relation to the properties of wood. Variations in the chemical nature of the wood are of the greatest importance in relation to certain qualities of wood, expecially those in which heartwood differs from sapwood. The cell wall itself varies in chemical nature, the proportions of cellulose, lignocellulose, and "lignin" differing greatly. Occasionally, the walls are even gelatinous in nature. Tannin compounds may be present in considerable quantities as substances infiltrating the cell wall; the lumina may contain various amount of gums, resins, and tannins.

Weight.—The wall substance of woods, either light or heavy, is of nearly the same specific gravity, about 1.53. Variations in weight are due to variations in the proportion of wall substance and of lumen space.

Where the latter is small in amount, that is, where the wood is dense, it is, of course, heavy. Hence, abundance of slender, thick-walled fibers makes a wood heavy, as in *Guaiacum* (lignum vitae), *Diospyros* (Fig. 103*A*), and *Malus pumila* (Fig. 102). Where the walls of all cells are unusually thick and the lumina small, especially where the proportion of fibers is high, the wood is heavy, as in *Guaiacum* and *Diospyros* (ebony). Where all the walls are thin and the lumina of parenchyma and fibers are large, the wood is light. The presence of numerous thin-walled vessels, as in *Populus* (Fig. 100) and *Tilia*, reduces the specific gravity greatly even though the fiber walls are not especially thin. Extremely light woods, such as those of the "cork-wood" type—for example, *Ochroma* (balsa) (Fig. 103*B*)—have a high proportion of large, thin-walled parenchyma cells. The specific gravity of wood ranges from about 0.04 (*Aeschynomene*) to about 1.4 in black ironwood (*Krugiodendron*). The majority of well-known, commercially important woods range from 0.35 to 0.65 in specific gravity. Species of *Quercus* range from 0.65 to 0.95; *Carya*, 0.74 to 0.84; *Fraxinus*, 0.65 to 0.72; *Acer*, 0.62 to 0.69; *Pinus*, 0.37 to 0.79; *Abies*, 0.35 to 0.47; *Populus*, 0.36 to 0.41; *Sequoia*, 0.29 to 0.42. Examples of heavy woods are *Guaiacum*, 1.1 to 1.4; *Eucalyptus*, 0.8 to 1.25; *Acacia*, 0.8 to 1.3. Among very light woods there are few of economic importance. Of these, balsa (*Ochroma*), with a specific gravity of 0.12 to 0.37, is used extensively in insulation, airplane manufacture, and for life rafts. The structural strength of this wood is high for its weight. Histologically there are two types of lightweight woods: those with alternating bands of thick-walled, lignified cells and thin-walled, unlignified cells; and those with elements homogeneously arranged. In some woods, such as those of the Caricaceae (papaya) and Phytolaccaceae (*Phytolacca dioica*), only the vessels are lignified; the wood resembles the flesh of a turnip.

Strength.—The presence of a large proportion of fibers or fiber-tracheids makes a wood strong. Hence, dense and heavy woods are usually strong woods. The length of the fibers and the extent to which the ends overlap are apparently features of minor importance as regards the strength of the wood.

Durability.—Resistance to decay by the action of fungi and bacteria is dependent largely upon the chemical nature of the wood—of the cell walls and of the cell contents. It is not correlated to any degree (except in some very light woods) with physical properties, such as weight and strength, or with structure. The presence of tyloses that block vessels does, of course, reduce the rapidity of entrance of fungal hyphae, and of water and oxygen, but it is the presence or absence of infiltrating, natural preservative substances, such as tannin, resin, and oils, that determines the durability of wood. Both light and heavy woods are durable—for example, those of *Sequoia, Catalpa, Castanea, Robinia, Maclura*; and other equally light and heavy woods decay rapidly—for example, those of *Populus, Tilia, Acer*, and *Carya*. Resistance to decay is not necessarily correlated with depth of color, though heartwood is generally more durable than sapwood, and, in a general way, the darker the color the greater the resistance. This condition results from the fact that depth of color often indicates roughly the amount of preservative substances. Durability as resistance to mechanical destruction depends upon hardness, density, and toughness.

Other Properties.—Flexible woods are fairly homogeneous and have long, straight, strongly overlapping fibers, and linear rays. This type of wood also cleaves readily. Toughness involves strength and pliability, and, to a large extent, the interlocking of the fibers. Where the fibers are strongly interlocked, the woods can be put to special uses, that of *Ulmus*, for example, for hubs and basket splints; of *Ostrya*, for mallets and tool handles. The interlocking of the grain may be due in part, as in *Platanus*, to the presence of low, proportionately very broad rays around which the fibers bend in their course. Woods with interlocked fibers and uneven texture are not readily "workable." In all properties the proportion of water present is of much importance.

Penetrability by Preservatives.—The rate of penetration of wood by preservatives, such as creosote, is dependent to a large extent upon the structure of the wood. Such open channels as vessels and resin ducts provide ready access to infiltrating fluids, but the preservation of regions about these openings alone is of little value. Penetration through thick cell walls can occur only slowly. More rapid penetration through closed cells must take place through the pit-pairs, from lumen to lumen, as does the passage of water in living tissue. Passage of fluids through the pit-pairs during translocation undoubtedly is chiefly through the marginal regions of the closing membranes. Upon this movement the structure and the behavior of bordered pits have an important bearing. When living sapwood is immersed in a preservative fluid under pressure, that is, where pressure is from all sides, there is little or no penetration; when the preservative is applied at one end of a piece of wood with light pressure, penetration is fairly rapid, but with increased pressure it is quickly cut down. The cessation of penetration under the higher pressure appears to be due to the closure of the pit-pairs by movement of the tori (Figs. 22C and 25C). Green timber is only slightly penetrable to preservatives. In certain processes such timber is treated under pressure, but the preservative is not forced into the wood until the green timber has had the water driven out in the treating retort, when it is, of course, seasoned timber. It is practicable to treat only seasoned timber with preservatives. The structural conditions responsible for the penetration of seasoned wood by preservatives are not well understood. It has been believed that microscopic checking of the cell walls is responsible for the passage of fluids from lumen to lumen, but such checking does not always occur, and the checking slits do not pass through the wall (Fig. 271,K). The pit membranes rupture only rarely in drying. It seems probable that the differences are due to changed conditions involving surface tension, capillarity, and the passage of fluids through very minute openings. In this case the openings involved are probably in the closing membranes of the bordered pit-pairs. Although the lumina and pits are larger in early wood than in late wood, the latter is more freely penetrated by preservative fluids, whether the wood is green or dry. This is doubtless due to the fact that valve action of the closing membranes of pit-pairs in late wood is weak or lacking. Sapwood is always much more penetrable than heartwood. This is doubtless because in the heartwood the tori have become fixed in the lateral position, and the cavities of the pits are often clogged with gummy or resinous material. Even the lumina may be more or less filled by these substances or by tyloses.

Grain of Wood.—Variations in size, shape, and orientation of cells, and in proportion and arrangement of cell types, determine variations in appearance of wood, known as *grain*. Coarse grain, fine grain, and cross-grain are largely self-explanatory terms. Spiral grain refers to a condition in which the cells of the vertical system lie parallel with one another, but the whole system winds spirally about the tree. The grosser features of wood structure, annual rings and wood rays, form the most conspicuous grain of wood. The alternation of layers of coarser and finer cells, early and late wood, produce the prominent markings in many woods, such as those of the hard pines and the ashes. Wood rays when cut longitudinally, as in radial section, or obliquely longitudinally in sections that approach the radial, form conspicuous markings which are the more prominent because of the dense structure of rays and their tendency to take a high polish. Large rays, as in *Quercus*, form the *silver grain* of wood. The presence of silver grain renders the wood of great value for furniture and cabinet work. *Curly grain* is the result of an undulate course of the cells. This occurs in individual trees or parts of trees, and is frequent in *Betula, Castanea, Acer,* and *Prunus,* for example. *Bird's-eye grain* is related to the presence of numerous dormant adventitious buds; these buds maintain their position in the bark of the tree, building up weak central cylinders, chiefly parenchymatous, as the tree trunk increases in diameter. The pith-like steles form the "eye" when cut in cross section (tangential section of the wood), and the fibers and other cells of the vertical system swing around them as they do about the bases of branches.

Compression Wood.—A type of wood somewhat darker than the sapwood, often reddish and resembling heartwood, is present in many conifers, as in *Pseudotsuga, Pinus,* and *Araucaria.* It occurs on the underside of branches, and on the lower side of tree trunks in trees that do not stand vertically. This wood is known as *redwood* or as *compression wood* because of its color and its location in the tree where compression seems to occur. In properties it is denser, more brittle, and shrinks longitudinally to a greater extent than the normal wood. Histologically it differs in only minor ways from normal wood: its cells are somewhat shorter and perhaps thicker walled; in cross-sectional shape, the tracheids are more rounded, and intercellular spaces are frequently present. Compression wood is apparently formed in response to geotropic stimuli, not as a result of compression during development.

Pith-ray Flecks.—Injury to the cambium region by insects known as "cambium-miners" produces ultimately areas of wound tissue in the mature xylem. These areas when seen in cross section appear as small patches of irregularly arranged, thick-walled parenchyma cells. They resemble wood rays ("medullary rays" or "pith rays") in that they consist of parenchyma; hence they are commonly known as *pith-ray flecks* and as *medullary spots.* But they have nothing to do with rays or with the pith. These injuries are common in rosaceous woods, and also in the wood of *Salix, Acer, Betula,* and other genera. The insects causing the injury apparently belong to different groups, but are chiefly Diptera. The larvae bore along the branches and trunk, forming tunnels downward through the cambium or the immature xylem. The cavity so formed is soon filled by the proliferation of surrounding cells. The cambium, if injured, is replaced, and growth is continued normally, burying the strands of wound tissue in normal wood. Such injury is of minor importance to the tree, but may greatly reduce the economic value of the wood, as in *Acer* and *Prunus.* (Pith-ray flecks are, unfortunately, sometimes explained as normal features of wood structure.)

Gummosis.—Another pathological condition, *gummosis,* deserves brief mention because of its frequent occurrence. As a result of injuries of various types which bring about exposure of tissue to slight drying, the wall of cells around the injured region are transformed into gum. The transformation appears to be the result of enzyme action, which first affects the middle lamella and may ultimately dissolve the entire wall. Tissues affected by gummosis may be only partly broken down, or become completely transformed into a mass of gum. The gum may fill the lumina of the affected cells, entering from the pits or may be exuded from the tissue. Gummosis is common in woody plants, being extensive after insect and other injury in such plants as cherry, peach, and acacia. It perhaps occurs in the xylem of many plants as a response to injury or diseased conditions, and may incidentally protect tissues from further injury.

References

(Further references for Xylem under Chap. IV)

AUCHTER, E. C.: Is there normally a cross-transfer of foods, water, and mineral nutrients in woody plants?, *Univ. Md. Agr. Exp. Sta. Bull.*, 257, 1923.
BAILEY, I. W.: The effect of the structure of wood upon its permeability. No. 1 The tracheids of coniferous timbers, *Amer. Ry. Eng. Assoc. Bull.*, **174**, 1–17, 1915.
———: The structure of the bordered pits of conifers and its bearing on the tension hypothesis of the ascent of sap in plants, *Bot. Gaz.*, **62**, 133–142, 1916.
———: The structure of the pit membranes in the tracheids of conifers, and its relation to the penetration of gases, liquids, and finely divided solids into green and seasoned wood, *For. Quart.*, **11**, 12–20, 1913.
BARGHOORN, E. S., JR.: The ontogenetic development and phylogenetic specialization of rays in the xylem of dicotyledons. I. The primitive ray structure, *Amer. Jour. Bot.*, **27**, 918–928, 1940. II. Modification of the multiseriate and uniseriate rays, *Ibid.*, **28**, 273–281, 1941.
———: Origin and development of the uniseriate ray in the Coniferae, *Bull. Torrey Bot. Club*, **67**, 303–328, 1940.
BROWN, H. P.: Pith-ray flecks in wood, *U.S. Dept. Agr. For. Serv. Circ.*, **215**, 1913.
——— and A. J. PANSHIN: "Commercial Timbers of the United States," New York, 1940.
CIESLAR, A.: Das Rotholz der Fichte, *Centralbl. Ges. Förstwesen*, **22**, 149–165, 1896.
GERRY, E.: Fiber measurement studies: length variations, where they occur and their relation to the strength and uses of wood, *Science*, new ser., **41**, 179, 1915.
———: Microscopic structure of wood in relation to properties and uses, *Proc. Soc. Amer. For.*, **8**, 159–174, 1913.
———: Tyloses: their occurrence and practical significance in some American woods, *Jour. Agr. Res.*, **1**, 445–469, 1914.
HIGGINS, B. B.: Gum formation with special reference to cankers and decays of woody plants, *Ga. Exp. Sta. Bull.*, **127**, 1919.
HYDE, K. C.: I. Tropical light weight woods, *Bot. Gaz.*, **79**, 380–411, 1925.
JEFFREY, E. C.: "The Anatomy of Woody Plants," Chicago, 1917.
KANEHIRA, R.: Anatomical characters and identification of Formosan woods with critical remarks from the climatic point of view, Taihoku, 1921.
———: On light-weight woods, *Jour. Soc. For.* Japan, **15**, 601–615, 1933. (Abstract in *Trop. Woods*, **37**, 52–53, 1934.)
KLEIN, G.: Zur Ätiologie der Thyllen, *Zeitschr. Bot.*, **15**, 417–439, 1923.
KOEHLER, A.: "The Properties and Uses of Wood," New York, 1924.
KRIBS, D. A.: Salient lines of structural specialization in the wood rays of the dicotyledons, *Bot. Gaz.*, **96**, 547–557, 1935.
KÜHNS, R.: Die Verdoppelung des Jahresringes durch künstliche Entlaubung, *Bibl Bot.*, **70**, 1–53, 1910.

Küster, E.: Secundäres Dickenwachstum: Holz und Rinde, *In* Linsbauer, K.: "Handbuch der Pflanzenanatomie," IX, 1939.

Muller, C.: Über die Balken in den Holzelementen der Coniferen, *Ber. Deut. Bot. Ges.*, **8** (sup.), 17–46, 1890.

Penhallow, D. P.: Anatomy of North American Coniferales together with certain exotic species of Japan and Australasia, *Am. Nat.*, **38**, 243–273, 331–359, 523–554, 691–723, 1904.

Pillow, M. Y., and R. F. Luxford: Structure, occurrence, and properties of compression wood, *U.S.D.A. Tech. Bull.*, 546, 1937.

Record, S. J.: "Identification of the Timbers of Temperate North America," New York, 1934.

────── and C. D. Mell: "Timbers of Tropical America," New Haven, 1924.

────── and R. W. Hess: "Timbers of the New World," New Haven, 1943.

Robinson, J.: The microscopic features of mechanical strains in timber and the bearing of these on the structure of the cell wall in plants, *Phil. Trans. Roy. Soc. London*, **210B**, 49–82, 1920.

Sifton, H. B.: The bar of Sanio and primordial pit in the gymnosperms, *Trans. Roy. Soc. Can. Sect.*, V, **16**, 83–99, 1922.

CHAPTER VIII

SECONDARY PHLOEM

In the preceding chapter consideration is given to those secondary tissues formed by the cambium toward the inside of the stele and known as secondary xylem, or wood. The present chapter deals with the analogous tissues formed by the cambium toward the outside of the stele, the *secondary phloem*. The terms "inner bark" and "bast" have been used by various authors to indicate these tissues, but these terms are not specific enough to be valuable in technical usage. The terminology used in the description of phloem structure is given in Chap. IV; part of that terminology and description is repeated here for completeness of discussion.

Extent and Amount of Secondary Phloem.—The extent and the amount of the secondary phloem depend upon the type of plant and upon the age of the part in question. The distribution of secondary phloem is, of course, controlled by that of the cambium. Hence, this tissue in its entirety may form a layer over all parts of the plant axis except the tips and may extend outward in the leaf traces into the larger veins of the leaf. In herbaceous stems, particularly those in which the vascular tissue is reduced and the stele broken, it may comprise bands or strips of tissue which, as seen in cross sections of the stele, appear as isolated groups of cells between the cambium and the pericycle (Fig. 111A,B). In such forms the tissue is radially thin, so that the total amount of such phloem is small. On the other hand, in a woody plant of considerable age, all the thick layer of tissue outside the cambium may be phloem, living and dead, combined with varying amounts of periderm. Between these extremes are all gradations. Normally, the amount of secondary phloem is less than the amount of secondary xylem, both in the space occupied and in the number of cells formed. In woody plants this difference in amount is exaggerated, not only because the older phloem becomes crushed, but also because the dead outer layers of phloem are actually lost either by weathering or by abscission.

Importance of Secondary Phloem.—The special importance of the secondary phloem is that, in the majority of cases in the dicotyledons and the gymnosperms, it very soon replaces the primary phloem which becomes crushed and functionless. This is particularly true of the woody plants in which secondary thickening is initiated close to the growing point and soon crushes the delicate primary phloem tissue. In fact, in such plants, the existence of the conducting cells of the primary phloem as functioning tissue is of such short duration that it is difficult to study. It appears in transverse sections of young twigs as obscure lines of crushed, thin walls and wall substance that have lost all semblance of their original cell form and structure. In some herbaceous dicotyledons this obliteration of the primary phloem is not so complete or, as in the potato plant, does not occur. In monocotyledons that have secondary thickening, the primary tissues remain intact. In general, however, in plants where secondary thickening takes place, the secondary phloem is the only phloem which is functionally important over any considerable length of time. The development of this tissue by the cambium provides for its constant renewal, a necessary feature since even secondary phloem is of short duration as a functioning tissue.

Structure of Secondary Phloem.—Considered as a whole, secondary phloem is a complex tissue made up of a number of cell types, all of which have a common origin in the cambium. It does not differ fundamentally from the primary phloem as the same types of cells occur in both. The secondary phloem, however, in comparison with the primary phloem of the same species, usually has a more regular arrangement of cells in radial rows, a higher proportion of sieve tubes which are larger and have thicker walls, shorter sieve-tube elements and fibers, and a longer functioning life. The elements which are normally present in secondary phloem are sieve cells or sieve tubes; companion cells, which accompany sieve tubes in the angiosperms; parenchyma cells of one or more types; phloem-ray cells. Usually some type of sclerenchyma is also present, and, less commonly, secretory cells, laticiferous ducts, and resin canals. The arrangement and the proportion of cells of the different types vary greatly in different species.

Sieve Cells and Sieve Tubes.—*Sieve cells* and *sieve tubes* (Fig. 52) are the characteristic elements in all secondary phloem from the standpoint both of structure and of function. *Sieve cells* are characteristic of gymnosperms. They are separate and distinct cells, in this respect resembling tracheids. *Sieve tubes* are characteristic of angiosperms. They are series of sieve-tube elements attached end to end with certain sieve areas more highly specialized than others. The sieve tubes of the secondary phloem of dicotyledons are of many different types with respect to the shape and nature of the end and side walls. In many woody species, such as *Carya cordiformis* (Fig. 109*B*), the oblique end walls of the sieve-tube elements frequently extend for nearly half the length of the element. These oblique walls have many specialized sieve areas which together form *compound sieve plates*. Sieve plates of this type are mostly confined

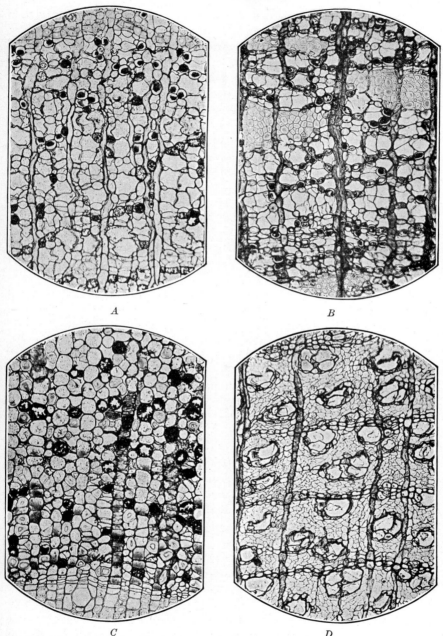

Fig. 107.—Secondary phloem in transverse section. *A, Populus deltoides; B, Juglans nigra; C, Cephalanthus occidentalis; D, Carya cordiformis.* All × 100. (*After MacDaniels.*)

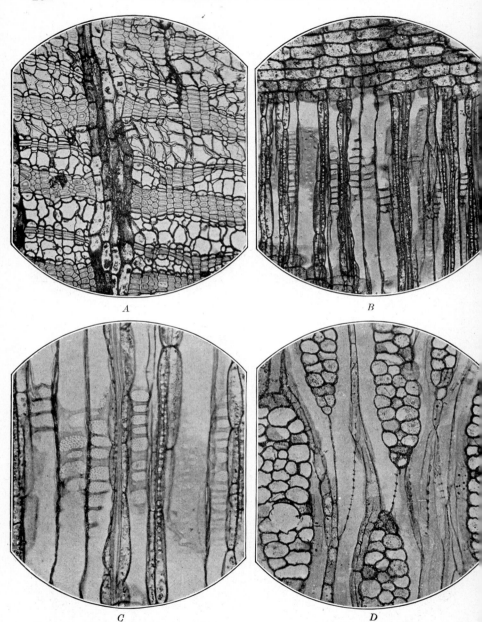

Fig. 108.—Secondary phloem of *Liriodendron*. *A*, transverse, × 100; *B*, radial, × 100; *C*, radial, × 200; *D*, tangential, × 100. (*After MacDaniels.*)

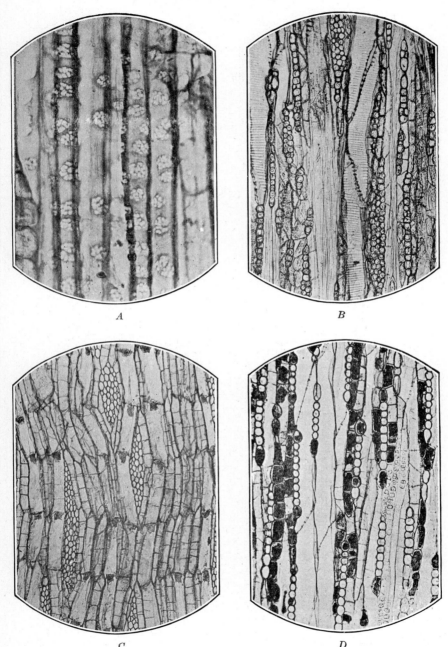

Fig. 109.—Secondary phloem in tangential section. *A, Pinus Strobus*, × 250; *B, Carya cordiformis*, × 100; *C, Robinia Pseudo-Acacia*, × 100; *D, Salix nigra*, × 100.

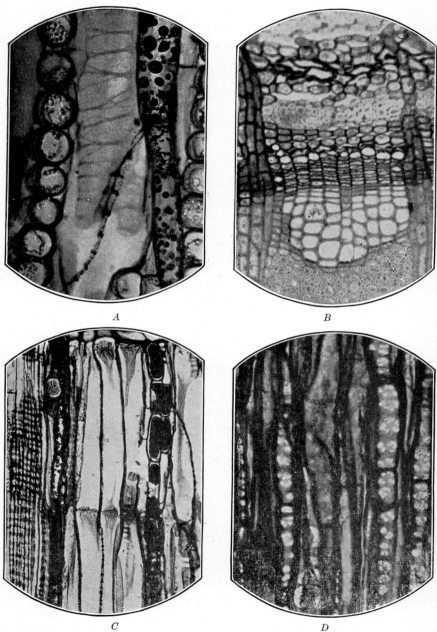

Fig. 110.—*A*, lattice, face view, and sieve plate, in section, of *Populus*, × 450; *B*, cross section of cambium region of *Robinia*, dormant state, showing the collapsed condition of all mature sieve tubes, × 100; *C*, radial section of cambium and phloem of *Ulmus*, dormant state; the cambium at the left; sieve tubes with transverse end walls and slime plugs in the center, × 100; *D*, phloem parenchyma of *Cornus*, showing sieve-plate-like pits, × 450.

to the radial walls and thus connect cells of the same functioning stage. The other extreme of variation is the type found in *Robinia* (Fig. 109*C*), *Maclura* and some species of *Ulmus* (Fig. 110*C*), in which the terminal walls of the sieve-tube elements are transverse. Here there is a single specialized sieve area. Such an end wall forms a *simple sieve plate*. Between these two extreme types is a series of intergrading forms. No one type is predominant in the woody angiosperms; even closely related species, as those of *Fraxinus*, may have markedly different types. In the majority of species, the sieve-tube elements of the secondary phloem of herbaceous plants have simple sieve plates (Fig. 111*C*); where compound sieve plates occur, they are not of the extreme type found in the Juglandaceae. The sieve tubes of seedlings and of the phloem formed in the first few years in woody twigs are of relatively small size with few compound sieve plates. In a given species, the type of sieve plate is usually constant in tissues of the same age.

The proportion and arrangement of sieve cells and sieve tubes in secondary phloem vary greatly in different species. In the gymnosperms, the sieve cells may make up the greater part of the phloem; parenchyma and sclerenchyma occur in small amounts. On the other hand, in most herbaceous plants, in seedlings, in young twigs of woody plants, and in the mature secondary phloem of some woody plants, for example, *Carya cordiformis* (Fig. 107*D*) and *Dirca palustris*, the proportion of sieve tubes may be much smaller than that of the other cell types. Sieve tubes also show great variety in arrangement with respect to other cell types in secondary phloem. In some plants, such as *Liriodendron* (Fig. 108*A*), *Juglans*, and *Tilia*, they occur in more or less definite tangential rows or bands; in others, such as *Carya* (Fig. 107*D*), they are in somewhat isolated groups of three or four surrounded by fibers or other types of cells. In *Cephalanthus* (Fig. 107*C*), the sieve tubes form, in cross sections of the phloem, radial rows which are frequently interrupted by fibers, a condition not uncommon in shrubs which approach the herbaceous type. In fact, in woody plants, almost any arrangement of sieve tubes or sieve cells with parenchyma and sclerenchyma may occur. In the secondary phloem of herbaceous plants generally, sieve tubes are without regularity of arrangement (Fig. 111*A,B*), although in certain groups, such as the Ranales, there may be a definite plan.

The lateral walls of sieve-tube elements that abut upon other sieve tubes in secondary phloem are frequently covered with pronounced lattices (Fig. 52*H,S*). These are abundant in some woody plants, such as *Ulmus* and *Populus* (Fig. 110*A*) but may be rare or absent in other genera. The presence of well-marked lattices is not correlated with sieve-tube type as classified on the basis of the angle of the end wall;

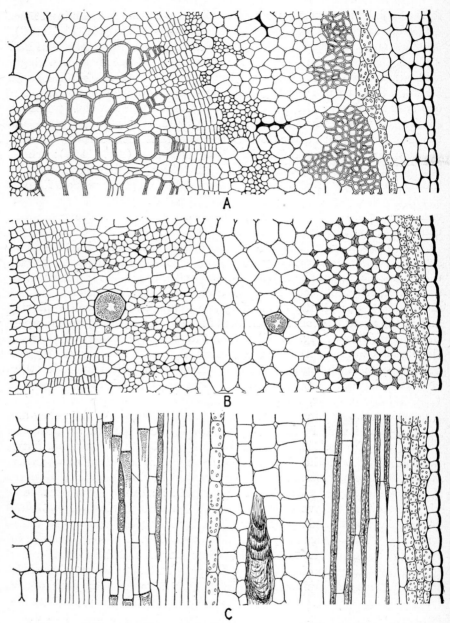

Fig. 111.—Secondary phloem in herbaceous stems. *A, Pisum,* transverse section showing primary phloem crushed; *B, C, Hibiscus esculentus: B,* transverse section showing companion cells (with protoplasmic contents); *C,* radial section showing sieve tubes with transverse end walls and several companion cells associated with each sieve tube. (*Courtesy of E. F. and E Artschwager.*)

lattices may occur in any type. The sieve tubes of the secondary phloem of herbaceous plants are usually without prominent lattices.

Companion Cells.—Companion cells (Fig. 52H,I,S,T) are absent in gymnosperms but are probably present in greater or less abundance in angiosperms of all types. In some plants, such as *Juglans* and *Solanum*, they accompany only a part of the sieve tubes; in others, for example, *Tilia*, from one to three companion cells are seen in cross section with every sieve tube. A companion cell may extend the entire length of the sieve-tube element to which it is adjacent or only part way, or several shorter cells may extend along the sieve-tube element, the series occupying the full length of the element or only a part of it (Fig. 111C). As seen in transverse section, therefore, a sieve tube may have no companion cell or one to several, each of which in turn may be a member of a vertical series of companion cells. Ordinarily, companion cells can be most readily identified in transverse sections of the phloem by their appearance as small, triangular or rounded cells apparently located in the corners of the sieve tubes (Fig. 107A), the two kinds of cells together outlining the mother cell from which they have been derived. In some plants, companion cells may extend as narrow cells across the entire width of the sieve tube. In the secondary phloem of herbaceous plants, where the vascular cylinder is well developed, as in *Lobelia* and *Solanum*, they are frequently difficult to recognize because of the small size of all phloem elements in this type of plant. In plants with less well-developed vascular cylinders, such as *Cucurbita* and *Ranunculus*, the companion cells are distinct and large enough to be easily recognized. Companion cells show no great diversity of type, varying chiefly in length.

Phloem Parenchyma.—Parenchyma cells are found in the secondary phloem of all plants (except extreme herbaceous types where secondary phloem is small in amount), some vines, and woody forms of some genera and families which are predominantly herbaceous. The proportion of parenchyma varies through wide limits. In the gymnosperms, parenchyma cells are relatively few in comparison with the number of sieve cells, a condition fairly constant in all species. In the angiosperms, however, there is great diversity. The secondary phloem of the seedlings and of the younger twigs of woody dicotyledons is largely composed of parenchyma, but in the more mature condition the proportion of these cells is much smaller. In some plants, for example, *Carya* and *Robinia*, there are few parenchyma cells in the secondary phloem of the older stems; in others, such as species of *Cornus*, parenchyma may predominate. In herbaceous dicotyledons, the proportion of parenchyma is usually smaller—frequently much smaller—than that of sieve tubes and companion cells. In many of the Compositae, the proportion of parenchyma is larger. In some families, such as the Ranunculaceae, and in certain genera in other families, no parenchyma is present, the phloem consisting wholly of sieve tubes and companion cells.

The arrangement of the parenchyma in secondary phloem varies as does that of the sieve tubes. As seen in cross sections of phloem, parenchyma may occur in tangential bands alternating with bands of sieve tubes and fibers, as in *Liriodendron* (Fig. 108A) and *Tilia;* in radial rows, as in *Cornus* and *Sambucus;* singly or in groups of a few cells, as in *Pinus;* or more or less definitely clustered about the sieve tubes, as in *Carya* (Fig. 107D). In longitudinal section, the cells are seen to form vertical series parallel with the sieve tubes (Fig. 109C, D).

The types of cells constituting the parenchyma of secondary phloem show considerable diversity. This can be most readily understood by a consideration of the development of these elements from the cambium (Chap. VI). Briefly stated, phloem parenchyma cells are formed directly from parenchyma mother cells, which, in turn, are formed from cambial cells. The cambial derivative may develop directly into a parenchyma cell or, as is more commonly the case, divides transversely, forming two or more cells. Phloem parenchyma cells, therefore, as seen in longitudinal section, may be elongated and pointed at both ends, resembling the cambial cell from which they were derived, or they may be rectangular or cylindrical, ranging from very elongate to short cylindrical or nearly cubical. The former, sometimes called *cambiform cells*, are not commonly found in the secondary phloem of woody plants, but rather generally in primary phloem, in herbaceous types, and especially in vines such as *Cucurbita*. Parenchyma cells formed by the transverse division of the parenchyma mother cell are common in the secondary phloem of all types of plants. The end cells resulting from such division remain pointed at one end.

In many woody angiosperms two or more types of phloem parenchyma, distinct in form and function, are often found in the same tissue. In *Tilia americana*, for example, one type of parenchyma cell is elongate, heavily pitted, and usually associated with the sieve tubes; the other is short and broad, apparently without an active protoplast, and usually contains large crystals. In other woody plants, such as *Robinia*, the parenchyma cells are uniformly short and broad with thin walls and abundant pits. The secondary phloem parenchyma of herbaceous plants does not show the diversity of type shown in the woody plants. For the most part, its cells are thin-walled and elongate, rectangular or rounded in cross section (Fig. 111A,C).

Phloem Fibers and Sclereids.—Sclerenchyma of one type or another is a characteristic feature of the secondary phloem of many plants. Fibers are common and occur in a variety of form and arrangement. Frequently, they occur in definite tangential bands, for example, in *Liriodendron* (Fig. 108*A*) and *Populus*. In other genera they are found singly, as in *Cephalanthus* (Fig. 107*C*). In some woody plants which have a hard, tough bark, as in *Carya* (Fig. 107*D*), the fibers make up the greater part of the secondary phloem and may nearly surround the groups of softer tissues. When abundant, they provide considerable mechanical support for the stem. In *Dirca*, the twigs owe their toughness largely to the phloem fibers.

In the gymnosperms, all conditions are found from a complete lack of sclerenchyma, as in the phloem of *Pinus Strobus*, to well-developed tangential bands of fibers, as in *Juniperus*, and large masses of sclereids as in *Tsuga*. In *Thuja occidentalis* the fibers are arranged in uniseriate tangential rows which alternate with rows of sieve cells and parenchyma.

Sclereids frequently occur in secondary phloem either alone or in combination with fibers. In some plants, for example, *Platanus* and *Fagus*, sclereids are the only type of sclerenchyma present in the phloem. In active phloem, sclereids are not generally so abundant as fibers, but, as the phloem loses its conducting function, the sclereids often increase in number by transformation of parenchyma cells. In the older, living, but probably nonconducting phloem of the woody plants, they may be abundant and present a great variety of form and arrangement. The occurrence of sclereids in phloem is further discussed in the consideration of the obliteration of this tissue.

Phloem Rays.—In vascular tissues formed by the cambium, rays are usually present. Only in reduced herbaceous types and in some specialized vines, such as *Clematis*, where secondary tissues are small in amount, are true vascular rays lacking. Vascular rays are initiated in the cambium and develop on the inside and the outside with the secondary xylem and the secondary phloem of which they are a part. The diversity of type of vascular rays in secondary phloem is, therefore, as great as is that of the rays in secondary xylem. The phloem rays vary in width and height, as do the xylem rays, and make up the same proportion of the tissue. They may be one cell wide, as in *Castanea* and *Salix* (Fig. 109*D*); two or three cells wide, as in *Malus pumila;* or several to many cells wide, as in *Robinia* and *Liriodendron* (Fig. 108*D*); or rays of various widths may exist in a single species. In the oaks there are two types of ray, one very broad and the other uniseriate. Usually, the phloem rays are of uniform width throughout their extent. They may increase in width outwardly, the increase being due to the multiplication of the cells or to the increase in size of cells toward the outer end of the ray. This provides in part for the necessary adjustment to the increase in circumferential extent of the phloem caused by the increase in the diameter of the axis. Phloem rays enlarging distally are especially prominent in twigs of certain genera, such as *Tilia*. The vertical extent of the ray in the phloem is as varied as its width, ranging from that of two or three cells, as in *Thuja*, to 8 to 10 cm, as in the broad rays of the oaks. Rays consisting of a single row of cells occasionally occur. Marginal cells, different in form, pitting, and content from the other ray cells, but parenchymatous, are found in some plants, as in *Salix* and *Nyssa*.

The abundance of rays is, in part, related to size. Broad rays are spaced farther apart than are those of the narrow type. In woody plants that in structure approach the herbaceous type, as, for example, *Cephalanthus*, the rays as seen in transverse section of the phloem (Fig. 107C) are separated by only one or two rows of sieve tubes or parenchyma. With the exception of the large rays found in some families, herbaceous plants are characterized by uniseriate rays spaced at short intervals. In herbs with reduced vascular tissue, the vascular rays have in evolutionary development been reduced and have disappeared in some forms.

Phloem ray cells are for the most part of uniform type. In woody plants, the common form, as seen in cross section, is rectangular and radially elongated. In semiherbaceous and herbaceous plants the cells tend to become cubical or globose. Transitional forms are frequently seen in shrubs, such as *Cephalanthus*, and in "woody" herbs, such as *Agrimonia* and *Potentilla*. In such plants phloem ray cells closely resemble phloem parenchyma and can only be distinguished from the latter by a study of the series of cells back to the cambium and to the rays of the xylem. All phloem ray cells are parenchymatous with active protoplasts. With age, many of them become sclereids.

A special type of ray cell is the so-called *albuminous cell* of the gymnosperms. These albuminous cells are situated at the upper and lower margins of the phloem rays and differ from the ordinary ray cells both structurally and functionally. Structurally they differ from ordinary ray cells in that they are joined directly with the sieve cells by sieve areas, are of much greater vertical diameter than the normal ray cells and do not contain starch. They are closely related to the sieve cells in their development and retain their protoplasts only as long as the sieve cells with which they are connected function. (Ray cells adjacent to them remain alive as long as does the surrounding tissue.) At the time that callus pads form within the albuminous cells over the sievelike connections with sieve cells, similar pads cover the sieve areas in the sieve cells themselves. Functionally, they are apparently intimately connected with the sieve cells; it is believed that they may function much as do companion cells in the angiosperms. In similarity of position and difference in function from other ray cells albuminous cells are like the marginal ray tracheids of the xylem. Cells suggestive of albuminous cells in appearance and position are found in the phloem rays of some woody dicotyledons, such as *Nyssa* and *Cornus;* these are not, however, the functional equivalents of such cells.

Seasonal Rings in Secondary Phloem.—As has been previously stated, the secondary phloem tissues are frequently arranged in definite tangential bands. Such layers of tissue often have the appearance of annual rings. These ring-like bands do not, however, have such definite seasonal limits as do those of the secondary xylem, because there is no sharp distinction between the phloem cells formed in the early part and those in the late part of the growing season comparable to the difference between early wood and late wood. Further, the last xylem cells formed in a season from the cambium become fully mature and lignified before growth ceases. On the phloem side of the cambium, on the other hand, at the end of the growing season there are usually several rows of cells that have not completely differentiated, but which remain dormant until growth is renewed, when they mature to form normal tissue. Therefore, no line can usually be drawn between the cells formed in two consecutive seasons. Seasonal formation of sclerenchyma bands may exist, but it is not known that there is any constancy in such growth. In fact, the number and width of these bands of sclerenchyma is apparently dependent upon environmental factors and vigor of growth. In tropical plants, new layers of phloem as well as of xylem are formed with each "flush" or period of new growth; these have, in general, no distinctness strictly comparable with the annual rings of temperate-zone plants.

Function of Secondary Phloem.—The function of the secondary phloem is, in general, that already discussed in Chap. IV as that of phloem as a whole. From the standpoint of function, the secondary phloem is a complex tissue with most of its parts interrelated in a definite manner. The sieve tubes, sieve cells, companion cells, and some parenchyma cells are structurally adapted to vertical conduction, whereas the phloem rays provide a means of horizontal conduction to and from the xylem and cambium. In gymnosperms, sieve cells are ordinarily pitted heavily only with other sieve cells and albuminous cells; in dicotyledons, sieve tubes are ordinarily pitted heavily only with other sieve tubes and with companion cells. The different types of parenchyma concerned with conduction frequently lie adjacent to the sieve tubes, from which they are separated by only a thin wall. These parenchyma cells are not, however, conspicuously pitted with the sieve tubes or with the companion cells, although they are usually heavily pitted with each other upon both radial and transverse walls. In some woody plants, for example, *Cas-*

tanea and *Cornus*, the pits in these cells are clustered in a way that strongly suggests sieve areas (Fig. 110*D*). Phloem ray cells are frequently pitted with parenchyma of this type. Taken as a whole, phloem has to do with the conduction of elaborated food products, both protein and carbohydrate, and possibly with mineral nutrients as well. Just which substances, however, move in the sieve tubes and which in the parenchyma is not known. From the fact that the proportion of such parenchyma to sieve tubes and sieve cells shows such great variation, that its pitting may be sieve-area-like, and that sieve tubes are few or sometimes possibly lacking, as in slender vascular bundles in leaves and in primary phloem of some woody plants, it may be inferred that parenchyma under some conditions may perform the functions of sieve tubes or that the functions of the two types of cells may be interchangeable.

Another function of the phloem parenchyma is storage of crystals, starch, and various organic materials. Such storage occurs in specialized parenchyma cells which differ from the heavily pitted parenchyma cells in being usually of rather large diameter with inconspicuous pits. Phloem ray cells are often packed with starch during the dormant season. This is especially true in roots where the phloem rays are usually relatively large. The abundance of this type of parenchyma in roots seems to be correlated with a small amount or absence of sclerenchyma. Parenchyma cells storing crystals lie usually close to fibers, where they may form extensive rows and even ensheathe the strands of fibers. Specialized parenchyma cells containing various secretions occur abundantly in the secondary phloem of some plants (Fig. 111*B*), both in the vertical system and in the rays.

Cessation of Function of the Phloem.—The length of the functioning life of the secondary phloem, at least that of the sieve cells, sieve tubes, and companion cells, is brief as compared with that of secondary xylem. In many woody plants the sieve tubes and sieve cells function for a single season or less, and in some tropical trees for a single "flush" of growth. In other species, and frequently in roots, these cells may be active over a longer period, though probably rarely more than two years. The phloem parenchyma and the phloem ray cells—other than the marginal albuminous cells of the gymnosperms—commonly remain alive and apparently in normal condition long after the sieve tubes have lost their protoplasts. The cessation of function in the secondary phloem is in some ways comparable to the formation of heartwood in the xylem except that the phloem is being continually stretched, crushed, and torn by circumferential increase and the pressure of cambial growth, and that its outer layers are lost by weathering and exfoliation. The cessation of activity of the entire tissue appears to be gradual, becoming complete only when periderm layers have formed within it, depriving all cells of food and water supply from the tissues within.

The exact time of loss of function by the sieve tubes and sieve cells is in doubt. It is generally believed that these cells function from the time of disappearance of the nucleus until the protoplast is lost. The opinion has been expressed that the functional life of the sieve tube or sieve cell ends with the disintegration of the nucleus, but the preponderance of evidence does not support this view.

Closely tied up with the functioning life of the sieve tube is the formation of the callus pads over the sieve plates. In many plants these are deposited at the end of the growing season, thus occluding the sieve tube while the plant is dormant; with the resumption of growth in the spring, the pads are dissolved, and the sieve tubes again become functional. In other plants, particularly in herbaceous forms, however, the plugging of the sieve pores with callus marks the permanent cessation of function of the cell.

The disappearance of the protoplasts of the sieve tubes and companion cells certainly indicates loss of function in these cells. In many genera, for example, in *Robinia*, *Quercus*, and *Pyrus*, this is accompanied by, or followed by, the crushing and flattening of the sieve tubes. This cell destruction—collapse of radial walls and obliteration of lumina—is caused by the pressure of the growing tissues beneath upon unlignified, empty cells. The crushing of the sieve tubes may be so complete that a group or layer of such cells is represented only by an irregular band of structureless wall substance. In some genera even this is quickly removed by absorption. This crushing and removal of the sieve tubes is commonly spoken of as the *obliteration* of these cells.

In the dormant phloem of many plants, such as *Pinus Strobus*, *Robinia Pseudo-Acacia*, and *Clematis virginiana*, there is only a narrow band of tissue with intact sieve tubes. Outside of this region all sieve tubes are functionless and crushed. There may even be no mature sieve tubes that are uncrushed (Fig. 110*B*). In other plants, such as *Tilia* and *Populus*, the sieve tubes may not be crushed for a considerable distance from the cambium, but all those not close to the cambium have lost their contents and thus apparently have ceased to function. In other plants, for example, *Salix*, the sieve tubes are not crushed at any time but remain normal in size and shape even after the formation of periderm layers which cut them off from the living tissues beneath. It is doubtful that the sieve tubes in woody dicotyledons become lignified as the phloem grows old as do many of the surrounding parenchyma cells.

The changes which take place in phloem as it ages show great diversity in minor details in different species but, in general, involve approximately the same phenomena. Simultaneously with, or following, the death of the sieve tubes, lignification of many or all phloem parenchyma and ray cells may occur. Druses and other crystals are formed in great numbers, both in parenchyma and in the newly formed sclerenchyma. In many plants the crystals form in thin-walled parenchyma, which lies beside and often ensheathes the fibers (Fig. 9N). Additional gums, tannin, and resins may be deposited in the parenchyma at this time. In annual herbaceous stems, the secondary phloem probably remains functional throughout the life of the stem or, at least, until the maturation of the seed. Few changes occur in such tissue, although in certain Compositae the phloem parenchyma becomes lignified toward the end of the growing season.

The manner and the extent of sclerification in the outer secondary phloem are varied. In such woody plants as *Fagus* and *Platanus*, nearly all cells, with the exception of sieve tubes and companion cells, become transformed into sclereids, and the bark is, consequently, very hard but not tough. Fibers in these genera are few or wanting. Many trees, for example, *Quercus*, show a mixture of sclereids and fibers in old secondary phloem. Phloem rays frequently become lignified at their outer ends, that is, as the cells become old. In the oak, the cells of the broad phloem rays become lignified when still young and close to the cambium. In *Juglans* and many other genera, no sclereids are formed; the outer phloem consists largely of fibers. The increase in the proportion of sclerenchyma in the outer phloem of some species is in part only apparent and is due to the collapse of the softer tissues, rather than to the lignification of additional cells.

Economic Value of Secondary Phloem.—Fibers of secondary phloem were formerly used commercially to a considerable extent and are still of some importance. The secondary phloem of various trees and shrubs of the Malvaceae, Tiliaceae, Moraceae, etc., has provided "bast" or "bast fibers" (Chap. IV) for economic purposes for centuries. The tapa cloth of the tropical Pacific Islands is composed chiefly of phloem fibers. Other uses of phloem are as a source of tannin, as in oak, chestnut, and hemlock bark, and, together with the cortex, as a source of some spices and drugs, for example, cinnamon and quinine. Secretory canals are often abundant in phloem, and the secretions may be of much economic value, as are rubber, which is obtained from the latex of *Hevea* and other genera; and various resins, such as kauri gum, obtained from *Agathis*, and spruce gum, from *Picea*.

References

ABBE, L. B., and A. S. CRAFTS: Phloem of white pine and other coniferous species, *Bot. Gaz.*, **100**, 695–722, 1939.

CHAUVEAUD, G. L.: L'appareil conducteur des plantes vasculaires et les phases principales de son évolution, *Ann. Sci. Nat. Bot.*, 9 sér., **13**, 113–438, 1911.

———: Recherches sur le mode de formation des tubes criblés dans la racine des dicotylédones, *Ann. Sci. Nat. Bot.*, 8 sér., **12**, 333–394, 1900.

CHEADLE, V. I., and N. B. WHITFORD: Observations on the phloem in the Monocotyledoneae, I. The occurrence and phylogenetic specialization in the structure of the sieve tubes in the metaphloem, *Amer. Jour. Bot.*, **28**, 623–628, 1941.

CURTIS, O. F.: "The Translocation of Solutes in Plants," New York, 1935.

DE BARY, A.: "Comparative Anatomy of the Vegetative Organs of the Phanerogams and Ferns," 1877. Engl. ed. 1884.

ESAU, K.: Ontogeny and structure of the phloem of tobacco, *Hilgardia*, **2**, 343–424, 1938.

———: Development and structure of the phloem tissue, *Bot. Rev.*, **5**, 373–432, 1939.

HEMENWAY, A. F.: Studies on the phloem of the dicotyledons, I. Phloem of the Juglandaceae, *Bot. Gaz.*, **51**, 131–135, 1911.

———: Studies on the phloem of the dicotyledons, II. The evolution of the sieve tube, *Bot. Gaz.*, **55**, 236–243, 1913.

HILL, A. W.: Histology of the sieve tubes of Pinus, *Ann. Bot.*, **15**, 576–611, 1901.

———: The histology of the sieve tubes of angiosperms, *Ann. Bot.*, **22**, 245–290, 1908.

HUBER, B.: Das Siebröhrensystem unserer Bäume und seine jahreszeitlichen Veränderungen, *Jahrb. Wiss. Bot.*, **88**, 176–242, 1939.

JANCZEWSKI, DE, E.: Études comparées sur les tubes cribreux, *Ann. Sci. Nat. Bot.*, 6 sér., **14**, 50–166, 1882.

LECOMTE, H.: Contribution a l'étude du liber des angiospermes, *Ann. Sci. Nat. Bot.*, 7 sér., **10**, 193–324, 1889.

LÉGER, L. J.: Recherches sur l'origine et les transformations des éléments libériens, *Mém. Soc. Linn. Normandie*, **19**, 51–182, 1897.

MACDANIELS, L. H.: The histology of the phloem in certain woody angiosperms, *Amer. Jour. Bot.*, **5**, 347–378, 1918.

PERROT, É.: "Le Tissue Criblé," Paris, 1899.

POIRAULT, G.: Recherches anatomiques sur les cryptogames vasculaires, *Ann. Sci. Nat. Bot.*, 7 sér., **18**, 113–256, 1893.

RUSSOW, E.: Sur la structure et le développement des tubes cribreux, *Ann. Sci. Nat. Bot.*, 6 sér., **14**, 167–215, 1882.

SCHMIDT, E. W.: "Bau und Funktion der Siebröhre der Angiospermen," Jena, 1917.

STRASBURGER, E.: "Ueber den Bau und die Verrichtungen der Leitungsbahnen in den Pflanzen," Histologische Beiträge III, Jena, 1891.

WILHELM, K.: "Beiträge zur Kenntniss des Siebröhrenapparates dicotyler Pflanzen," Leipzig, 1890.

WILSON, C. L.: Lignification of mature phloem in herbaceous types, *Amer. Jour. Bot.*, **9**, 239–244, 1922.

CHAPTER IX

PERIDERM AND ABSCISSION

PERIDERM

The protection of living tissues from desiccation is characteristic of all plants except submersed aquatics. In plants with primary growth only, such protection is usually given by the cuticle and the cutinization of the epidermis or possibly the hypodermis or outer cortex, as described in Chap. II. In plants with secondary growth there is an adjustment of the epidermal, cortical, and phloem tissues to the resulting increase in diameter, or, when any of these outer layers are ruptured, new protective layers arise that prevent desiccation. In herbaceous types, generally, and in some woody plants, the formation of new cells by radial division, and the increase in size of the cells are the means of adjustment to an increasing diameter. The cells of the outer primary regions in herbaceous plants often do not lose their capacity for division until the axis has attained approximately mature size, and cell division in these regions in woody plants may continue for a year or longer, for example, in species of *Acer* and *Cornus*. In the older stems and roots of most woody plants, however, secondary growth is followed by the rupture and death of the outer tissues and the formation of a new protective layer known as the *periderm*.

Structure of the Periderm.—The periderm commonly consists of three layers of tissue: the initiating layer of meristematic cells known as the *phellogen*, or *cork cambium;* the layer of cells formed by this meristem toward the outside, the *phellem*, or *cork;* and usually a layer formed toward the inside, the *phelloderm* (Fig. 112A).

Phellogen.—The phellogen is an excellent example of a secondary meristem in that it arises from living cells that have become permanent, that is, they have differentiated and are mature epidermal, cortical, or phloem cells. It is a lateral meristem, in that it increases the diameter of the axis, cutting off cells on the tangential face for the most part in much the same way as does the true cambium. In shape, the phellogen cells show little variation. As seen in tangential section, they appear polygonal and more or less isodiametric, except in a few special plants; in transverse section they appear rectangular and radially flattened (Fig. 112A,B,D). Except in the lenticels, intercellular spaces are lacking.

In the formation of a phellogen layer, mature living cells become meristematic and form a continuous uniseriate layer of initials. This meristematic activity always occurs in epidermal, cortical, or other

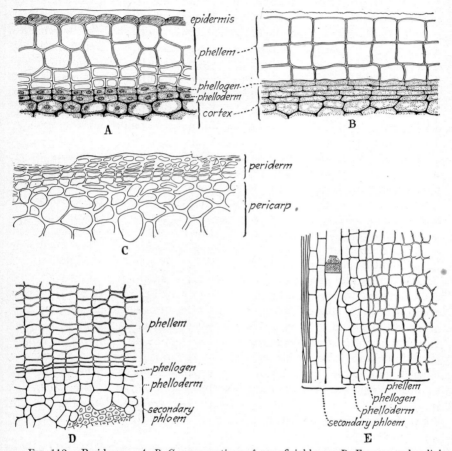

Fig. 112.—Periderm. *A, B, C,* cross sections of superficial layer; *D, E,* cross and radial sections of a deep layer. *A,* in twig of *Populus deltoides:* the phellogen arose in the outermost cortical cells, and has formed four layers of phellem cells and one of phelloderm; the phellem is capped by the dead, tannin-filled epidermal cells. *B,* in twig of *Solanum Dulcamara:* the phellogen arose in the epidermis, the outer halves of the epidermal cells have become typical phellem cells; they are capped by the cuticle; no phelloderm has been formed. *C,* in fruit of *Malus pumila:* the outer cells are readily loosened, the periderm becoming "scurfy." *D, E,* in secondary phloem of *Salix alba,* var. *vitellina.* The phelloderm cells are somewhat irregularly arranged.

parenchymatous tissue where living cells are sufficiently abundant to permit the formation of a layer of considerable vertical and tangential extent. When the formation of a phellogen is about to take place in epidermal cells, the protoplasts lose their central vacuoles and the cytoplasm increases in amount and becomes more richly granular. After the formation of this initial layer, tangential, and, to a lesser extent, radial division takes place, much in the same manner as does division in the true cambium. The derivative cells are normally arranged in radial rows, the cells of the phelloderm being less markedly so than those of the phellem. In periderm, the number of tangential rows of undifferentiated cells present at any one time during the period of activity of the phellogen is usually much less than the number of similar rows in the cambium zone during the growth period of the cambium. In fact, in periderm the only immature cells to be found are frequently the row of initials, the cells formed by this meristem having all matured before further division of the phellogen.

In the ratio of the number of cork cells to phelloderm cells formed by the phellogen, considerable diversity is shown in different plants. Generally, several to many times as many cells are cut off toward the outside (phellem) as toward the inside (phelloderm). Phelloderm cells may be few or none; rarely, phelloderm is greater in amount than the phellem. In the cork oak, *Quercus suber*, and in plants with corky-winged stems, such as species of *Euonymus*, *Liquidambar*, and *Ulmus*, the phellogen forms great numbers of soft cork cells and relatively little phelloderm.

Phellem.—The cells constituting phellem, commonly known as cork cells, are like the cork cambium cells from which they are cut off. They are uniform in shape—polygonal, as seen in tangential section, and often radially thin as seen in transverse sections of the stem (Fig. 113). In the maturing of phellogen derivatives there is no complicated differentiation comparable with that which occurs in the formation of vascular tissues. In shape, the cells of some types of thin-walled cork may be radially elongated, as in commercial cork (Fig. 113A,B); in the superficial persistent periderm of plants like *Betula* and *Prunus*, the cork cells are conspicuously elongated tangentially (Fig. 113D,F). Cork cells lack intercellular spaces.

Mature cork cells are nonliving and usually appear to be without pits. Where pits have been reported, they are described as in the inner cellulose layer (that next the lumen) only, a condition possibly related to the translocation of materials in the formation of the suberin lamella.

A number of different types of cork cells occur. Of the two common types, in one the cells are thin-walled, empty, and radially elongate, making up a light tissue of the bottle-cork type; in the other the cells are thick-walled and radially thin with the lumen filled with dark-staining material of a resinous or tanniferous nature. These two types may be found in different plants or in alternating bands in the same plant,

PERIDERM AND ABSCISSION 251

for example, in *Betula* (Fig. 113*D,E*). It is because of this alternate formation of thick-walled and thin-walled cells and the ease of rupture of the latter, that birch periderm splits into thin papery sheets. Rarely,

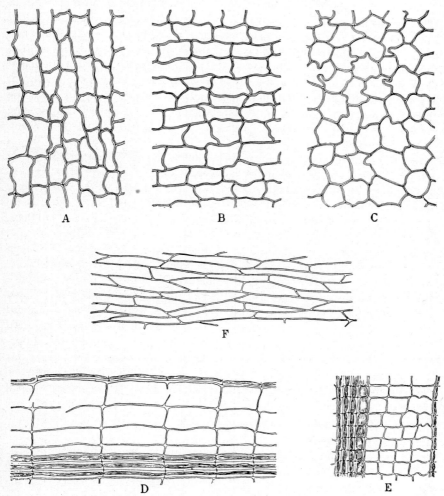

FIG. 113.—Phellem. *A, B, C*, commercial cork, from *Quercus suber*, transverse, radial, and tangential sections respectively; *D, E, F*, birch "bark," from *Betula alba*, transverse, radial, and tangential sections respectively. *D* and *E* show a complete (probably annual) layer, of two kinds of cells, the larger, thinner cells showing the tendency to rupture which causes birch "bark" to peel in thin, papery sheets.

sclereids and cells with crystals form part of the phellem.

The cell wall of cork cells is characteristically made up of the primary wall which is of cellulose, or in some plants lignified, and in others slightly suberized; next to the primary wall, a thick median suberized layer of the secondary wall called the *suberin lamella;* and next to the lumen the thin inner cellulosic layer of the secondary wall, which in some plants is lignified. In thin-walled cork the inner, cellulose layer is absent. Often the various layers are not readily seen. The substance suberin, which is believed to make up the suberin lamella, is similar in its properties to cutin, in that both are highly impervious to gases and water, are highly refractive as seen under the microscope, and resist the action of acids. Cork tissue in some few species, notably, *Quercus suber,* is elastic to a high degree, this quality depending in part upon the elasticity of the cell walls themselves and in part upon a change in the shape of the cells as the

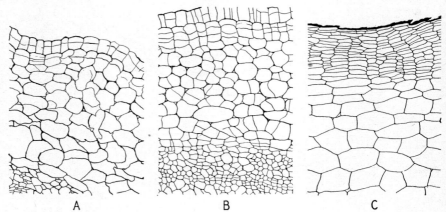

Fig. 114.—Periderm in the potato tuber (var. Irish cobbler). *A, B*, cross sections of outer part of young tuber, showing origin of periderm: *A*, two phellogen layers arising, one in the epidermis, one in the subepidermal layer; *B*, the inner phellogen dividing, the outer not developing. *C*, cross section of outer part of mature tuber showing thick phellem. (*After Artschwager.*)

stretching takes place. In the majority of plants, however, the cork tissue is both inextensible and inelastic. The prevention of water loss by the periderm is secured by the suberization of cell walls and compact arrangement of the cork cells.

Phelloderm.—The cells of the phelloderm are living cells with cellulose walls. They are more or less loosely arranged and in most plants, do not differ from adjacent cortical cells except in their more or less definite arrangement in radial rows (Fig. 112*A*). In some species they function in photosynthesis and in starch storage. They are pitted like other parenchyma cells. Sclereids and other specialized cells occasionally occur in phelloderm. The term "secondary cortex" is sometimes applied to phelloderm, but is more often applied to periderm as a whole. The basis for the latter use is that as a protective layer it is a cortex in the physiological sense. Since the periderm commonly develops secondarily in phloem, however, the term "cortex" cannot logically be applied to it. The use of the term "secondary cortex" in either sense is clearly inaccurate and undesirable.

Origin of the Periderm.—The origin of the first phellogen layer to be formed on the young stem is always in mature living cells in the tissues

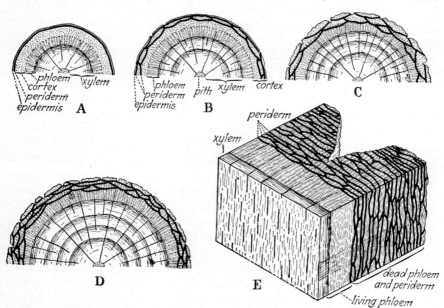

FIG. 115.—Diagrams, showing the position and extent of successively formed periderm layers in a typical woody stem. *A*, a one-year-old twig, the first periderm layer, a complete cylinder, formed beneath the epidermis; *B*, a two-year-old twig, the epidermis and first periderm ruptured and new, shell-shaped layers formed deeper in the cortex; *C*, a three-year-old stem, the outer tissues weathered away and more periderm layers formed still more deeply in the stem, invading the secondary phloem; *D*, a four-year-old stem, the cortex and outer secondary phloem with their periderm layers largely weathered away, the new cork layers invading the younger phloem; *E*, the outer tissues of an old tree trunk, showing the narrow band of young, living secondary phloem, and the thick, deeply fissured rhytidome made up of dead phloem and its many shell-shaped periderm layers; a considerable amount of similar tissue has weathered away.

outside the phloem. In many species, for example, *Malus pumila, Viburnum Lantana, Quercus suber*, and *Solanum Dulcamara* (Fig. 112*B*), the first layer of cork cambium is differentiated in the epidermis itself. More frequently, the phellogen layer is formed in the layer of cells just beneath the epidermis, as in *Populus* (Fig. 112*A*), *Magnolia* (Fig. 77), *Castanea, Ulmus,* and *Juglans*. When origin is in this position the epidermis is soon ruptured and disintegrates. In the potato tuber a phellogen arises in both the epidermis and the subepidermal layer (Fig. 114), but the outer meristem does not function. Rarely, the first-formed cork cambium of stems has its origin deeper in the cortical tissues, or even in the pericycle, as in *Ribes* and *Thuja*. Under these conditions the outer tissues, deprived of food and water, soon die, are ruptured, and sloughed off.

With the increase in diameter of the stem and the consequent deep splitting of the outer tissue, including the first periderm bands, additional periderm layers are formed, progressively deeper in the stem. The newer layers are formed successively in the inner cortex, the pericycle, and the phloem (Fig. 115). In old stems all layers, after the first few, are formed in the secondary phloem; the majority of periderm layers in large branches and tree trunks have formed in secondary phloem. The formation of each periderm layer shuts off all tissues to the outside from food and water supply, so that soon after the initiation of a layer, all tissues outside of it die. As a result there is formed an outer "crust" of successive and overlapping layers of cork enclosing pockets of dead cortical or phloem tissue. These tissues make up the *rhytidome*, often known as *shell bark* and *scale bark* (Fig. 115D,E).

The many uses of the term *bark* have led to considerable confusion. In a nontechnical sense, bark is applied to the tissues which are readily removed when logs or twigs are peeled, that is, to those outside the cambium. Both "bark" and "outer bark" have been used to designate the superficial layers of dead tissue made up of periderm layers and enclosed tissues, and "inner bark" to designate the living phloem tissues next to the cambium. The term "inner bark" has also been applied to the cambium zone itself. Periderm alone is sometimes called "outer bark," as in *Betula*, and sometimes "the bark." Some authors have used "bark" as synonymous with "cortex" in the technical sense, and still others use "outer bark" in this sense. The further use of the term "bark" in a technical sense is obviously inadvisable. The term may best be restricted to nontechnical usage as designating all tissues outside the cambium, a usage already long-established. The more specific terms "cortex," "pericycle," "phloem," and "periderm" can well be employed in anatomical usage, and the term "rhytidome" may be used to indicate alternate layers of periderm and dead cortical or phloem tissues.

The age of the stem at which the formation of periderm starts varies with different species and with different environmental conditions. In woody twigs the first (outermost) periderm forms usually in the first season, the epidermis rupturing at that time. In some genera deeper periderm formation may also begin the first year. The first layer often suffices for a few seasons, after which the deeper layers form. The apple tree and the pear tree begin to form internal periderm in the sixth to eighth year. Some species of *Populus* and *Prunus* retain their superficial smooth bark for twenty or thirty years or more, and *Fagus* and some other genera do not form internal periderm throughout the life of the plant.

In roots deep periderm formation is characteristic; the phellogen forms just beneath the endodermis. The first-formed periderm may persist as a continuous layer over the entire root system except near the tips. Adjustment to the increase in diameter of the root is secured by radial division of the phellogen cells and the living cells underneath. Underground roots of many species because of favorable conditions for decay, soon lose the dead outer tissues and have a continuous smooth outer surface of periderm, but roots exposed to the air develop a rough bark like that on the stem. In many herbaceous roots no periderm is formed, but the superficial layers become suberized.

Extent of the Periderm.—Considered in its entirety, the periderm of woody plants covers the plant axis except slender, "fibrous" roots and the younger parts of roots and stems near the growing points. In twigs, the first-formed layer makes a continuous cylinder. In a few genera, for example, *Fagus* and some species of *Betula*, this first phellogen layer covers the trunks until they are old. In *Vitis*, the first and also later formed periderm layers form continuous sheets of tissue parallel with the outer surface, and concentric cylinders of cork result. But in most woody species the extent of later formed layers is only rarely as great as the first and usually is much less, sometimes only a few square centimeters. The longitudinal and circumferential extent of any one phellogen layer varies greatly with the species and the age of the axis. The central portion of such a restricted layer of periderm may be parallel with the outer surface of the stem, but its edges curve outward and meet the older, outer cork layers; in this way a lenticular scale of cortical or phloem tissue is cut off. The periderm layers formed at different times overlap or join each other so that together they form a continuous corky covering over the stem (Fig. 116). The innermost last-formed layers have a living phellogen. The outer periderm layers are lost through weathering or exfoliation or remain and build up the rhytidome (Fig. 115).

The periderm layer of roots is commonly a continuous layer covering the entire surface except the growing tips.

Duration of Periderm.—The length of time that the first-formed periderm layers persist as functioning tissue varies greatly. In trees with smooth bark, for example, *Fagus, Carpinus,* and *Betula*—the original periderm may persist for a great many years or for the life of the tree. In these plants the increase in circumference of the periderm is accomplished by the radial division and subsequent enlargement of the phellogen cells. In the majority of woody plants, however, the first-formed periderm layer is replaced sooner or later by other periderm layers which arise successively deeper and deeper in the cortical, and ultimately in the phloem tissues (Fig. 115). These later formed phellogen layers are active for only a brief period of time; their component cells then mature to form cork cells. In *Vitis*, the successive periderm layers of the first few years are complete cylinders, and the outer cylinder is ruptured and sloughed away immediately, so that usually but one or two layers of phellem are present in the younger stems.

Fig. 116.—Inner (tangential) surface of rhytidome of *Pinus Strobus* with living phloem tissues weathered away exposing (face view) the overlapping shells or scales of the innermost periderm layers.

In plants in which a given periderm layer is active for only a brief period, there is no seasonal regularity of activity. On the other hand, in species with persistent, superficial periderm layers, for example, species of *Betula*, *Prunus* (Fig. 113D,E), and the cork oaks, *Quercus suber* and *Q. occidentalis*, there is seasonal variation in type of phellem cells formed by the cork cambium with the resultant formation of rather conspicuous bands or layers of cork which probably represent "annual rings" (Fig. 117B). This condition is readily visible in bottle cork, where the layers resemble the annual rings of xylem. In *Betula* also, the layering is prominent (Fig. 113D,E).

Commercial Cork.—The development of the periderm layers in the cork oaks is of special interest. Here the phellogen first arises in the epidermis, forming externally masses of cork tissue and internally relatively few phelloderm cells. This phellogen persists indefinitely. At the age of about twenty years, when the tree is about 40 cm in circumference, this outer layer, known as virgin cork, is removed by stripping, the separation being made probably through the phellogen layer, though possibly through the phelloderm. The underlying exposed phelloderm and cortical cells die, and a new phellogen layer is formed several millimeters deeper in the cortex. By this phellogen new cork is formed more rapidly than by the first layer, and after nine or ten years the new cork layer has attained sufficient thickness to be commercially valuable and is in turn

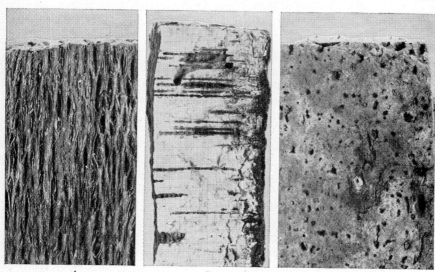

A   *B*   *C*

FIG. 117.—Phellem of *Quercus suber*, commercial cork as it comes from the tree. *A*, the weathered outer (tangential) surface, showing fibers and rays of the secondary phloem, tissues cut off and killed by the formation of the underlying phellem; *B*, cross section, showing "annual rings," and lenticels in longitudinal section, the torn phloem fibers of the outer surface; *C*, the smooth inner (tangential) surface, along which the layer was peeled from the trunk, showing cross sections of the lenticels.

removed. This cork is of better quality than the nearly useless virgin cork, but is not so good as that obtained at the third and subsequent strippings which take place at intervals of about nine years until the tree is 150 or more years old. After the successive strippings the new phellogen layers develop at greater and greater depth in the living tissue. The cortex is lost in the first few strippings and the subsequent cork layers form in the secondary phloem. A strip of "cork bark," therefore, has one smooth surface, the inner, where the phellogen was split (Fig. 117*C*), and one rough surface which has been cracked by weathering. Upon the rough surface there is apparent the remains of the secondary phloem which was killed by exposure after the stripping of the previous layer (Fig. 117A). The cross section (Fig. 117B) shows the bands which are probably annual layers. This section and the tangential surface (Fig. 117C) show longitudinal and cross sections respectively of lenticels (page 262). If the trees are not stripped, the initial periderm layer probably persists indefinitely, forming cork of considerable thickness.

Most commercial cork probably is obtained from the cork oak, *Quercus suber*. The thin, tough sheets of cork from the canoe birch, *Betula papyrifera*, and related species, were important in the North American Indian civilization but are little used today. The properties which make commercial cork valuable are its imperviousness, its lightness, toughness, and elasticity.

Morphology of the Periderm.—All stages in the formation of periderm can be observed in a tree of considerable age. In a pear tree, for example, early in the growing season the new shoots are covered with epidermis only. Extending down the stem from these young shoots to a region six or eight years old is the zone covered by the first-formed superficial periderm, of epidermal origin, which gives a smooth, tan or gray-green surface. On the upper part of this zone the scaling remains of the epidermis can be seen. Below this is a comparatively short zone where the internal periderm has been formed in small areas, but where the outer layer has not yet been broken. The discoloration of the dead tissue cut out locally by the internal periderm layers gives the stems of this zone a blotchy appearance. The surface of the dead patches may be somewhat sunken because of the shrinkage of the dead tissue (Fig. 118A). This zone shades into a region of scaly bark where the outer scales have broken loose and are dropping off (Fig. 118B). Below this on large trunks the bark persists in weakly defined ridges, formed of periderm and dead phloem tissues.

The character of the older bark of different plants is dependent partly upon the number, extent, and nature of the periderm layers and partly upon the nature of the cortical and phloem tissues which are cut off by the successively formed phellogen layers. In many plants the outer phloem and the cortical tissues undergo extensive sclerification in which all parenchyma cells are converted into sclereids. This may take place in part before the tissues are cut off by the phellogen and in part afterward as the cells are dying. It is to these masses of sclerenchyma, together sometimes with firm or stony periderm cells, that the bark of species of *Quercus*, *Carya*, and of *Acer saccharum* owe their extreme hardness. Soft bark such as that of *Magnolia acuminata* and *Ulmus americana* has soft cork layers and few sclerenchyma masses.

Variations in the extent and firmness of the periderm layers account in large part for the different methods of bark exfoliation. The "scale bark" of the younger parts of the trunk and branches of many trees, for example, *Acer rubrum* and *Pyrus communis* is the result of the formation of small, isolated periderm layers which cut off scales of tissue. The breaking of the bark of the shagbark hickory, *Carya ovata*, into narrow, persistent strips attached by their upper ends is due to the method of

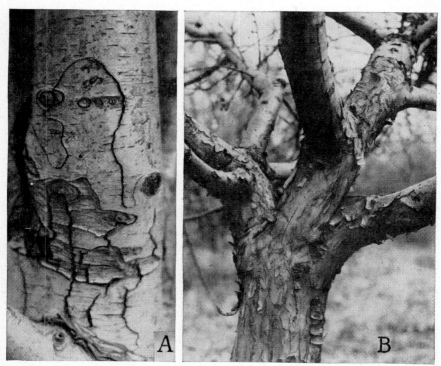

Fig. 118.—The trunk of young trees, *A*, *Pyrus communis*; *B*, *Malus pumila*, showing the cracking and scaling of the outer tissues. In *A*, the smooth, continuous, outer periderm is cracking, and small, restricted, inner periderm layers are beginning to form, two of which, just below the center of the figure, have "cut out" and killed patches of outer tissue; *B* shows stages ranging from that seen in *A* to that where the trunk is freely scaling over its entire surface.

splitting of long and narrow, vertical strips of periderm and to the abundance of fibers in the phloem. In some of the red oaks and many other trees, the periderm layers are firm throughout, and, as a result, the bark adheres tightly to the trunk without exfoliation, gradually disintegrating on the surface.

In some species the periderm forms definite abscission layers which bring about the shedding of the outer bark in sheets. In *Platanus occidentalis* the outer bark of the trunk and larger limbs is cut off every spring in this way. The "ring bark" of *Vitis* is shed in a somewhat similar manner, but the outer tissues usually hang to the old vine in shreds for some time. Such exfoliation has been ascribed to two types of periderm formation: one in which a layer of thin-walled unsuberized cells is laid down by the phellogen between layers of firmer cork, and the other in which thick-walled lignified cells are formed between layers of thin-walled cork tissue. In both types the periderm layer splits under the action of moisture upon the unsuberized tissue and the strain set up by the increase in the diameter of the stem.

Protective Layers in the Monocotyledons.—In herbaceous monocotyledons a persisting epidermis with its cuticle, and often cutinized walls, is usually the only protective layer. When this is weakened or ruptured, the primary cortical cells beneath become secondarily suberized by the formation of a suberin lamella on the cellulose wall, as in typical cork cells. This occurs commonly in the Gramineae, Juncaceae, Typhaceae, and other families.

The long-lived stems of woody monocotyledons show a variety of structure in their external layers. Typical periderm formation is infrequent. In the royal palms (*Roystonia*), which have a smooth white trunk, a persistent periderm, with a layer of hard phellem, covers the entire surface. In species of *Cocos* and some other genera, successive continuous layers of periderm cut off the outer layers of the stem, including ends of leaf traces and some of the stem bundles, forming a rhytidome-like layer, somewhat resembling that of dicotyledons. Some woody genera, such as species of *Aloë*, that have specialized secondary thickening, have continuous outer layers of periderm, the phellem made up of layers of thin-walled cells alternating with thick-walled cells (Fig. 119C).

Storied Cork.—In many monocotyledons with secondary thickening—for example, *Dracaena, Cordyline, Curcuma*, some species of *Aloë*—the protective layer consists chiefly of *storied cork*. Such cork differs from the phellem of typical periderm in origin and in arrangement of its radial rows of cells. In typical periderm, the phellogen cells form continuous rows of closely similar phellem cells (Fig. 112A,B); in storied cork, the initials—primary cortical cells—lie in an irregular, often broken line, and each cell by tangential divisions forms a limited number (three to eight) of phellem cells. The cells of a row so formed differ in size and shape and show by their size and arrangement their origin from a single mother cell (Fig. 119A). Tangential bands of these cell rows form "stories." The bands are irregular in outline and indefinite in extent. They enclose undivided cortical cells, which become suberized, and may merge with other similar bands laterally or radially (Fig. 119B,C). Radial series of bands formed in this manner constitute a type of cork probably restricted to monocotyledons.

Layers of storied cork form progressively deeper and deeper in cortical tissue but the margins of these do not join according to a characteristic pattern as in dicotyledons and conifers. The rhytidome formed in monocotyledons consists of alternating and overlapping, irregular layers of storied cork, suberized cortical cells, and unsuberized, crushed cortical cells (Fig. 119C). Such a layer is therefore a complex mass of protective tissue, lacking the completeness of phellem layers characteristic of the stems of woody plants of other groups.

The Function of Periderm.—The principal function of the periderm is protection of the underlying tissues from drying out, effected by

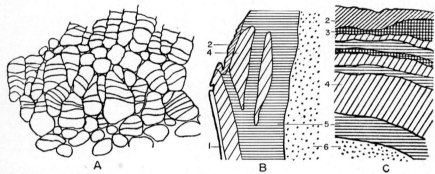

Fig. 119.—Protective layers in monocotyledons. A, *Curcuma longa*, transverse section of part of cortex showing storied cork—successive tangential divisions have occurred in an irregular layer of these cells; B, *Cordyline australis*, radial section of stem showing layer of storied cork (5) enclosing clusters of suberized, undivided cortical cells (4), 1, epidermis; C, *Cordyline indivisa*, transverse section of stem showing superficial layer of crushed cells (2), alternating tangential layers of cortical cells (3), suberized, undivided cortical cells (4), storied cork (5), and cortex (6). (*After Philipp.*)

maintenance of corky and suberized layers. Cork layers, when of considerable thickness, may also afford a certain amount of protection against mechanical injury to the tissues beneath. In addition to these functions, the periderm layers serve for the protection of various specialized structures or plant parts. Thus in fruits and tubers a periderm layer frequently takes the place of a heavily cutinized or cuticularized epidermis, as in the russet apple (Fig. 112C) and the common potato (Fig. 114). In some tropical fruits—for example, the sapodilla (*Achras zapota*) or the sapote (*Calocarpum mammosum*)—surface cork layers are well developed and give the fruit a gray-brown, somewhat rough appearance. Cork layers also occur on the dorsal side of bud scales of many woody plants, and as wings and ridges on many dry fruits (Fig. 171). They also occur rarely on leaves, as on the petioles of some species of *Ficus*.

Wound Cork.—One of the special functions of the periderm is the protection of wounds by the production of *wound cork*. Dead tissue is usually shut off from that which is healthy by a suberized layer formed from preexistent cells that become chemically changed. This change may be followed by formation of a phellogen layer in the layers of uninjured living parenchymatous tissues adjacent to the wound. This layer forms phellem and phelloderm in the normal way, thus sealing the wound. Such a layer not only prevents water loss from the wound, but also protects the healthy tissues against infection by fungi and bacteria for cork is particularly resistant to the action of microorganisms.

Wound cork may occur in any part of the plant though the readiness with which it is formed varies with the species, the organ or tissues concerned, and the environmental conditions. In general, wound cork forms more readily in woody plants and dicotyledons than herbaceous plants and monocotyledons. It develops only occasionally in leaves and rarely in sclerenchymatous tissue. Even where wound cork forms readily, as in the potato, a low temperature and low humidity may greatly retard its formation.

Lenticels.—In the periderm of nearly all plants, small restricted areas have loosely arranged cells with abundant, small intercellular spaces. These areas, known as *lenticels*, are thicker radially than the rest of the periderm layer because of the loose arrangement, the larger size, and possibly greater number of the cells.

Distribution of Lenticels.—Lenticels are conspicuous on twigs, and on other smooth-surfaced organs, as more or less raised, often somewhat corky spots where underlying tissues break through the epidermis. They are of almost universal occurrence on the stems of woody plants. Only a comparatively small number of these have been reported to be without them, among them *Philadelphus, Tecoma radicans, Vitis vinifera,* and some others, mostly vines, which shed the outer layers of bark annually, and thus maintain new tissues in close relation to the outer air. Lenticels are also found on many roots. The large lenticels of *Morus alba* are conspicuous features of the orange-colored surface. The so-called "dots" on apples and plums are familiar examples of lenticels on fruit.

The number of lenticels per unit of surface area varies greatly with different species. In some plants a lenticel forms beneath each stoma or group of stomata and the number of lenticels is therefore dependent upon the number of stomata or groups of stomata. In other plants lenticels may form between the stomata, if stomata are few or if stomata are abundant, lenticels may form under only a small proportion of them. In stems, lenticels are usually scattered, but are occasionally in vertical or horizontal rows. Where there are large multiseriate vascular rays, rows of lenticels may be found opposite these bands of tissue, suggesting that the ray and the lenticel together may form radial passageways for gaseous interchange between the inner tissues and the external atmosphere. In roots, lenticels occur in pairs, one on each side of a lateral

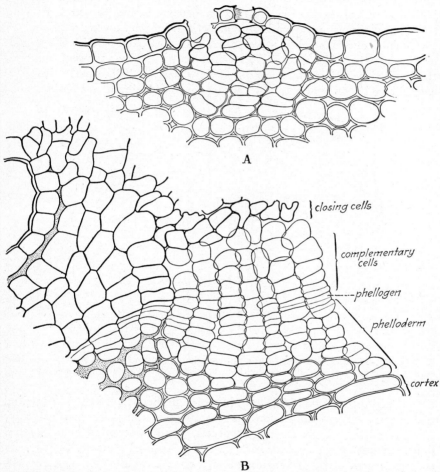

Fig. 120.—Early stages in lenticel formation. *A*, the lenticel phellogen has just arisen beneath a stoma, and the development of the first complementary cells has broken the epidermis; *B*, the lenticel (only one-half shown) is well developed. (*After Devaux.*)

rootlet. In storage roots, for example, the carrot, the lenticels occur in vertical rows in the position of the rows of secondary roots.

The Origin of Lenticels.—Lenticels on young stems commonly originate beneath stomata (Fig. 120) just previous to, or coincident with, the formation of the first periderm layer. Inasmuch as the lenticel is a part of the periderm, and the formation of the normal periderm usually spreads outward from the edges of the lenticel, the initiation of the periderm may be said to start with lenticel formation. The time of lenticel formation varies in different species with the duration of the epidermis. In the majority of plants, lenticel formation occurs during the first growing season and sometimes even before growth in length has ceased.

In lenticel formation in stems, the cells immediately below the stoma, or group of stomata, divide in different planes forming a mass of rounded, thin-walled cells known as *complementary tissue*. These early divisions are followed sooner or later by the differentiation in the adjoining inner

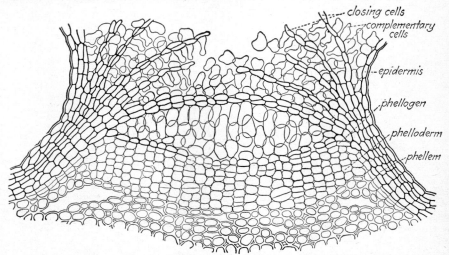

Fig. 121.—Lenticel of *Prunus avium* in transverse section of stem. A number of successive layers of complementary and closing tissue have been formed, and the large layer of phelloderm dips inward into the cortex. (*After Devaux.*)

tissue of a normal phellogen layer in which divisions are in the tangential plane only. The young complementary cells enlarge in size, lose their chlorophyll, and later their protoplasts, finally becoming colorless and light. Additional complementary cells are formed beneath the first by the phellogen. The formation of any considerable mass of such cells ruptures the epidermis through the stoma over the growing tissues (Fig. 120A) and exposes the mass of complementary tissue. With the continued growth of the phellogen, the epidermis around the opening of the lenticels is thrust back as flaps of tissue allowing pale masses of complementary cells to protrude (Fig. 120B).

Structure of Lenticels.—The lenticel, as a complete structure, is usually a lens-shaped mass of tissue bulging somewhat into the cortical parenchyma on the inside and raising the surface of the organ on the other (Figs. 121, 122). According to the orientation on the axis of the breaks in the epidermis, lenticels are called transverse or longitudinal. In roots, lenticels are always transverse; in stems, they may be transverse or longitudinal. Orientation of lenticels is related to the type of vascular ray within—transverse lenticels tend to be associated with uniseriate rays and vertically short rays; longitudinal lenticels tend to be related to elongated rays and so-called "compound rays."

In the development of lenticels, the phellogen does not form normal suberized phellem cells but gives rise to large numbers of unsuberized cells. All the cells formed to the outside may be of the same type—the thin-walled, more or less rounded, cells of complementary tissue—or the

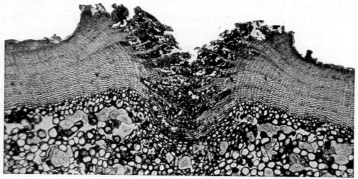

FIG. 122.—Lenticel, several years old, of a young stem of *Prunus serotina* in transverse section of the stem, showing relation to periderm, alternate layers of closing and complementary cells, and the layering of the periderm.

masses of complementary tissue may alternate with bands of denser, more compact tissue known as *closing layers*.

The cells of the complementary tissue are only loosely attached to each other, so that abundant radial air passageways are present. Two types of this tissue are recognized: one in which the cells are more or less firmly united, forming a fairly compact tissue, for example, in *Salix* and *Ginkgo;* the other in which the complementary tissue is composed of almost wholly unattached cells, which give the tissue a powdery consistency. Examples of this latter type are found in the stems of *Betula* and *Prunus* (Figs. 121, 122), and in the roots of *Morus*. Here the masses of powdery complementary tissue are held in place by the closing layers which although sufficiently dense to hold the looser complementary tissues in place are traversed by radial air passages as is also the phellogen layer itself. With the continued formation of new masses of complementary cells, the closing layers are ruptured. During the growing season the lenticel is filled chiefly with complementary cells; all closing layers are ruptured. At the end of the growing season the lenticel is "closed" by the formation of a closing layer which closes the passageways except for the minute intercellular spaces between these cells. In the spring the rapid development of complementary tissue breaks open the closing layer.

Duration of Lenticels.—The duration of a lenticel depends upon the formation of internal periderm. In those plants which form internal periderm early, the lenticel may be cut off early and lost by the exfoliation of the outer tissues. On the other hand, in plants with persistent superficial periderm layers, such as *Betula* and *Prunus avium*, the lenticel may persist for a great many years. In such species the lenticels become greatly elongated tangentially, due to the increase in circumference of the periderm layer accompanying secondary growth. The phellogen layer of the lenticel is increased in extent by radial division of its cells at about the same rate as is that of the surrounding periderm. The elongated lenticels form conspicuous markings on the smooth bark of birch and cherry trees and the roots of *Morus*.

With the formation of internal periderm, new lenticels are formed by the specialized functioning of special areas in the phellogen. The lenticels in the deeper layers occur beneath the cracks in the outer bark and gaseous interchange takes place through them. Lenticels on rough bark surfaces are not readily seen. In *Quercus suber*, where the layer of phellem may be several centimeters thick, the lenticels persist, forming cylindrical masses of loose complementary tissue reaching to the surface (Fig. 117*B,C*). This lenticel tissue forms the spots of dark, porous, crumbling tissue in commercial cork. Because of the presence of these pervious radial cylinders of tissue, bottle corks are cut vertically from the cork sheet so that the lenticel cylinders extend transversely through them. Because sheets of cork as they come from the tree are rarely more than 3 cm thick, cork with a diameter greater than that cannot be obtained by cutting in the usual manner. Corks of large diameter are sometimes cut radially from the sheet but in these the lenticels run longitudinally in the cork and the cork is not "tight." Large corks are usually cut from sheets of ground and compressed cork or from "multiple sheets" composed of layers cemented together. Such corks are of low quality.

ABSCISSION

In vascular plants the loss of various plant parts is a common occurrence. In annual herbaceous plants, this may be no more than the loss of bracts or floral parts. In perennials in which there is seasonal renewal of growth, particularly in woody plants, there is continual death and loss of many of the older parts. These parts may remain attached to the plant until they decay or weather away, or they may be cut off by the process of *abscission*. Leaves, floral parts, fruits, and foliage branches are commonly cut off from the parent plant, and protection of the exposed surfaces usually includes the formation of a periderm layer.

Abscission of Leaves.—The leaves of most pteridophytes and herbaceous angiosperms are not shed after their death but gradually decay in position or are torn away. The leaves of gymnosperms and woody dicotyledons generally—less commonly those of herbaceous angiosperms—are abscised, usually before their death, by structural modifications in the tissues at the base of the leaf. The seedlings of some herbs whose leaves are not deciduous show incomplete abscission, which is probably vestigial in nature.

Although the details of abscission vary, there is present at the base of all mature deciduous leaves, a narrow transverse zone, the *abscission zone*, which differs structurally from the parts above and below. Within this zone, days or even weeks before leaf fall, there is developed a well-defined *separation layer*, which is the immediate structural cause of leaf fall. When the tissues underlying the separation layer are exposed at leaf fall, they are protected against desiccation and infection by one or more protective layers, at least one of which lies within the abscission zone. These layers are of two types: a *primary protective layer*, of primary origin, and a *periderm*, of secondary origin.

The abscission zone, which is structurally the weakest part of the petiole, is evident as soon as the leaf is mature. Its position may be visible externally by a shallow furrow or a difference in color or surface of the epidermis. In this zone, the vascular bundles are usually reduced in diameter, the sclerenchymatous tissues are weak or absent, collenchyma is lacking, and some of the parenchyma cells have denser cytoplasm than those of other parts of the petiole.

The separation layer consists of a few to several layers of cells which differ from those above and below usually by difference in shape, by smaller size, by the presence of abundant starch grains and dense cytoplasm, and by differences in the staining qualities of the cell wall. In and below this layer, part of the conducting elements of the vascular bundles, chiefly the primary cells, are to some extent blocked by tyloses and gums, but sufficient conduction is maintained through the secondary elements to keep the leaf turgid until its fall.

Shortly before leaf fall the middle lamella and outer walls of cells of this layer swell, then become gelatinous, and finally, just before abscission, break down and dissolve. In some plants, perhaps commonly, the inner wall is also largely dissolved and only a thin cellulosic wall remains about the protoplasts. The cells are then wholly free from one another. All parenchyma cells of the layer, including those of the vascular tissues, are involved in this change so that the leaf is finally supported only by the vascular elements and these are broken in time by the weight of the leaf and by wind action. Wet weather, adding the weight of water to the leaves and hastening the hydrolysis of the gelatinous cell walls, accelerates leaf fall. Frost is not an important element in the fall of the leaves in most trees. The formation of ice crystals in the separation layer has sometimes been called the important cause of leaf fall, but the leaves of most trees fall commonly before temperatures above the ground level drop to freezing. In some plants, for example, *Ailanthus, Catalpa, Acer Pseudo-Platanus, Prunus avium*, frost may in this way bring about the abscission of leaves if conditions have not been favorable for earlier fall. The fall of the leaf therefore results from structural separation followed by mechanical rupture and is hastened or delayed by external factors.

The protective layers may be of both primary and secondary origin or of secondary origin alone. The protective layer of primary origin is formed by the transformation into a lignosuberized layer of existing parenchyma cells in the abscission zone, or of cells formed by irregular division of such cells. The protective layer of secondary origin is a typical periderm layer. Whenever cell division occurs in the formation of the primary protective layer, the bulk of the tissue is not increased; the new cells are crowded into the space occupied by the original cells.

The anatomical modifications associated with leaf abscission vary with the species and apparently at times within the species. The four plants illustrated in Fig. 123 give some of the important variations in the anatomy of leaf abscission.

Castanea.—The separation layer is formed from existing parenchyma cells of the abscission zone just before leaf fall. After the leaf has fallen, the walls of cells below the separation layer become lignified, a thin suberin layer is added on the inside of each wall, and the protoplasts disappear. This primary protective layer forms over the unmodified inner cells. After leaf fall, the remnants of the separation layer form a thin layer over the surface of the scar. Through these layers extend the torn ends of the vascular bundles. The periderm layer of the stem does not extend to the protective layer of the leaf scar. Later in the autumn, as the formation of the lignosuberized layer is nearing completion, a typical periderm layer is developed in the cells immediately below. This layer runs through the vascular bundles, developing there by the activity of the living cells and even of the tyloses in the vessels. Like the separation layer, this leaf-scar periderm does not connect with the stem periderm. In the second year another periderm layer forms deep beneath the first and becomes continuous with the stem periderm. The tissue between the two leaf-scar periderm layers becomes lignosuberized and forms an inner primary protective layer. (Fig. 123A, 1, 2, 3, 4.)

This is a simple and apparently primitive type of abscission. Only a separation layer is formed before leaf fall and no cell divisions take place

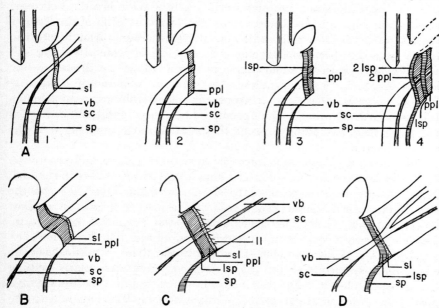

Fig. 123.—Diagrams of leaf abscission in woody dicotyledons. *A1, B–D*, radial sections of part of nodal region before leaf fall; *A2–A4*, after leaf fall. *A, Castanea: A1*, showing separation layer extending through vascular bundle which lacks sclerenchyma in the abscission zone; *A2*, soon after leaf fall, showing primary protective layer below the surface of the scar; *A3*, late autumn stage, showing leaf-scar periderm beneath the primary protective layer; *A4*, second year stage, showing same layers as in *A3*, and in addition, a primary protective layer and leaf-scar periderm of the second year, the latter connected with the stem periderm. *B, Catalpa*, showing separation layer and primary protective layer. *C, Betula*, showing lignified layer, separation layer, primary protective layer, and leaf-scar periderm in contact with the stem periderm which extends to the separation layer. *D, Populus*, showing leaf-scar periderm directly below separation layer and connecting with the stem periderm which extends to the separation layer. *sl*, separation layer; *vb*, vascular bundle; *sc*, sclerenchyma; *sp*, stem periderm; *ppl*, primary protective layer; *2ppl*, primary protective layer of second year; *lsp*, leaf-scar periderm; *2lsp*, leaf-scar periderm of second year; *ll*, lignified layer. (*Based on Lee.*)

in the formation of this layer or in the primary protective layer. No protective layer, either of primary or secondary origin, is formed until after the leaf has fallen. A small area around the leaf scar is unprotected by cork layers over winter.

Catalpa.—Abscission in *Catalpa* differs from that in *Castanea* in that, in addition to the separation layer, a primary protective layer is formed before the leaf falls. Here there is cell division in the formation of both the separation and primary protective layers. A periderm layer develops much later, usually not until the second year. (Fig. 123*B*.)

Betula.—The first change seen before leaf fall in the abscission zone in *Betula* is the irregular division in cells of a thick layer which very soon becomes lignosuberized, forming a primary protective layer. Following closely the beginning of this process, the adjacent cells above form a separation layer in which some cell divisions occur. Just before leaf fall, a very thin periderm develops below the first protective layer and makes contact with the stem periderm which extends to the separation layer. After the leaf has fallen, the periderm increases in thickness. The vascular bundles and accompanying sclerenchyma pass through this periderm layer. In the second year a new periderm is formed below the first, extending through the bundles and connecting with the stem periderm.

The structural conditions associated with abscission and protection in *Betula* are complex as compared with those of the others illustrated. New-cell formation occurs in both the primary protective layer and the separation layer. These two layers and a periderm are all present before leaf fall. A lignified layer distal to the separation layer is also present; it occurs frequently in other genera. (Fig. 123*C*.)

Populus.—At the approach of leaf fall two or three rows of cells in the abscission zone become active. The distal cells divide irregularly forming the separation layer. The proximal cells divide regularly, establishing a phellogen. The periderm layer so formed becomes continuous with the stem periderm which extends to the separation layer, although the point of union of the two layers remains clear. In the second year another periderm layer forms below the first.

In *Populus* irregular cell division occurs in the formation of the separation layer. A primary protective layer is absent, but a periderm is formed before abscission and this first periderm makes contact with the stem periderm. This immediate contact of leaf-scar and stem periderm suggests an advanced type of the protective structure. Only *Salix* and *Populus* are so far known to have this type of abscission. (Fig. 123*D*.)

In those herbaceous plants in which abscission of leaves occurs, for example, *Coleus*, the method of separation of the leaf is essentially the same as that in woody plants except that the separation layer is not formed well in advance of leaf fall as it is in many woody species.

Abscission of Floral Parts.—In flowering plants, floral parts, such as petals, stamens, and often other floral appendages, may be lost by abscission. The abscission of these appendages is not markedly different from the abscission of leaves of herbaceous plants. A separation layer is formed, but not long in advance of the fall of the floral appendage, nor is there the specialization and differentiation of layers associated with leaf abscission in woody dicotyledons.

The Abscission of Stems.—In addition to the abscission of leaves and other appendages, most plants lose parts of the stem by abscission. These parts may be immature or herbaceous stems in which there is little hard vascular tissue or sclerenchyma, as in flowers and young fruits, or they may be mature and woody with well-developed hard tissues such as the foliage branches of poplar and elm, the fruit-cluster stalks of horsechestnut, and the pedicels of mature fruits of apple, pear, plum, and walnut.

Conditions unfavorable to growth hasten or bring about abscission whereas favorable conditions tend to delay or prevent abscission. Lack of pollination and of fertilization may cause the fall of flowers; unfavorable growth conditions may cause young fruits to fall. (The factors causing early or late fall of mature fruit are not fully understood.) Some senile parts may be lost by abscission.

Abscission of Immature and Herbaceous Stems.—In stems in which the tissues are soft, no definite abscission zone is apparent, and the separation layer forms immediately before abscission. The location of this separation layer is variable, even in a given plant: in the apple flower, for example, the separation layer may form anywhere in the basal region of the pedicel. The cells of the separation layer undergo changes similar to those described in leaf abscission. Abscission occurs in this simple way by the formation of a separation layer just before the cutting off of flowers and young fruits and of the tips of leafy stems. The cutting off of the tips of leafy stems is characteristic of many genera that have indeterminate growth, for example *Ailanthus, Morus,* and *Ulmus*.

Abscission of Woody Stems.—In woody plants only a small proportion of the twigs and small branches persist in a living condition for more than a few years. In most species the dead twigs and small branches cling to the larger branches or trunks until they are mechanically broken off or decay. In other plants, for example, *Populus, Ulmus, Quercus,* and *Agathis,* the shedding of branches is brought about by abscission. In these plants a definite abscission zone is present in which the hard tissues are greatly reduced and modified and through which a separation layer is formed. In *Populus, Ulmus,* and some other genera, the majority of the branches, cut off by abscission toward the end of the growing season, are living and many have leaves attached. The age at which branches are shed varies: in *Populus alba,* abscised branches range in age from 1 to 20 years; in *Ulmus americana* abscised branches are rarely older than 7 or 8 years. The size of abscised branches also varies with the species. In *Populus* the largest are about two centimeters in diameter at the scar; in *Agathis* and *Castilloa*, up to five centimeters in diameter.

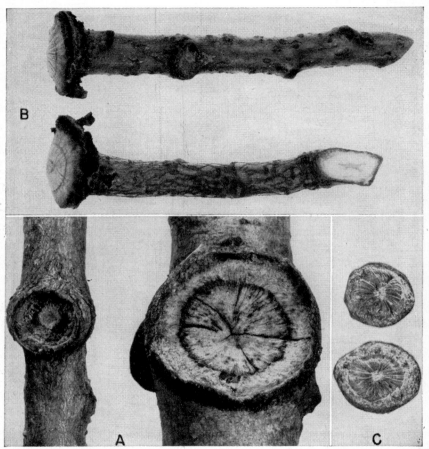

Fig. 124.—Abscission scars on branches of *Populus*, showing smooth surface. *A*, main branch showing concave ("socket") scar in face view; *B*, lateral branches showing swollen bases and convex ("ball") scar in side view; *C*, scars on abscised branches in face view. (Natural size.)

Abscission of woody stems is best described by examples. In *Populus grandidentata*, the smaller limbs are swollen at the base where they are attached to the trunk or main branches (Fig. 124). This swelling consists largely of thickened parenchymatous cortex in which stone cells, but no fibers, are present (Fig. 125*A*). In the xylem, the vessels of this region are reduced in number and are conspicuously modified, with abundant scalariform or reticulate pits instead of the usual circular pits (Fig. 125B). The vessels, fibers, and other cells are less strongly lignified, with their walls apparently mostly of cellulose as indicated by their staining reactions. Parenchyma cells are more abundant than elsewhere in the xylem. Through the living cells of this zone—both those of the xylem and those of the other regions—the separation layer

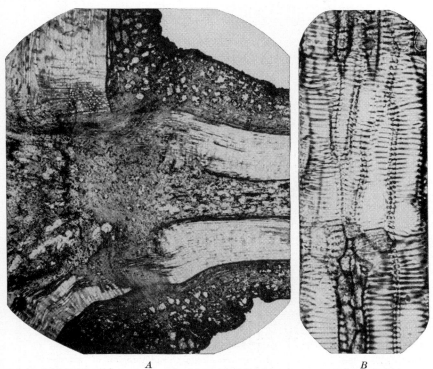

A   *B*

Fig. 125.—Branch abscission in *Populus*. *A*, radial section of lateral and main branch showing abscission zone with lateral branch swollen at base, parenchymatous cortex with sclereids but lacking fibers, and the modified xylem cylinder at the base of the lateral branch; *B*, scalariform-pitted vessels from the abscission zone in *A* at higher magnification.

is formed. Those xylem cells in which the separation layer does not form —the weakly lignified vessels and fibers—rupture easily because of their chemical nature and the transverse scalariform pitting.

When the branch falls, the surfaces of the two scars are smooth (Fig. 124) because the separation extends chiefly through living cells of the vascular tissue; and their surfaces follow the curved separation layer (Fig. 124*A*,*B*). The two surfaces are related to each other as those of a ball and socket joint—the scar of the abscised branch has a convex surface; that of the main branch a concave surface (Fig. 124A,B). Following the fall of the branch, or possibly even before its separation, a periderm layer, which is continuous with the periderm of the main branch, is formed in the living tissue just below the face of the scar.

The formation of an abscission zone in lateral branches does not necessarily result in their shedding. In *Populus*, although a well marked

Fig. 126.—Radial section of *Ulmus americana* through region of branch-abscission after development of protective layer of periderm. The periderm (light-colored tissue) extends over the entire scar area. On its external surface are the ragged remnants of vascular tissue and cortex cut off by the phellogen which arose some distance below the wound surface. The phellem is thick over the stele, filling the central concave area, and thin in the cortical regions.

abscission zone is formed in all the lateral branches while they are small, those branches that receive abundant light and make rapid growth are not shed, and this weak zone of soft tissue is buried and reinforced by later formed annual rings of normal xylem. When these older branches die, they are not cut off but remain until they are lost by decay.

In species of *Ulmus* and *Quercus*, the process of abscission is much the same as in *Populus*, but the position of periderm formation over the wound is deep in the abscission zone (Fig. 126). In *Dirca palustris*, the periderm, functioning in much the same manner as a separation layer, forms beneath the scar cutting off the ends of the severed vessels and fibers.

A breaking away of living twigs different from the abscission described above is present in some species of *Salix*, for example, *S. nigra* and *S. fragilis*. No separation layer is formed, but somewhat above the base of the small twigs, there is a weak zone. In this zone the cortex is considerably thicker than in the normal internode, and the cortex and phloem lack fibrous sclerenchyma, which is abundant in these tissues elsewhere in the twig. The xylem of this region is less strongly lignified than in the normal internode. Because of these structural weaknesses, mechanical stress causes rupture of the twig straight across the weak zone. The fall of the branch is not followed by the formation of a smooth periderm layer over the wound as in species whose branches are abscised. The base of the branch persists as a dead stub and is eventually buried by secondary growth of the main stem. Under favorable conditions twigs in which these weak zones are present—like those with abscission zones—persist indefinitely and with addition of xylem each year become strong branches.

In the pedicel of the mature apple fruit an abscission zone is present where the pedicel joins the fruit cluster base. The specialization of this zone is comparable to that of abscission zones of woody branches generally, with reduced fibrous and vascular tissue and increased parenchyma. At the point where the separation layer is formed the pedicel is constricted and here rupture occurs through the separation layer and the weak vascular tissues.

Entire mature inflorescences may be similarly cut off, as in chestnut, oak, willow, or they may remain attached to the parent plant until they are lost by weathering or decay, as in sumach and lilac.

References

ARTSCHWAGER, E. F.: Anatomy of the potato plant, with special reference to the ontogeny of the vascular system, *Jour. Agr. Res.*, **14**, 221–252, 1918.
———: Studies on the potato tuber, *Jour. Agr. Res.*, **27**, 809–835, 1924.
BLOCH, R.: Wound healing in higher plants, *Bot. Rev.*, **7**, 110–146, 1941.
CLEMENTS, H. F.: The morphology and physiology of the pome lenticels of *Pyrus Malus*, *Bot. Gaz.*, **97**, 101–117, 1935.
DEVAUX, H.: Recherches sur les lenticelles, *Ann. Sci. Nat. Bot.*, 8 sér., **12**, 1–240, 1900.
DOULIOT, H.: Recherches sur le periderm, *Ann. Sci. Nat. Bot.*, 7 sér., **10**, 325–395, 1889.
FOUILLOY, E.: Sur la chute des feuilles de certaines monocotylédones, *Rev. Gén. Bot.*, **11**, 304–309, 1899.
GILSON, E.: La subérine et les cellules du liège, *La Cellule*, **6**, 67–114, 1890.
HÖHNEL, VON, F. R.: Ueber den Ablösungsvorgang der Zweige einiger Holzgewächse und seine anatomischen Ursachen, *Mitt. Förstlich. Versuchswesen Oesterr.*, **1**, 255–282, 1878.

———: Ueber den Kork und verkorkte Gewebe überhaupt, *Sitzungsber. Kais. Akad. Wiss.*, **76**, 507–662, 1877.

KLEBAHN, H.: Die Rindenporen. Ein Beitrag zur Kenntniss des Baues und der Function der Lenticellen und der analogen Rindenbildungen, *Jenaische Zeitsch. Naturwiss.*, **17**, 537–592, 1884.

KUHLA, F.: Ueber Entstehung und Verbreitung des Phelloderms, *Bot. Centralbl.*, **71**, 81–87, 113–121, 161–170, 193–200, 225–230, 1897.

LEE, E.: The morphology of leaf fall, *Ann. Bot.*, **25**, 51–106, 1911.

LLOYD, F. E.: Abscission in *Mirabilis Jalapa*, *Bot. Gaz.*, **61**, 213–230, 1916.

MACDANIELS, L. H.: Some anatomical aspects of apple flower and fruit abscission, *Proc. Amer. Soc. Hort. Sci.*, **34**, 122–129, 1937.

MASSART, J.: La cicatrisation chez les végétaux, *Mém. Cour. et autres Mém. Acad. Roy. Belgique*, **57**, 3–68, 1898.

MOHL, VON, H.: Ueber die anatomische Veränderungen des Blattgelenkes welche das Abfallen der Blätter herbeiführen, *Bot. Zeit.*, **18**, 1–7, 9–17, 1860.

———: Einige nachträgliche Bemerkungen zu meinem Aufsatze über den Blattfall, *Bot. Zeit.*, **18**, 132–133, 1860.

———: Ueber den Ablösungsprocess saftiger Pflanzenorgane, *Bot. Zeit*, **18**, 273–277, 1860.

MYLIUS, G.: Das Polyderm. Eine vergleichende Untersuchung über die physiologischen Scheiden: Polyderm, Periderm, und Endodermis, *Bibl. Bot.*, **79**, 1–119, 1913.

NAMIKAWA, I.: Contributions to the knowledge of abscission and exfoliation of floral organs, *Jour. Coll. Agr. Hokkaido Imp. U.*, **17**, 63–131, 1926.

OLIVIER, L.: Recherches sur l'appareil tégumentaire des racines, *Ann. Sci. Nat. Bot.*, 6 sér., **11**, 5–133, 1881.

PARKIN, J.: On some points in the histology of monocotyledons, *Ann. Bot.*, **12**, 147–154, 1898.

PFEIFFER, H.: Die pflanzlichen Trennungsgewebe, *In* Linsbauer, K.: "Handbuch der Pflanzenanatomie," V, 1928.

PHILIPP, M.: Über die verkorkten Abschlussgewebe der Monokotylen, *Bibl. Bot.*, **92**, 1–28, 1924.

PRIESTLEY, J. H., and L. M. WOFFENDEN: Physiological studies in plant anatomy, V, Causal factors in cork formation, *New Phyt.*, **21**, 252–268, 1922.

SAMPSON, H. C.: Chemical changes accompanying abscission in *Coleus Blumei*, *Bot. Gaz.*, **66**, 32–53, 1918.

TISON, A.: Recherches sur la chute des feuilles chez les dicotyledonées, *Mém. Soc. Linn. de Normandie*, **20**, 121–327, 1900.

VAN TIEGHEM, P.: Sur les diverses sortes de méristèles corticales de la tige, *Ann. Sci. Nat. Bot.*, 9 sér., **1**, 33–44, 1905.

WEISSE, A.: Ueber Lenticellen und verwandte Durchlüftungseinrichtungen bei Monocotylen, *Ber. Deut. Bot. Ges.*, **15**, 303–320, 1897.

WETMORE, R. H.: Organization and significance of lenticels in dicotyledons, *Bot. Gaz.*, **82**, 71–88, 113–131, 1926.

CHAPTER X

THE ROOT

That part of the plant axis which normally develops beneath the surface of the soil is commonly called the "root" as distinguished from the aerial portion of the axis, the stem. There are, of course, roots that are aerial and stems that develop underground, but from the standpoint of anatomy, there are fundamental differences between root and stem in the arrangement and method of development of the primary tissues: the primary xylem in the root is exarch, as contrasted with the endarch condition in the stem (typical of gymnosperms and angiosperms); and the root has radial arrangement of vascular tissue, with alternating strands of xylem and phloem, whereas the bundles of stems are collateral, bicollateral, or concentric. Roots differ further from stems in that they do not have appendages comparable with leaves, do not give rise directly to flowers, lack stomata, and form lateral branches from relatively permanent tissue in the pericycle, rather than from the promeristem at the growing point. Other differences are the presence of the root cap, a structure wholly lacking in stems; the almost universal occurrence of an endodermis, which is frequently lacking in stems; and the usual pericyclic origin of the initial periderm layer in roots, a condition found only rarely in the normal aerial axis.

Function of the Root.—The function of the root is twofold. Physiologically the root is the absorbing organ of the plant, taking up water and mineral salts in solution and conducting them to the stem; it also serves as a storage organ for food materials translocated to the root from the leaves. Mechanically, the root anchors the plant and holds the stem in a position that makes possible the support of a large leaf surface. It is structurally effective for support in its tensile strength, flexibility, and extensive ramification through the soil. Absorption takes place for the most part by diffusion through the walls of the root hairs, although in some plants—for example, species of *Ranunculus*—root hairs are wanting, and water enters the root directly through the thin-walled epidermis. Water may be also absorbed through the epidermis of the root in the region distal to the zone of root hairs. Usually the older roots and those in which secondary thickening has taken place are incapable of absorption because of the presence of periderm and serve only for conduction, support, and storage. Storage may take place in the cortex, phloem, and xylem of typical roots or in specialized fleshy roots, such as the sweet potato and carrot. There is also abundant starch in the roots of herbaceous perennials during the season when the tops are dormant or wanting.

Gross Morphology of Roots.—Roots show great variation in form and structure. This variation may be directly related to function or may be characteristic of the species. Roots with obvious function or structure are described as storage, fleshy, fibrous, aquatic, brace, tap, aerial, and holdfast. Different species have in general root systems of a characteristic structure and habit but environment, particularly soil moisture and soil type, may cause considerable variation. For example, plants growing in dry soils usually have much more extensive root systems than plants of the same species growing in moist soils. In a natural environment in which are living various kinds of plants a variety of root systems occupy the soil—some species fill the surface soil with masses of fibrous roots, others have sparsely branched tap roots which occupy the deeper parts of the soil, and small herbs may fill in surface pockets with seasonal or transitory root systems.

The length of time that a root functions as an absorbing organ varies with the type of plant. In many herbaceous plants, particularly spring-flowering bulbs and herbs with storage roots, the fibrous roots make rapid growth for a short time and then die. The roots of tulips and narcissi start growth in the fall, continue to grow in the early spring, and die in early summer. With woody plants also there may be an annual renewal of small or "feeder" roots which at the end of a growing season die back to the more permanent part of the root system. Probably most perennial plants show renewal of the "feeder" roots.

Primary and Secondary Roots.—The part of the main axis of the plant growing down into the soil is called the *primary root;* its branches are *secondary roots.* Primary and secondary roots differ in their mode of origin. The primary root is present in early stages of the seedling, in some plants in the embryo within the seed. The apical growing region of the primary root is developed as a part of the differentiation of the embryo; one end of the axis possesses root structure, or at least root-tip meristem. The origin of secondary roots as lateral structures is discussed later in the chapter.

Between the root and the stem, a transitional region is formed in which the exarch xylem and the radial structure of the root pass over into the normal stem structure. The anatomy of this region is discussed in Chap. XI.

Ontogeny of the Root.—The early stages of root development are discussed in Chaps. III and V and the development of some of the primary tissues in Chap. IV. Development from promeristem to permanent tissue is similar to that of stems. The radial arrangement of the primary xylem and phloem and the centripetal development are evident as soon as procambium cells are distinguishable. The first mature vascular cells are elements of the protophloem (page 131). Cells of the dermatogen continue to divide anticlinally until the earlier formed vascular cells are mature. As division in these cells ceases, the development of root hairs begins. The mature primary tissues of roots—with the exception of arrangement and order of development of the vascular tissues—are similar to those of stems. A pith is typically absent in the roots of dicotyledons, except in the primary root of seedlings and some storage roots, but is commonly present in monocotyledons.

Root Cap.—The specialized structure known as the root cap is found on the tips of roots of nearly all plants. The manner of its origin is described in Chap. III. The cap functions as a constantly renewed protective sheath for the primary root meristem as it is forced through the soil. The cells of the root cap are simple and form a usually homogeneous, parenchymatous tissue of short-lived cells. They have no definite arrangement or may stand in rows radiating outward from the initials. In some species that have radiating rows the central mass may stand out by greater definiteness of arrangement as a so-called *columella*. This poorly defined structure has sometimes been considered an orienting region for the root tip, but an orienting function cannot be ascribed to this part of the cap because many plants have no such core, and others show all stages in its differentiation. (A new term, *stalace*, has been proposed to replace "columella," used otherwise in morphology.)

The cap is a continuously developing structure; its outer cells, dying and loosened from one another, are lost by attrition and decay and are replaced, as lost, by new cells formed by the initials. The loose outer cells of the tip are reported to be mucilaginous in nature. The cap is probably present on all roots except those of some parasites and in mycorrhizal roots. The roots of some aquatic plants have vestigial caps when young, but the initials soon die, and the cap disappears.

Root Hairs.—The most characteristic features of the epidermis of roots are the uncutinized cell walls, the lack of a cuticle, and the root hairs (Fig. 78), which are the specialized absorbing organs. In the great majority of plants, root hairs are confined to a part of the root somewhat back of the tip, where elongation has ceased. Generally, they persist for a short time only; the older root hairs wither or rot away and new ones form farther along the root as it increases in length. In this way the root hairs mature progressively farther and farther away from the base of the root and are constantly coming into contact with new soil. In certain plants, particularly some specialized herbs and plants of aquatic habitat, root hairs are wanting. Even roots on which root hairs would normally be present in soil may lack them when grown in water. On the other hand, a number of plants—for example, *Gleditsia triacanthos*, *Eupatorium purpureum* and *Schizaea rupestris*—have persistent root hairs, which may be generally distributed or may be confined largely to the proximal parts of the roots. The presence of persistent root hairs is correlated with little or no secondary growth and absence of periderm. Soon after the root hairs cease to function and wither away, the epidermis, or often the hypodermis, becomes suberized and forms a protective layer for the older root. In some plants, for example the Commelinaceae and some related monocotyledons, the hypodermis develops secondary root hairs in the older primary root-hair zone.

In the formation of a root hair the outside wall of the epidermal cells grows out to form a slender tubular structure somewhat resembling a pollen tube. The cell wall is very delicate and translucent, giving the root hair a white appearance. The protoplast of the epidermal cell extends into the tubular expansion as a lining of cytoplasm over the entire wall, surrounding the large interior vacuole. The nucleus is usually present near the middle or the distal end of the tube. In water or moist air, where no firm obstacles are encountered, root hairs are straight and assume a position at right angles to the root axis. In soil, they may take on any shape; their form depends upon the contacts with the soil particles (Fig. 78). Frequently, the root hairs become so firmly attached to the soil particles, that the particles cannot easily be removed, even by washing.

Cortex of the Root.—The cortex of most roots consists wholly or largely of relatively unspecialized parenchyma (Figs. 127, 128). The root cortex is usually thicker in proportion to the size of the axis than that of the stem, thereby serving better the function of storage. In some fleshy roots the thickness of the cortex is many times that of the stele, which appears as a slender thread or core extending through the spongy cortical tissue. Generally, the cortex of the root is not so firm as that of the stem of the same species, in that scelerenchyma is relatively small in amount or wanting. Also the cortical tissue of roots is not so dense because it has more intercellular spaces than that of stems. Secretory cells, resin ducts, and similar structures are normally present in the root cortex of many species. In roots that have no secondary thickening the cortex may persist for a number of years or throughout the life of the

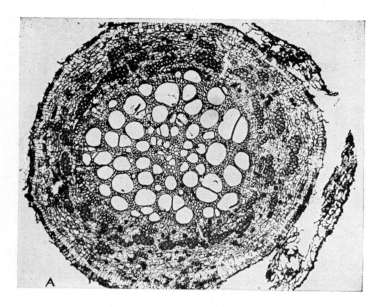
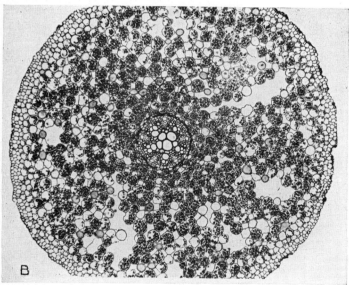

Fig. 127.—Root structure, as seen in cross section. *A*, a woody root of *Populus deltoides* with large amount of secondary tissues, showing a periderm formed in the pericycle, and the cortex sloughing off. *B*, a root of *Ranunculus acris* without secondary tissues, showing that the protostelic central cylinder is pentarch; the parenchymatous cortex is fleshy, with large intercellular spaces, its outer layers have become suberized and form a hypodermis; the epidermis is decaying. (The central parts of similar roots are shown enlarged in Figs. 129*A* and 128*D* respectively.)

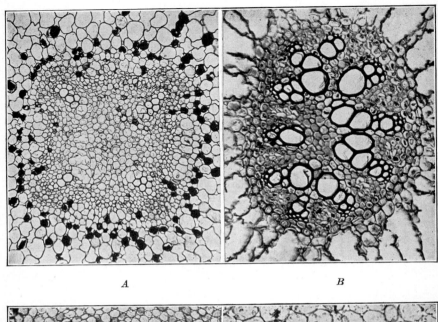
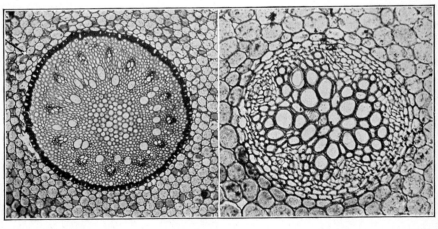

Fig. 128.—The primary structure of the stele of roots. *A*, young root of *Salix nigra*, tetrarch; only the outer parts of the xylem strands are mature; the phloem is hardly separable from the parenchyma and the procambium; the thin-walled endodermis is evident. *B*, polyarch root of *Polygonatum biflorum;* a small, irregular pith is present; the phloem and the endodermis are seen indistinctly set off from surrounding cells. *C*, polyarch root of *Smilax herbacea;* the phloem strands are dark-stained; the xylem strands are distinguishable with difficulty from the sclerenchyma in which they are embedded, but the darker-stained protoxylem region and the large innermost vessels limit the strands; a large pith is present; the thick-walled endodermis is prominent. *D*, pentarch root of *Ranunculus acris:* no pith is present.

root. Such is the condition in the roots of monocotyledons, pteridophytes, and many of the herbaceous dicotyledons. When secondary thickening takes place to any extent, as in woody dicotyledons and gymnosperms, and in many herbs, or when an internal periderm layer is formed, the cortex is soon destroyed.

Limiting the cortex on the inside, and frequently considered to be a part of it, is the endodermis (Fig. 128). This layer is present almost without exception in the primary body of roots. Its structure and its function have been described in Chap. V. The endodermis is destroyed soon after the incidence of secondary growth and its functioning life is short, therefore, in roots with secondary growth.

Pericycle of the Root.—The pericycle of roots, as compared with the cortex, is a relatively narrow zone of tissue. Its component cells are largely parenchymatous, and, although they become permanent early in the development of the root, they readily become meristematic and initiate new structures by the formation of secondary meristems. In the pericycle lateral roots are initiated and, in the majority of plants, the first periderm layer also is formed. In many roots the pericycle as seen in transverse section, is an unbroken ring of tissue but in a few plants, the bands of primary xylem abut upon the endodermis and the pericycle is divided into as many segments as there are arcs of xylem. The pericycle is commonly a persistent structure in roots even where secondary thickening is fairly well developed because new cells are added by the division of the existing parenchyma so that the layer is not ruptured by the growth beneath it. Structurally it is like the cortex of stems after growth under similar conditions. With continued secondary growth in the older roots of woody plants, periderm layers are formed in the phloem, and the pericycle is lost.

Primary Vascular Tissues of the Root.—Because of the radial arrangement of the strands of primary xylem and phloem which is characteristic of roots (Fig. 128), there are no concentric zones or rings of these primary tissues in roots similar to those found in stems. The xylem, as seen in transverse section, forms radially extended groups of cells. In the development of these strands of primary xylem, the first cells of the procambium to mature into xylem are those in the pericyclic region next to the endodermis. From these points of beginning, xylem cells mature progressively toward the center of the stele (Fig. 128A,B) until in most plants they abut upon those of the other xylem groups (Figs. 128D, 129A). Because of this sequence in development, the first-formed protoxylem cells are at the outer ends of the rows (Fig. 128A,B). No pith is present in most roots, but the roots of the monocotyledons and some other plants, especially herbs, frequently possess pith (Fig. 128C) even when other roots of the same plant are pithless.

The number of arcs of xylem varies greatly in different groups of plants (Fig. 128). In the roots of monocotyledons the number of radial plates of primary xylem is often fifteen or twenty. In most dicotyledons, both woody and herbaceous, and in the gymnosperms, relatively few primary xylem strands are present. Pteridophytes also have few strands. Roots are known as *monarch, diarch, triarch, tetrarch, pentarch,* and *polyarch,* when the number of xylem groups is one, two, three, four, five, and

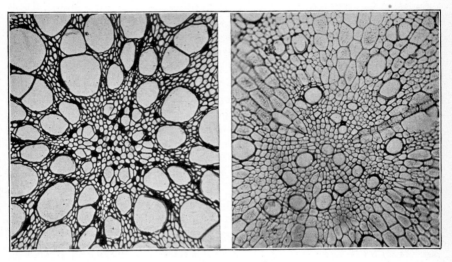

A *B*

FIG. 129.—Central portions of protostelic roots with secondary thickening. *A*, a woody root, *Populus deltoides,* tetrarch; the small-celled, tapering protoxylem tips of the primary xylem ridges project far into the secondary xylem, which closely surrounds the primary xylem, no definite line of demarcation being visible except about the ridges. *B*, a storage root, *Tephrosia virginica,* triarch; the primary xylem is small in amount, with a median vessel; the protoxylem points lie at the ends of the wide parenchymatous wood rays; the secondary xylem consists largely of parenchyma.

many, respectively (Figs. 128, 129). The number of xylem and phloem strands is constant for some species but most species show considerable variation, for example, being either diarch or tetrarch, or triarch or hexarch; less commonly, triarch or tetrarch, pentarch or hexarch. Often the different roots of an individual vary in structure—in some pines, for instance, the vigorous and main roots are tetrarch, the others diarch; in individuals of other species of pine, hexarch and tetarch roots occur.

Spiral and annular elements are much fewer in the protoxylem of roots than in that of stems, probably because very few xylem elements mature during the elongation stage, developing rather in the zone just behind this, where there is no elongation. The region of elongation of rapidly growing stems may be 8 or 10 cm or more in length, whereas that in roots may be only 1 cm or less. The region where greatest absorption takes place, namely, that normally covered with root hairs, is behind the zone where growth in length takes place. It is evident that in the stem the function of ringed and spiral elements is to conduct water and nutrients to the growing tip. The root tip, on the other hand, is surrounded by moist soil and absorbs water and nutrients directly.

The primary phloem of roots occurs as strands of tissue lying between or alternate with the radially extending plates of xylem (Figs. 128A, B, 130A). As a rule, the groups of primary phloem cells as seen in transverse section are more or less rounded or somewhat triangular in shape

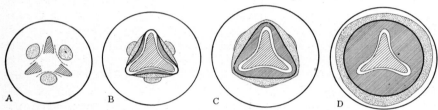

FIG. 130.—Diagrams to show origin of secondary growth in roots, the primary xylem lightly cross-hatched, the secondary closely; the primary phloem finely stippled, the secondary coarsely; the cambium represented by a heavy line; the endodermis and other layers not shown. A, young root, the inner xylem not mature, the cambium arising in the positions shown by the dotted lines; B, the cambium has formed secondary xylem and phloem beneath the primary phloem, where it first appeared, and has developed laterally, surrounding the tips of the xylem ridges, where no tissues have yet been formed, the primary phloem is crushed; C, secondary growth has continued, tissues forming about the ridges completing the three-angled cylinder of secondary xylem and phloem, the primary phloem is still further crushed; D, with continued secondary growth the cambium cylinder has become round in cross section, the primary phloem has disappeared. (A band of parenchyma is shown between primary and secondary xylem; this is often lacking.)

and are separated from the xylem groups by parenchymatous tissue. Frequently, groups of phloem cells are not well defined but merge into the surrounding parenchyma which they resemble. The first-formed cells are very slender sieve tubes which are soon crushed and absorbed. The direction in which the protophloem cells mature is usually, perhaps always, centripetal. The order of development apparently has no morphological significance, and no types are recognized as in the xylem.

The primary phloem of angiosperm roots consists of sieve tubes, companion cells, and parenchyma, and differs in no fundamental way from that of stems of the same species. As with all primary phloem in plants where secondary tissues are formed, the size of the elements is small compared with that of normal secondary phloem cells. The phloem of monocotyledonous roots is ordinarily more distinctly set off from the surrounding parenchyma or sclerenchyma than is that of other groups.

Secondary Growth in Roots.—In roots that have secondary thickening, the cambium originates as bands or strips of meristem in the procambial or parenchymatous tissues between the groups of primary phloem and the center of the stele (Fig. 130A). Here short tangential rows of cambium initials are formed which cut off secondary xylem cells toward the inside and phloem cells toward the outside. From the borders of these first-formed cambium strips, the layer of initials is extended laterally by differentiation of new initials in the parenchyma between the primary phloem and xylem strands until the segments of cambium meet in the pericycle between the xylem and the endodermis (Fig. 130B). Thus a continuous cambium sheet or cylinder is formed, which is undulate in cross section because it bends out around the primary xylem groups and dips inward beneath the primary phloem. However, because secondary tissues are formed earlier and perhaps more rapidly by the segments of cambium inside the primary phloem, the cambium soon becomes cylindrical (Fig. 130D). With the formation of the secondary tissues, the primary phloem is crushed (Fig. 130B,C), and the endodermis ruptured. The crushed cells are usually soon absorbed. The primary xylem remains intact, being readily visible in older roots, surrounded by secondary xylem (Fig. 129A).

The secondary vascular tissues of the root do not differ fundamentally from those in the stem. Such differences as occur are apparently adaptations to differences in function; the stem serves as the support of a large leaf surface in dry air and the root as anchorage in moist earth and as storage. The xylem of roots, as compared with that of stems, has larger and more numerous vessels with thinner walls, fewer fibers, more parenchyma, and larger or more abundant rays; the phloem, less sclerenchyma and more storage parenchyma. The arrangement of xylem and phloem elements in the secondary tissues of roots is fundamentally the same as that found in the stems of the same species.

Formation of Lateral Roots.—As previously stated, one of the characteristic features of roots as distinguished from stems is the method of formation of lateral appendages of the axis. In stems, the primordia of the branches and leaves are laid down in the apical meristem at the growing point, according to a definite plan. In roots, on the other hand, no branches or appendages of any kind are laid down in the meristem at the apex, and when lateral roots are developed, they are initiated in relatively permanent tissues and without regular order with reference to each other, except that of general acropetal succession. Lateral roots are commonly formed most abundantly in the region just behind the zone of root-hairs.

Exceptions are found in aquatic plants such as *Eichornia,* in which lateral roots arise from the immature pericycle anterior to the zone of elongation. This development would obviously be impossible in a root surrounded by earth since lateral roots extend at right angles to the mother root. Lateral roots in turn produce root hairs, and later, lateral roots. A branching system thus formed reaches all parts of the surrounding soil.

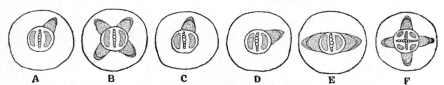

Fig. 131.—Diagrams to show position of origin of lateral roots. *A–E,* various positions found in gymnosperms and angiosperms whenever root has less than three xylem "rays"; *F,* position (opposite xylem "rays") found in roots of all vascular plants whenever there are three or more xylem "rays." (*After Van Tieghem and Douliot.*)

Lateral roots are endogenous, that is, the root meristems are formed in the inner tissues of the mother root, and the lateral root appears externally only after its growth is well begun. In angiosperms and gymnosperms the meristems arise in the pericycle just beneath the

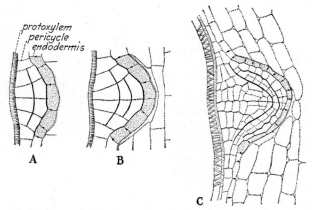

Fig. 132.—Early stages of lateral root development, radial section; *Hypericum.* *A,* a group of cells of the uniseriate pericycle have enlarged radially and divided tangentially; *B,* further enlargement and division have occurred, the endodermis and inner cortex are stretched; *C,* the root-tip meristem is well established, the endodermis will soon be ruptured. (*After Van Tieghem and Douliot.*)

endodermis; in pteridophytes growth begins in the cells of the endodermis. In all groups the originating cells are mature or nearly so. In gymnosperms and angiosperms the points of origin of lateral roots are opposite the xylem "rays" whenever there are three or more of these strands (Fig. 131*F*); in the pteridophytes they are always opposite the xylem "rays." In the seed plants where the root is diarch, the lateral roots depart at an angle with the xylem "rays" (Fig. 131A–E). The location of these points varies but is usually between the xylem and the phloem strands. With the usual restriction of lateral-root formation to the

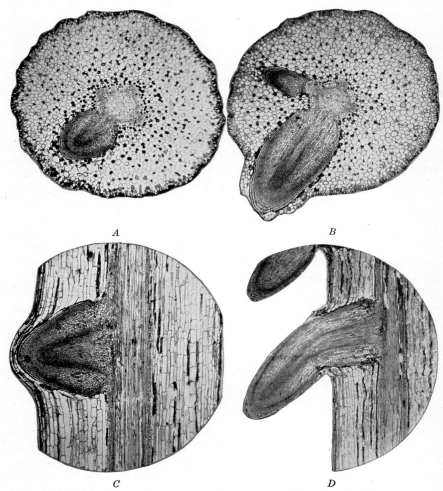

Fig. 133.—Late stages of lateral-root development, *Salix nigra*. A, B, transverse sections; C, D, radial sections. A, the tip of the young lateral root is passing through the cortex; in B, it has forced outward the outer cortical layers and the epidermis, and broken through; C, the lateral root tip is about to break through the outer cortex; D, the lateral root is free from the mother root, its attachment to the central cylinder is obvious.

regions opposite the xylem, these roots appear in vertical rows, as many rows of laterals as there are rays of xylem. A tetrarch root, for example, has four rows of laterals. Lateral roots not located opposite rays of xylem are likewise in rows. This longitudinal-row condition is often readily visible in such storage roots as those of the radish, parsnip, and turnip; it can also be seen in water-grown roots of willow and other plants. Regularity of arrangement is distorted in soil, unless the diameter of the main root is large.

In the formation of a lateral root, the cells of the pericycle of an area (circular in tangential section), at least two cells in diameter, become meristematic. These cells all divide tangentially (Fig. 132*A*) either before or after increase in radial diameter. Succeeding divisions are in any plane. A definite growing point with its initial cells, root cap, and other characteristic structures is quickly formed. As this meristem develops, the outer tissues are stretched (Figs. 132*B,C*, 133*C*) and soon ruptured (Fig. 133*B*). The root forces its way, apparently partly by absorption of the surrounding tissue, but in large part by mechanical pressure, through the endodermis, cortex, and epidermis, and continues growth in the normal way. In many plants there is said to be a partial chemical dissolution of the cortical tissues by the root cap as it forces its way out. There is no connection between the epidermal cells of the lateral root and the disrupted tissue of the cortex of the main root through which it passes, but at the proximal end of the lateral root where the meristem was initiated, parenchyma and vascular elements of the lateral root and the tissues of the mother root are in close contact, providing a passageway for conduction. With the formation of secondary growth the cambia of the two roots are connected, and annual rings of tissue are laid down which are continuous around the point of insertion of the lateral. Not all lateral roots grow with the same rapidity; some persist and grow rapidly, forming part of the main root system of the plant, whereas others remain small or are lost altogether. In some plants, and especially in storage roots, there is a seasonal renewal of secondary rootlets.

Adventitious Roots.—The term adventitious roots, used loosely, covers roots that arise on stems and those that arise in regions of the main root other than the pericycle behind the zone of elongation. Such roots are important in plant propagation, particularly in the rooting of stem cuttings and the generation of root systems on root cuttings or root grafts. Roots that have their origin in callus tissues near a cut surface are called *wound roots* as distinguished from roots which arise from preformed meristematic *root germs* or cushions, sometimes inappropriately called "morphological roots," or from definite groups of cells that are by position and structure capable of root formation, though they may not be distinguishable from the surrounding tissue.

Typical wound roots arise endogenously, after formation of vascular tissue in the basal callus, from a meristem that may be regarded as a continuation of the cambium. Roots that develop from "root germs" arise in most plants in the pericyclic region or, in older axes where the pericycle is no longer active, in the secondary phloem. Root germs are associated with the vascular rays in many woody plants; they are often in the vicinity of the node, especially in the leaf gaps, as in *Lonicera* and *Ribes*, or in the branch gaps, as in *Salix* and *Cotoneaster*, but they may be internodal in origin as in *Salix* and *Populus*. In herbaceous plants, they are most frequently associated with the interfascicular cambium (*Begonia*). Roots of root-germ origin are more important in some species than wound roots, in the successful propagation of woody cuttings. In other species, wound roots predominate but wound roots may occur in species where root germs predominate, and roots that develop from root germs may appear in species that usually lack them. In some of the pteridophytes and rarely in the angiosperms, adventitious roots are said to arise from cells of the outer cortex.

Different species vary greatly in the ease with which rooting can be induced on either stems or roots. Some plants, notably *Salix* and *Ribes*, which have preformed root germs, root very easily from stem cuttings, whereas others, such as stem cuttings of most forms of *Pyrus*, *Malus*, or *Carya* in which root germs are absent, root with great difficulty or not at all. Some root cuttings, for example those of *Rubus* and sea kale, form roots readily and others, such as those of *Malus* and *Carya*, with difficulty. Leaves of some plants, for example *Begonia*, *Saintpaulia*, and *Bryophyllum*, generate roots from the leaf margins or petioles. Ability to form roots changes with age—seedlings and juvenile forms root more readily than older plants.

Periderm of Roots.—Sooner or later in the development of the majority of perennial roots, especially in the dicotyledons and gymnosperms where secondary thickening takes place, a periderm layer is formed. In stems the first-formed periderm usually arises in the epidermis or in the cortical layers immediately beneath, whereas in roots, though suberization of the outer layers may take place, the first true periderm commonly arises in the outer layers of the pericycle and in woody plants persists as a continually expanding layer for a considerable number of years. The endodermis and cortex are ruptured and soon decay, so that the root has a smooth, brownish covering of cork cells broken only by the lenticels. Lenticels do not occur conspicuously in all species but in some, for example, *Morus* and *Gleditsia*, appear as transversely elongated roughenings of the root surface. The structure of root lenticels is not fundamentally different from that of the lenticels of stems. In some species, the complementary cells accumulate to such an extent that they are conspicuous as a powdery mass in the lenticellular cavity.

In the large roots of many trees, the first periderm persists indefinitely and the bark is smooth, commonly showing conspicuous lenticels. In some trees, other periderm layers arise successively deeper and deeper in the phloem, as in stems, and the outer tissues are lost. There is no extensive accumulation of rhytidome on roots because of the rapid decay of dead tissues beneath the soil. When roots are exposed at the base of a tree, bark is formed resembling that on the trunk.

There are many exceptions to the usual course of periderm development in roots. In those roots without secondary thickening, the epidermis may persist intact, often becoming cutinized, or it may be lost by decay and be replaced as the limiting layer by an outer primary cortical layer which becomes cutinized, as in many monocotyledons. In such plants the true epidermis may shrivel and disappear as soon as the root hairs cease to function. Such an accessory protective layer, serving as an epidermis, is sometimes called an *exodermis* but it is merely a type of hypodermis. Further, the term "exodermis" is also applied to a uniseriate layer in certain specialized roots, for example, the innermost layer of the velamen of orchid roots. A persistent epidermis is especially common among the roots of the monocotyledons. In this group periderm forms the protective layer only in the larger, older roots, as in the Araceae and Liliaceae. In herbaceous dicotyledons a hypodermis is often the protective layer (Fig. 127B). Periderm layers also frequently occur in the roots of the Primulaceae, Gentianaceae, and others; these layers occur in the outer cortex, but not superficially, as in the monocotyledons. In the pteridophytes, periderm does not form in the roots, but the epidermis and the outer cortical layers become cutinized or lignified without change in size or shape of the cells. The chemical nature of these lignified or cutinized walls is apparently unlike that in similar cells in most of the seed plants.

Unusual types of structure are found in roots as in stems, but in roots these are associated chiefly with secondary tissues as in some of the Centrospermae. Anomalous structure in roots is similar to that in stems which is discussed in Chap. XI.

Roots show great specialization related to environmental conditions and to particular functions. A discussion of the modifications found in some of these specialized forms are found in Chap. XIV.

References

ALTEN, H.: Beiträge zur vergleichenden Anatomie der Wurzeln nebst Bemerkungen über Wurzelthyllen, Heterorhizie, Lenticellen, Göttingen, 1908.

ESAU, K.: Developmental anatomy of the fleshy storage organ of *Daucus Carota*, *Hilgardia*, **13**, 175–209, 1940.

———: Vascular differentiation in the pear root, *Hilgardia*, **15**, 299–311, 1943.

FLAHAULT, C.: Recherches sur l'accroissement terminal de la racine chez les phanérogames, *Ann. Sci. Nat. Bot.*, 6 sér., **6**, 1–168, 1878.

FRIEDENFELT, T.: Der anatomische Bau der Wurzel in seinem Zuzammenhänge mit dem Wassergehalt des Bodens, *Bibl. Bot.*, **61**, 1–118, 1904.

JANCZEWSKI, DE, E.: Recherches sur le développement des radicelles dans les phanérogames, *Ann. Sci. Nat. Bot.*, 5 sér., **20**, 208–233, 1874.

KROEMER, K.: Wurzelhaut, Hypodermis, und Endodermis der Angiospermenwurzel, *Bibl. Bot.*, **59**, 1–151, 1903.

LEMAIRE, A.: Recherches sur l'origine et le développement des racines laterales chez les dicotylédones, *Ann. Sci. Nat. Bot.*, 7 sér., **3**, 163–274, 1886.

MAXWELL, F. B.: A comparative study of the roots of the Ranunculaceae, *Bot. Gaz.*, **18**, 8–16, 41–47, 97–102, 1893.

NÄGELI, C.: Ueber das Wachsthum des Stammes und der Wurzel bei den Gefässpflanzen, *Beit. Wiss. Bot.* (Nägeli), **1**, 1–156, Leipzig, 1858.

—— and H. LEITGEB: Entstehung und Wachsthum der Wurzeln, *Beit. Wiss. Bot.* (Nägeli), **4**, 73–160, Leipzig, 1868.

OLIVIER, L.: Recherches sur l'appareil tégumentaire des racines, *Ann. Sci. Nat. Bot.*, 6 sér., **11**, 5–133, 1881.

PRIESTLEY, J. H., and R. M. TUPPER-CAREY: Physiological studies in plant anatomy, IV. The water relations of the plant growing point, *New Phyt.*, **21**, 210–229, 1922.

SACHS, J.: Ueber das Wachsthum der Haupt und Nebenwurzeln, *Arb. Bot. Inst. Würzburg*, **1**, 384–474, 585–634, 1874.

TUBEUF, C.: Die Haarbildungen der Coniferen, V. Die Wurzelhaare der Coniferen, *Förstlich. Naturwiss. Zeitsch.*, **5**, 173–193, 1896.

VAN DER LEK, H. A. A.: Onderzoekingen over de vegetatieve vermenigvuldiging van houtige gewassen, 1. Over de wortelvorming van houtige stekken. Laboratorium voor Tuinbouwplantenteelt, No. 1. Overdruk uit Deel 28 der Mededeelingen van de Landbouwhoogeschool te Wageningen (Nederland).

VAN TIEGHEM, P.: Recherches sur la symétrie de structure des plantes vasculaires, *Ann. Sci. Nat. Bot.*, 5 sér., **13**, 5–314, 1870.

—— and H. DOULIOT: Recherches comparatives sur l'origine des membres endogènes dans les plantes vasculaires, *Ann. Sci. Nat. Bot.*, 7 sér., **8**, 1–660, 1888.

WHITAKER, E. S.: Root hairs and secondary thickening in the Compositae, *Bot. Gaz.*, **76**, 30–59, 1923.

CHAPTER XI

THE STEM

The part of the axis of a plant that bears the leaves and reproductive structures and is commonly aerial and ascending is called the stem. Stems and roots are alike in general structure, each possessing a stele with xylem and phloem, pericycle, endodermis, and cortex with an epidermis. Stems differ from roots in fundamental vascular structure and in the possession of appendages borne at definitely fixed positions known as nodes. The difference in vascular structure lies chiefly in the arrangement of the xylem and phloem: in the root the strands of primary xylem and phloem lie in different radii, separated from one another; in the stem the strands lie side by side in the same radius. Further, the xylem of the root is always exarch, whereas that of the stem is exarch, endarch, or mesarch, being endarch most commonly in present-day plants. In the development and structure of secondary vascular tissues, roots and stems are very much alike.

Origin of the Stem.—The first stem meristem develops during the specialization of the embryo. Lateral stems normally arise by the development of new apical meristems laterally in the terminal meristem of the mother stem. Adventitious branches develop, on both stems and roots, by the formation of similar meristems secondarily in the pericycle, phloem, or even in the cambium. In a few plants, they are said to arise in epidermal or subepidermal cells. Branch meristems also develop freely in wound tissue of some species. Such meristems arise by the division of meristematic cells, or of more or less permanent parenchyma, in several planes, forming an apical growing point like that of the normal stem tip. If this meristem is deeply buried in the stem tissues, it breaks its way through to the surface much as the lateral root meristem breaks through the root cortex. (The course of development of the mature stem from meristem is discussed in Chap. III.)

Root-stem Transition.—Root and stem form a continuous structure, the axis. There is, therefore, a transition region where root and stem meet, and where the various parts of each organ merge into those of the other. The epidermis, cortex, endodermis, pericycle, and secondary vascular tissues are directly continuous from the root into the stem. The primary vascular tissues are also continuous, but not directly so, since the

Fig. 134.—Root-stem transition; diagram of four types, A, B, C, D (see text). Figures A-1, B-1, etc., root; A-5, B-5, etc., stem; intermediate figures represent stages in transition found at successive levels, showing splitting, rotation, and fusion of the vascular strands. Xylem cross-hatched; phloem stippled.

types of bundles and of arrangement are markedly different in the two organs—the radially arranged, independent strands of xylem and phloem, the xylem exarch, of the root pass into bundles containing collaterally placed xylem and phloem, the xylem usually endarch. In the xylem this change of position involves a twisting and inversion of the strands (Fig. 134). The change from one type of vascular structure to the other takes place in a part of the axis called the *transition region*. The transition occurs either abruptly or gradually; the length of the transition region, which is commonly short, varies from less than a millimeter to two or three, rarely several, centimeters. It may occur in the top of the radicle, at the very base of the hypocotyl, near its center, or in the upper part. The hypocotyl may, therefore, possess root structure or stem structure through most of its length, or may be largely given over to transition region. Often the transition region coincides with that of the origin and departure of the cotyledonary traces, and the structure of the region is especially complex. Whenever the inversion of the bundles has not been accomplished at the level at which the cotyledonary traces depart, these outgoing strands are inverted during their passage into the cotyledons. Rarely the transition region extends to the first or even the third or fourth node above the cotyledons; a part of the stem then possesses bundles which are partly inverted. In the monocotyledons and the pteridophytes the region is usually short.

Externally, this region which limits root and stem may be visible as a zone of depression, or of change in diameter, but in many plants such an external line does not correspond exactly with the transition region.

In the change from root to stem there is commonly a considerable increase in the diameter of the stele. Accompanying this, go multiplication of vascular tissues and a forking, rotation, and fusion of strands. These changes take place according to a definite plan, of which four somewhat different types have been recognized.

Type a. (Fig. 134A).—Each xylem strand of the root forks by radial division; the branches, as they pass upward, swing laterally, one to the right and one to the left, turning at the same time through 180 degrees, and join the phloem strands on the inside. The latter have meanwhile remained unchanged in position and orientation; they pass as straight strands from the root into the stem. In this type there are formed in the stem as many primary bundles as there are phloem strands in the root. Examples of plants in which this type occurs are *Dipsacus*, *Mirabilis* and *Fumaria*.

Type b. (Fig. 134B).—This type differs from the first type in that the strands of phloem, as well as those of xylem, fork, the branches of each swinging laterally as they pass upward to meet in pairs in positions alternating with those of the strands in the root. The xylem strands become inverted as before; the phloem strands retain their orientation. In this way there are formed in the stem twice as many bundles as there are phoem strands in the root. This type of transition is more common than the first. It occurs, for instance, in *Acer, Cucurbita, Phaseolus,* and *Tropaeolum.*

Type c. (Fig. 134C).—In the third type, the xylem strands do not divide, but continue their direct course into the stem, twisting, however, through 180 degrees. The phloem strands meanwhile divide, and the halves swing laterally to the position of the xylem, joining the xylem strands on the outside. In this type, as in the first, as many bundles are formed in the stem as there are phloem strands in the root. Examples of plants in which this type occurs are *Medicago, Lathyrus,* and *Phoenix.*

Type d. (Fig. 134D).—In the fourth type, half of the xylem strands divide and the branches swing laterally to join the other unforking strands, which meanwhile become inverted. The phloem strands do not divide, but unite in pairs with the triple strands. Thus a bundle of the stem is made up of five united strands, and there are but half as many bundles in the stem as there are phloem strands in the root. This type of transition is rare, and apparently is known only in a few monocotyledons, such as *Anemarrhena.*

Where internal phloem is present in the stem, branches depart from the phloem strands of the root at the level at which the root structure begins to change. These branches pass inward and come to lie inside the new xylem strands, establishing bicollateral bundles. Where the xylem strands of the root are united laterally to form a hollow cylinder (a pith is probably always present in the very top of the seedling root adjacent to the transition region), the strands become separated before the forking or the change in position and orientation occurs.

In some of the monocotyledons the transition region is very short and its structure difficult of determination, owing to the development of a ring of vascular tissue in connection with the attachment of numerous strong lateral roots which develop at that point. In the cycads, the transition region contains a plate or a ring of vascular tissue into which all the strands of the stem and root pass. In these plants the bundles of the stem and root do not, therefore, pass directly into one another.

TYPES OF STEMS

The fundamental structure of stems has already been discussed (Chap. V). In review, it may be said that stems are commonly siphonostelic—protosteles occurring in living plants only among the ferns and some other pteridophytes—and usually possess secondary growth. Stems differ greatly in the amount and arrangement of primary vascular tissue and in the amount of secondary tissues. The primary vascular tissue ranges in amount from that of a solid cylinder of considerable thickness to that of a few small strands forming isolated bundles. The various conditions apparently represent stages in evolutionary progress where there has been a thinning of the cylinder of primary vascular tissue, a breaking up of the cylinder into longitudinal strands, and changes in the arrangement of the resulting bundles so that these no longer form a cylindrical series. Whatever the amount and the arrangement of the primary xylem, the secondary vascular tissues may form a solid cylinder enclosing it and in this way develop an unbroken vascular cylinder even from a series of unevenly placed strands. The amount and the arrangement of secondary xylem also vary from that of the complete cylinder of indefinite thickness formed in typical perennial woody axes to the isolated thin strands of certain types of annual herbaceous stems, and to the condition in other plants where secondary growth is absent. Secondary vascular tissues, like primary, have been reduced in evolutionary modification, the cylinder being first thinned radially and then broken up tangentially. In forms most specialized in this respect, no secondary vascular tissues are formed. All stages in these changes in primary and secondary vascular tissues are represented among living plants. Great variety is found in stem structure.

The Woody Stem.—Perennial woody plants present apparently simple stem structure. In these an unbroken layer of secondary vascular tissue sheaths a more or less continuous cylinder of primary xylem. Variations in the structure of this cylinder (already discussed under *Primary Vascular Skeleton* in Chap. V), range from cylinders that are unbroken except by leaf and branch gaps to those consisting of discrete bundles often complex in arrangement. The simplicity of the secondary cylinder, concealing the basic primary structure, seems to make the entire structure simple. Even where the primary cylinder is broken only by leaf and branch gaps (Figs. 72A, *Thuja;* 71A, *Populus*) prominent "bundles" stand out as protoxylem-containing ridges in the otherwise very thin cylinder. These ridges represent the downward continuation of the leaf traces and other early maturing parts of the cylinder (Fig. 71A). In other plants the cylinder consists of more or less closely placed bundles of varying size (Fig. 71B)—with proportionate amounts of protoxylem—without connecting primary xylem. The areas between these primary strands are ray-like projections of the pith on the inside and pericycle on the outside. They are soon closed by the cambium which builds secondary vascular tissue in these spaces.

Woody, perennial stems, such as those of some palms and other large monocotyledons, may be developed without secondary growth.

The Herbaceous Stem.—Herbaceous plants have no distinctive anatomical structure. Annual stems are commonly supposed to have their vascular tissues—whether it is primary only, or primary plus second-

298   AN INTRODUCTION TO PLANT ANATOMY

ary—characteristically in the form of discrete bundles arranged in a cylinder (Fig. 135F), but this condition is not typical of herbaceous plants. Among dicotyledons, the majority of herbaceous forms possess

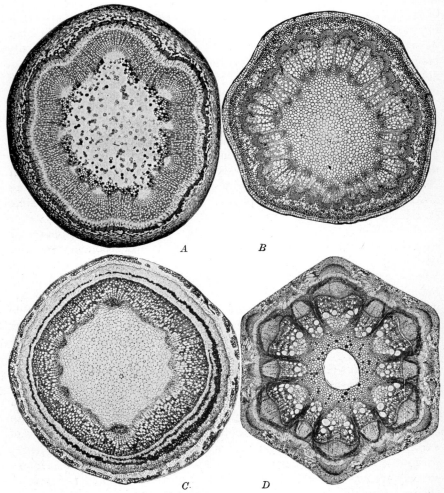

Fig. 135.—Types of dicotyledonous stems. A, *Liquidambar*, tree, stele continuous; B, *Platanus*, tree, stele dissected; C, *Lonicera*, woody vine, stele continuous; D, *Clematis*, woody vine, stele dissected. (Figure continued on opposite page.)

cylinders of vascular tissue that are complete, except for the presence of leaf and branch gaps (Fig. 135E). Such conditions obtain not only in coarse and stout stems, such as those of many composites, mints, and legumes, but even in slender delicate stems, such as those of species of

Veronica and *Stellaria*. The herbaceous stem with discrete bundles is not to be considered typical of annual stems. It represents, clearly, the extreme stage of reduction of vascular tissues. Although infrequent in

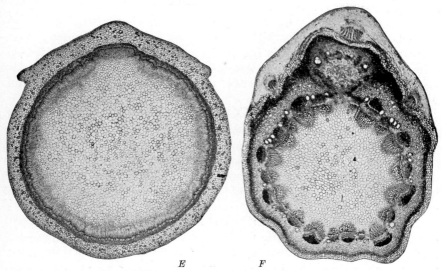

Fig. 135 (continued from opposite page).—Types of dicotyledonous stems. *E*, *Digitalis*, herb, stele continuous; *F*, *Artemisia*, herb, stele dissected. (*After Sinnott and Bailey.*)

occurrence, it is found in various families and in both stout and slender plants. In some forms, for example, in species of *Trifolium*, *Geum*, and *Agrimonia*, the lower part of the stem may have a complete cylinder of

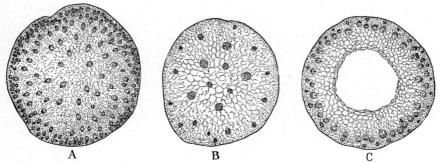

Fig. 136.—Diagrams of monocotyledonous stems. *A*, *Zea mays*, many bundles, no pith cavity; *B*, *Trillium grandiflorum*, few scattered bundles; *C*, *Triticum sativum*, many bundles about the periphery, central pith cavity.

vascular tissue, and the upper part separate bundles. The condition in monocotyledons is essentially the same as that in extreme dicotyledonous herbs; here the bundles may lie in a ring, but more frequently they are distributed through the stele according to a plan determined by the number of leaf traces, leaf arrangement, and other factors (Fig. 136).

Herbaceous stems, except those of the extreme type where the cambium fails to unite the primary skeleton, are structurally like woody stems; differences are those of period of persistence, not of basic structure. The herbaceous stem is one in which cambial growth is limited to one season or part of one season, or is lacking. The amount of vascular tissue, primary or secondary, is not necessarily less in a mature herbaceous stem than in the one-year-old stem of a closely related woody plant of the same type.

The type of stem in a given herb depends upon the type of stele present in the woody ancestral forms. Where the vascular cylinder of the woody type possesses primary tissues in the form of a continuous cylinder, related herbaceous forms also possess unbroken primary cylinders; where the primary cylinder of woody forms is discontinuous, related herbaceous plants have a type of stele in which the cylinder consists of discrete bundles.

In some details, a mature herbaceous stem with well-developed secondary vascular tissue may differ from a similar woody stem. Commonly, in annuals, as seasonal growth of the stem ceases, the cambium disappears, all its cells maturing as xylem and phloem elements. All tissues outside the cambium become strongly compressed—cortical parenchyma, especially delicate, photosynthetic types, may be completely flattened; and, even before flowering and seed-formation is completed, the softer cells of the phloem, including sieve tubes, are crushed. The phloem fibers of such plants show structural evidence of this lateral compression in their "jointed," or "broken" structure. In herbaceous stems the epidermis rarely is ruptured with increase in diameter. It is maintained at first by slow division; later, the compacting and crushing of the outer tissues provide space for the developing secondary xylem and phloem.

The morphological problem underlying the evolutionary changes in structure from the typical woody stem to the extreme herbaceous stem lies outside the scope of this book. From the evidence of comparative morphology and of the fossil record, it is clear that in angiosperms the herbaceous type of stem has been derived from the woody, undoubtedly independently in many families. In a few families, such as the Berberidaceae, woody types have been derived from herbs. Woody plants of this type have many features in common with herbs, especially structure of the primary skeleton and histological structure of the xylem.

The Monocotyledonous Stem.—Secondary growth of the usual type is typically lacking throughout the body of monocotyledons but vestiges of cambial activity in the bundles both of the stem and the leaves have been found in nearly all groups. The stele is broken up into bundles which are commonly distributed throughout the axis; the endodermis is lacking, and the limits of cortex, pericycle, and pith are often indistinguishable, since bundles are scattered throughout (Fig. 136*A,B*). In many forms, as in most grasses, there is present a central region (Figs. 136*C*, 181*A*)—which may or may not represent the pith morphologically —in which no bundles occur; in other forms a more or less readily separable cortical region may also be seen.

The vascular system of most monocotyledons is highly complex. The leaf-trace bundles are numerous and follow various types of courses in their descent, uniting in different ways with other strands (Fig. 72*E–M*). A common condition is that where all bundles are common bundles. The traces, upon entering the stem from the leaf, penetrate deeply, the median traces more deeply than the lateral, and then in their descent return toward the periphery. The downward course may be vertical, or the traces may swing laterally and become oriented in various ways. Each common bundle sooner or later fuses with other similar bundles. The anastomoses occur chiefly at the nodes, and in some groups, such as the grasses, are abundant and largely restricted to that region.

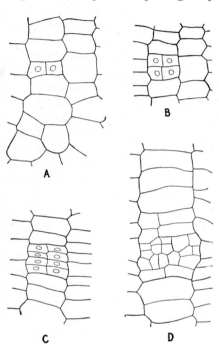

FIG. 137.—Diagrams illustrating early development of bundle in secondary growth of monocotyledons; *Aloë pleuridens*. *A–D*, successive stages. (*After Chamberlain*.)

Secondary Growth in Monocotyledons.—The type of secondary tissue formed in the monocotyledons is very different from that formed in other groups. The cambium does not produce phloem on the outside and xylem toward the inside in the normal way but forms on the inner side amphivasal or collateral bundles in parenchymatous ground tissue commonly called *conjunctive tissue*. The bundles are usually without definite arrangement but may lie to some extent in radial rows. Occasional anastomoses occur. The extracambial tissues formed are small in amount and parenchymatous in nature. In the cambium layer proper, the divisions are largely tangential, and the cells so formed are consequently radially arranged. This order is usually evident in the conjunctive tissue and aids in the separation of the secondary from the primary tissues in which the interfascicular parenchyma has no definite arrangement (Fig. 97). It is not evident in the tissues of the bundles because of the method of development of these strands. A very small group of cambium derivatives divide longitudinally—at first anticlinally and periclinally, then haphazardly (Fig. 137)—forming a strand of cells that mature as xylem and phloem. The cells that form the tracheids become 15 to 40 times their original length; the other cells elongate little or not at all. The tracheids may be scalariform, a type rare in secondary tissue. The mature bundles differ from the primary bundles in the small amount of phloem and the absence of annular and spiral protoxylem cells. In some genera differences in relative abundance of bundles and in wall thickness of the conjunctive cells set off weakly limited growth rings. The relation of these rings to annual growth is unknown.

Secondary tissue of this type is indefinite in amount, as is that of the normal type, but usually develops slowly, and large trunks are not commonly formed.

Increase in diameter of this type occurs in the arborescent Liliiflorae, for example, *Dracaena, Yucca, Aloë, Cordyline;* some of the palms; rarely in herbs, as in *Veratrum;* and in fleshy parts of some of the Dioscoreae.

The thickening which takes place in the bases of some palm stems is not due to the activity of a definite cambium layer but is, rather, the result of gradual increase in size of cells and of intercellular spaces, and, rarely, of the proliferation of strands of tissue to form new fibers; it represents long-continuing primary growth.

The Vine Type of Stem.—The stems of vines are of both structural types. Many vines, such as *Vitis, Celastrus,* and *Solanum Dulcamara,* have vascular steles in the form of woody cylinders (Fig. 135C); others, such as *Clematis, Humulus,* and *Pisum,* have vascular tissues arranged in the form of a ring of bundles (Fig. 135D). These bundles are separated by rays of parenchyma, which in many forms are increased, like the bundles, by cambial growth. In such forms, as in the type of herbaceous stem with discrete bundles, the xylem and phloem are highly specialized in structure—vascular rays are commonly absent; the vessels are porous and of large diameter and great length; tracheids and fibers are proportionately few; sieve tubes are of the highest type; and phloem parenchyma and fibers are scarce or lacking. Many vines show not merely these tissue specializations, but also possess anomalous general structure. In this group fall *Aristolochia* and *Menispermum,* which, unfortunately, are often cited as examples of typical stem structure, and used to demonstrate the origin of a woody from an herbaceous stem.

In the stem of the young herbaceous vine, such as that of *Pisum*, *Apios*, and *Adlumia*, the vascular bundles are separated by wide segments of parenchyma. In these stems the cambium may be restricted to the bundles, as in *Adlumia*, or may form a complete cylinder by extension across the interfascicular rays of parenchyma. This interfascicular cambium may be vestigial, as in the upper parts of the pea vine, forming no secondary tissue, or a very few vascular cells, as at the base of the pea vine; or it may be very active, increasing the parenchymatous rays at the same rate as the fascicular cambium builds up the bundles, as in *Clematis*. In such plants as the last, the type of structure found in the one-year stem is maintained as the stem becomes woody and perennial. Yet, since the bundles increase in tangential extent as secondary growth continues, the stem apparently becomes a woody stem. The bundles are, however, still separated, though by wedges of secondary rather than primary parenchyma. These radial plates or rays of tissue commonly extend unbroken from node to node, often for several internodes and constitute an important feature of vine structure. Such rays, either of primary structure, or of both primary and secondary structure, are prominent structural features of many vines, both annual and perennial. Similar parenchymatous segments are present also in some vines with complete woody cylinders, and constitute one of the prominent modifications of such steles found in lianas (Fig. 141). Since they consist of soft tissue, they are sometimes crushed as the stem becomes older, perhaps by the "play" of the bundles upon one another during the periods of lateral stresses to which a vine stem is peculiarly subject. Apparently, this type of structure represents one type of modification related to the mechanical requirements of vines.

"Medullary Rays" of Vines and Herbs.—In vines and herbs with dissected vascular cylinders, the vascular bundles are in early stages separated by sheets of parenchyma which merge internally with the pith and externally with the cortex (Fig. 135*D*, *F*). In these early stages there is usually little or no evidence of limitation of pith or cortex in the region of these bands, nor is there histological evidence that they belong morphologically, even in part, to the vascular cylinder. Since they appear as radiating portions of the pith, they are commonly called *medullary rays*. When these primary structures are later increased in radial extent by secondary growth, as they are in many vines, such as *Clematis* (Fig. 135 *D*), they closely resemble broad "medullary rays" (vascular rays) of secondary wood and phloem. For this reason both these rays and true vascular rays have been termed "medullary rays," and believed to be homologous. The theory that the woody stem is developed from the herbaceous stem is based in large part upon this misconception. The broad rays of vines and herbs extend from node to node and often through several internodes and in this respect differ much from the vascular rays which are of very limited vertical extent. Further, they clearly represent, morphologically, entire segments of the vascular cylinder, being equivalent to many vascular rays plus the tissue surrounding and included between these rays. It is obvious that the same term should not be applied to both types of rays. The term "medullary ray" as applied to rays of secondary vascular tissue is inapt, and has been supplanted by the term "vascular ray." Since the broad rays of vines and herbs, when only primary, resemble projections of the pith, they may perhaps aptly be called medullary rays, especially if this term is not applied to vascular rays. That they are not homologous with the structures hitherto known commonly as medullary rays is obvious.

Internal (or Intraxylary) Phloem.—In the siphonosteles of ferns the amphiphloic condition is usually prominent. Here the internal phloem forms a continuous layer and is closely similar to the external phloem. In the angiosperms, where internal phloem is also frequently present, this tissue is less conspicuous. It occurs usually as strands, large or small, more or less closely associated with the primary xylem (Fig. 138). In amphiphloic ferns, where the cylinder is broken by leaf gaps, the internal phloem unites with the external through the gaps, and amphicribral bundles are formed (Fig. 139*B*). In angiosperms with internal phloem and broken vascular cylinder, the internal phloem forms the innermost part of bicollateral bundles, as in *Cucurbita;* where the primary xylem forms a more or less complete cylinder, strands of internal phloem also form a fairly continuous layer. Internal phloem is in most plants only primary; rarely a cambium arises inside the primary xylem and a small amount of internal secondary phloem is formed, as in *Tecoma*.

The cells of internal phloem are like those of external phloem, except that fibers are few or lacking, and the sieve tubes and companion cells occur in small, restricted groups, surrounded by parenchyma (Fig. 138). This parenchyma, together with that of the protoxylem region, forms the perimedullary zone of the pith (Chap. V). In ontogeny internal phloem develops later than external primary phloem. Such a layer is, of course, morphologically, a part of the vascular cylinder and not the outer layer of the pith. Where an inner endodermis is present, this layer separates the inner phloem from the pith. The internal phloem and the internal endodermis are often continuous through leaf and branch gaps with the outer phloem and the outer endodermis.

Internal phloem occurs in many families among the angiosperms, chiefly in the more highly specialized groups, such as the Solanaceae, Gentianaceae, Myrtaceae, Cucurbitaceae, Convolvulaceae, Apocynaceae, Asclepiadaceae, Onagraceae, Campanulaceae, and Compositae. In some of these plants the internal phloem is apparently degenerate; in others it probably plays a prominent part in conduction. The latter is the condition in the tomato and potato plants, especially in the rhizomes and tubers of the potato. In the potato tuber, except for a slender central core, strands of internal phloem are present throughout most of the region inside the thin xylem cylinder. Like the pith, the cortex is narrow. The storage tissue of the potato is, therefore, to a large extent a par-

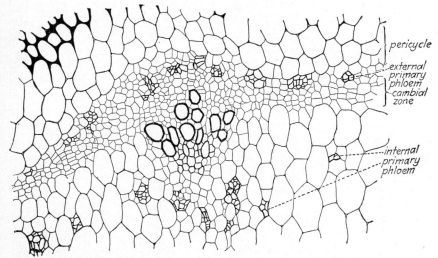

FIG. 138.—Internal (intraxylary) phloem in young stem of *Solanum tuberosum*. The primary phloem cells are in small groups external and internal to the cambial zone. (*After Artschwager.*)

enchymatous vascular cylinder; the pith makes up only a small part of the tuber.

The Vascular Bundle.[1]—The vascular tissues of a plant form a continuous system. In the axes of the majority of plants these tissues are arranged in the form of solid or hollow cylinders. But in the axes of other plants, and in appendages generally, the vascular tissues occur in more or less distinct strands, which are united proximally and sometimes also distally with other similar bundles or with the central vascular

[1] A general discussion of the vascular bundle is placed in this chapter for convenience alone. Bundles are not, of course, restricted to stems. Further discussion of bundle structure will be found in other chapters.

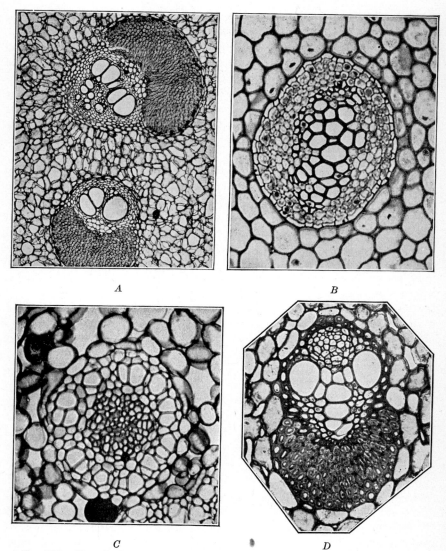

Fig. 139.—Vascular bundles. *A, Sabal Palmetto*, collateral monocotyledonous bundles with heavy bundle cap; *B, Polypodium vulgare*, amphicribral bundle; *C, Acorus Calamus*, amphivasal bundle; *D, Scirpus lineatus*, collateral monocotyledonous bundle, with sclerenchyma sheath, two prominent vessels flanking the circular phloem mass. (*A, Courtesy of the U.S. Forest Products Laboratory.*)

tissues. To these more or less free parts or extensions of the vascular body the term *vascular bundle* is applied (Chap. V). The bundle is, obviously, an important structural feature of the plant; but as such it has doubtless received undue attention. The study of the stele and of the relationship of axis and appendages shows clearly that the bundle is merely a segregated portion of the conducting system. In the axis the bundle is not the fundamental unit of structure from which the vascular cylinder has been built up. The disposition of the tissues in a cylinder is the primitive condition, and the discrete bundles of axes are the portions of such a cylinder broken up in specialization (Chap. V).

Size and Shape of the Vascular Bundle.—Vascular bundles vary greatly in structure and shape, in course taken in the plant body, and in relation to other bundles and the central vascular body. The various types of bundles, in so far as they depend upon the relationship of xylem and phloem to each other in the bundle (concentric, collateral, etc.), and upon the course of bundles in the axis (cauline and common), have been considered in Chaps. IV and V. Since any essentially free strand of the vascular system, large or small, can be called a "bundle," a bundle, as seen in cross section, may consist of an indefinite number of cells, or of very few, even two (Fig. 150). Bundle tips in leaves and fruits often consist of a single cell. In cross-sectional shape, the bundle is most commonly oblong, or elliptical, but linear, forked, and irregularly shaped bundles also occur.

Histological Structure of the Vascular Bundle.—The proportionate amount of xylem and phloem varies much in both collateral and concentric types of bundles. In both types the bundle may consist largely of one tissue, and collateral bundles are sometimes (reduced or vestigial bundles, and bundle tips) made up of xylem or phloem alone. In collateral bundles, the xylem, especially when small in amount, may extend laterally about the phloem to a greater or less extent, a condition found in *Equisetum*, in some monocotyledons, and in other herbaceous angiosperms. In bundles of this type the xylem may even be separated into groups.

In some specialized bundles, the cells of both xylem and phloem are reduced in number; fibers and parenchyma cells are few or lacking in each tissue. The phloem of such bundles may consist wholly of sieve tubes and companion cells; the xylem, of vessels and tracheids, or of tracheids or vessels alone. This is the condition in many monocotyledons and in some dicotyledons, such as certain genera of the Ranunculaceae. Under these conditions, the sieve tubes and companion cells are often symmetrically arranged, and of the few vessels one, two, or three are of large diameter. In the bundle of many monocotyledons there are two huge vessels, one on each "shoulder" of the xylem group, as in *Zea;*

in fewer forms, there is one large central vessel, as in *Musa*. In still other genera, there are several vessels much alike in size.

Small bundles are, of course, largely or wholly of primary tissues, and commonly have their cells arranged irregularly. In many dicotyledons even the primary vascular tissues are arranged in radial rows, for example, in *Trifolium* (Fig. 140A) and *Asclepias*. Where the bundle has a small amount of secondary tissues, the central part usually shows weak radial arrangement and the inner and outer cells are irregularly arranged.

Just as the proportion of primary and secondary tissues varies, so

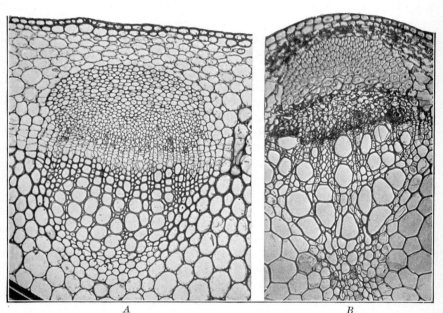

FIG. 140.—Collateral dicotyledonous vascular bundles. *A*, *Trifolium pratense*, young bundle, secondary tissues in early stages, primary xylem radially arranged, bundle cap immature; *B*, *Pisum sativum*, mature bundle, cell arrangement distorted by vessel development.

does that of protoxylem and metaxylem, the former being commonly large in amount in small stem and leaf bundles. The extent to which protoxylem lacunae are formed likewise varies; such spaces are especially large in many of the monocotyledons.

Strands or sheaths of fibers are often associated with conducting cells of xylem and phloem to form a complex conducting and supporting structure that was long aptly called a "fibrovascular bundle." The term was extended to bundles without caps or sheaths and came to be applied loosely to all vascular strands, simple or complex. Morphologically the fibrous part of the bundle may constitute a part of the vascular tissues—bundle-cap fibers in some plants are a part of the protophloem—or may lie outside those tissues. The term *"vascular bundle"* has supplanted the older term.

The association of fibers morphologically external to the bundle with the vascular tissues is apparently connected with the mechanical relations of support for the organ concerned, and support and protection for the softer conducting cells. The fibrous cells associated with vascular bundles—whether a part of the phloem, pericycle, or cortex—commonly form "caps" or crescent-shaped masses (in cross section) external, or both external and internal, to the bundle (Figs. 135*D*,*F*, 139*A*,*D*). These caps may be united laterally by flanges of similar cells, so that a complete supporting and protecting cylinder is formed. Where the bundles are rather widely separated, and especially where they are much reduced and specialized, they are often completely sheathed by fibers (Fig. 139*D*). This is the condition in the stems of many monocotyledons and other herbaceous forms.

The term "vascular bundle" is used in studies of physiological anatomy to denote bundles without fibrous sheaths and in which the xylem and phloem themselves contain no fibers. In the physiological sense the term "fibrovascular bundle" indicates a bundle whose xylem and phloem contain fibers as a constituent part of these tissues, whether or not fibers external to these conducting tissues are also present.

ANOMALOUS STRUCTURE IN STEMS

Though the great majority of plants possess stelar structure of the type considered normal, many have unusual structure. This unusual structure is of many different types, which, however, fall into two groups: (*a*) those in which a cambium of normal type and persistence, by peculiarity or irregularity in its activity, develops vascular tissues of unusual arrangement and proportion of xylem and phloem; (*b*) those in which the cambium, and, consequently, the secondary xylem and phloem, is abnormally arranged, or in which the original cambium is replaced by other cambium layers secondarily formed. These additional cambia also may be of unusual extent and arrangement. To the many combinations of unusual structure that are produced by these modifications is added the anomaly of the presence of medullary and cortical bundles. As a result, extremely complex structures are formed. The interpretation of all such conditions can usually readily be made by ontogenetic studies, either of the developing axis itself or of the seedling. Descriptions of a few of the many varieties of anomalous structure will give an idea of the principal types of departure from the normal.

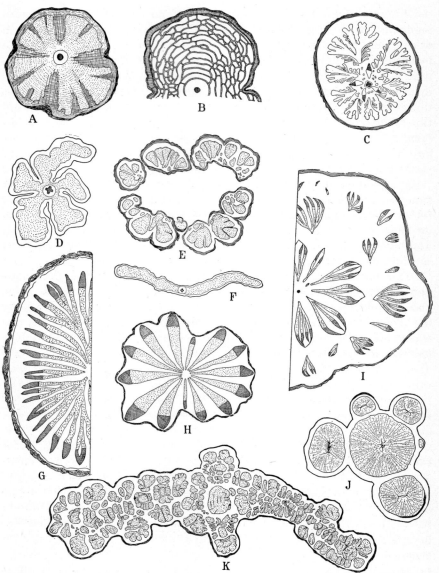

Fig. 141.—Anomalous stem structure—liana types. *A*, A Bignoniaceous genus; *B*, *Securidaca lanceolata*; *C*, *Bignonia* sp.; *D*, *Bauhinia rubiginosa*; *E*, *Serjania ichthyoctona*; *F*, *Bauhinia* sp.; *G*, *Aristolochia triangularis*; *H*, *Begonia fruticosa*; *I*, *Piper fluminense*; *J*, *Thinouia scandens*; *K*, *Bauhinia Langsdorffiana*. (*After Schenck.*)

Where a cambium, normal in position and activity, forms in some segments much larger proportions of xylem than of phloem, and in others more phloem than xylem, a ridged and furrowed xylem cylinder is formed. This may be of simple structure, as in Fig. 141A, or very complex, as in Fig. 141C. In some genera, as in *Aristolochia* (Fig. 141G), segments of the cambium form only ray-like parenchyma and, with increase in diameter, new areas of cambium are constantly given over to the formation of rays of parenchyma. A strongly fluted vascular cylinder is thus formed. Restriction of the activity of the cambium to certain regions also results in the formation of ridged stems (Fig. 141D). Strap-like stems (Fig. 141F) are formed in the same manner.

Stems of other peculiar shapes or types are formed by unusual position of the cambium. This layer, while the stem is young, is thrown into folds or ridges; the tips of the ridges are pinched off and after separation develop "steles" by themselves (Fig. 141J). In some plants the cambium appears originally in several separate strips, each of which surrounds portions, even individual strands, of the primary tissue. A stem with this structure appears to be made up of several fused stems. This apparently compound condition becomes more marked as the stem becomes older and the parts separate as the outer layers of each strand die because of the development of periderm layers (Fig. 141E). In this way there is formed a stem composed of strands lying together more or less like the strands of a rope. A somewhat similar structure is brought about by the breaking into strips of the original cambium cylinder, and even of the vascular cylinder formed by this meristem, by the proliferation of xylem parenchyma (Fig. 141K). Excessive increase of parenchyma in the xylem and phloem ruptures the first-formed, original tissues and the cambium sheet which formed them.

Interxylary Phloem.—Variations of another type in the activity of the cambium produce *interxylary phloem*. Phloem of this type is secondary phloem in the form of strands embedded in secondary xylem. There appear to be two methods by which interxylary phloem comes to lie embedded in secondary xylem. Although it is possible that there is only one method (the second described below), since behavior of the cambium in such growth types has been studied in detail in but few plants. In genera like *Combretum* and *Entada*, small segments of the cambium are said to produce phloem cells toward the inside for a brief period, in place of the xylem cells which are normally produced. After a brief period of such activity these cambium segments return to normal function, and thus bury the inwardly formed phloem with xylem. In other forms, such as *Strychnos*, the interxylary phloem strands are formed by the cambium toward the outside as a part of the normal external phloem, but the strands later become embedded in the xylem in the following manner. Small segments of the cambium cease to function, their cells becoming transformed into mature conducting tissue. New segments of cambium then arise, as secondary meristems, in the phloem some few rows of cells out from the original cambium, or in the pericycle. These unite with the edges of the segments of the general cambium cylinder—which have continued meanwhile their normal activity—and thus enclose a strand of phloem cells. This is repeated in other segments of the cambium so that the secondary xylem possesses numerous scattered strands of embedded phloem.

The formation of phloem embedded in other secondary tissues occurs also in secondary growth in monocotyledons (Chap. VI). Other types of phloem burial are brought about by the development of accessory cambium layers external to the phloem, as discussed in the following paragraphs. In some genera, a combination of these methods occurs.

Accessory Cambium Formation and Activity.—The formation of secondary cambial zones is responsible for many of the unusual types of stems. These meristems commonly develop in the pericycle and function as does a normal cambium, or, when the first cambium has functioned in an unusual manner, repeat this peculiar behavior. Such secondary cambial activity follows the cessation of function of the first layer, one or even many additional layers successively appearing and ceasing to function. Thus a cylinder of alternate concentric layers of xylem and phloem is formed (Fig. 141*B*). The restriction of the extent of the secondary cambium layers to certain narrow parts of the circumference results in the formation of much ridged or flattened stems, and, where the secondary layers form on but one side, or on two opposite sides, of a strap-like stem (Fig. 141*F*).

Of these types of modified stele, the majority are found in plants of special growth habit, many of them lianas; the modifications appear to be largely associated with the habit of the stem and with the mechanical demands upon its structure.

In the Chenopodiaceae, Amaranthaceae, and allied families a somewhat different type of unusual growth is present. Here there is first formed a hollow cylinder of vascular tissue or a ring of irregularly arranged bundles. These bundles are partly of secondary nature, but cambial activity soon ceases and a new, secondary cambium arises in the pericycle just outside the bundles. In some species the cambium forms tissues centripetally, consisting of bundles (similar to those already formed) embedded in nonvascular tissue. This embedding tissue, which has been variously termed *conjunctive tissue, interfascicular tissue,* and *intermediate tissue,* consists of elongate, lignified cells, which in some shrubby forms make a very hard "wood." Centrifugally, the cambium forms a very little parenchyma or no cells at all. The "bundles" formed in this way may be arranged irregularly or in definite concentric rings. In *Chenopodium* (Fig. 142), the phloem is formed centrifugally and later buried by the development of an arc of new cambium formed without it. In this genus this secondary cambium persists, continuing indefinitely to form this complex tissue of embedded bundles; in other genera this first secondary cambium is quickly replaced by others, which in succession form rings of embedded bundles.

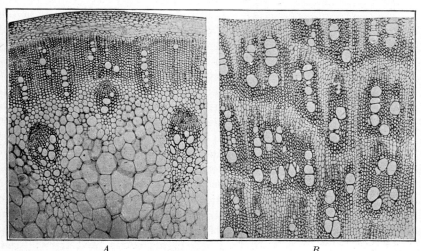

A   *B*

FIG. 142.—Anomalous secondary growth; *Chenopodium album*. *A*, portion of young stem, showing "medullary" bundles (with primary and secondary tissues), within the secondary cylinder; *B*, portion of secondary cylinder, showing alternate bands of xylem, phloem, and "conjunctive tissue," the xylem dark, the phloem in restricted patches, other tissue light.

The activity of the secondary cambium layers in other members of this group of families is more nearly like that of the normal cambium but it is complicated by prolonged primary growth. The development of the common beet root (*Beta vulgaris*) serves as an example of this type (Fig. 143). (The beet "root" consists of root, transition region, and several internodes of stem.) The first cambium forms a ring of bundles close about the primary xylem. A secondary cambium soon arises in the pericycle and this is followed in rapid succession by others originating similarly. All layers continue to function, perhaps indefinitely, though more slowly after an early period of activity. The cambium arises apparently as a continuous band, but forms more or less separate bundles, bands of conjunctive parenchyma developing between the vascular strips. The position of each new cambium, as it arises in the pericycle, is such that it encloses a few layers of pericyclic cells. These multiply and build up a parenchymatous layer as rapidly, or even more rapidly, than the cambium builds up the vascular layer. Alternate bands of proliferated pericycle and of vascular bundles are thus formed. The former constitute the dark-colored, the latter the light-colored, rings in the beet root. The bundles are themselves largely parenchymatous with only a few lignified cells in the xylem. Growth continues throughout all layers, in the bundles apparently both by cambial activity and by proliferation of the parenchyma of the xylem and the phloem. In this way the beet root increases in diameter by growth throughout its layers. The layers are not always complete cylinders and are united irregularly with other layers, so that a complex, asymmetrical structure is formed.

Fig. 143.—Anomalous secondary growth; *Beta vulgaris*, root, cross section. Alternate layers of vascular bundles and proliferated pericycle (cross-hatched); phloem stippled, lignified xylem cells in radial rows, xylem parenchyma and secondary interfascicular tissue unshaded; all layers growing.

Anomalous structure is sometimes due to the presence of medullary and cortical bundles. Such bundles may occur with other structural peculiarities, or in stems otherwise typical in structure. Medullary bundles are infrequent in ferns, as in *Pteridium*.[1] In the dicotyledons, medullary bundles occur in a considerable number of families, such as the Piperaceae, Ranunculaceae, Amaranthaceae, Berberidaceae, Cucurbitaceae. Cortical bundles are of less frequent occurrence; they are known in the Calycanthaceae and Melastomaceae, rarely elsewhere. Many so-called "cortical" bundles are leaf-trace bundles which run downward through the cortex for some distance before entering the stele. This is the condition in *Begonia* and *Casuarina*. In plants with a fleshy cortex, such as many of the Cactaceae, where the leaves are reduced and photosynthesis is carried on largely by the cortex, branches from the base of the leaf traces penetrate the cortical tissues.

[1] This fern, so commonly used to illustrate stem structure, is, because of this and other features of unusual structure, an unfortunate choice. It should be supplanted by typical forms.

The various types of stem and root structure that are commonly described as "anomalous" occur in many families of vascular plants—ferns, cycads, and angiosperms. In the last group, they occur in many families, in some of which all forms are of peculiar structure.

References

ARBER, A.: Studies in the Gramineae, IX. The nodal plexus, *Ann. Bot.*, **44**, 593–620, 1930.

BANCROFT, H.: The arborescent habit in angiosperms, A review, *New Phyt.*, **29**, 153–169, 227–275, 1930.

CHAMBERLAIN, C. J.: Growth rings in a monocotyl, *Bot. Gaz.*, **72**, 293–304, 1921.

CHAUVEAUD, G.: L'appareil conducteur des plantes vasculaires et les phases principales de son évolution, *Ann. Sci. Nat. Bot.*, 9 sér., **13**, 113–438, 1911.

COL, A.: Recherches sur la disposition des faisceaux dans la tige et les feuilles de quelques dicotylédones, *Ann. Sci. Nat. Bot.*, 8 sér., **20**, 1–288, 1904.

COMPTON, R. H.: Theories of the anatomical transition from root to stem, *New Phyt.*, **11**, 13–25, 1912.

DORMER, K. J.: An investigation of the taxonomic value of shoot structure in angiosperms with especial reference to Leguminosae, *Ann. Bot. N.S.*, **9**, 143–152, 1945.

GÉRARD, R.: Recherches sur le passage de la racine à la tige, *Ann. Sci. Nat. Bot.*, 6 sér., **11**, 279–430, 1881.

GWYNNE-VAUGHAN, D. T.: Observations on the anatomy of solenostelic ferns, Part I, *Ann. Bot.*, **15**, 71–98, 1901. Part II, *Ann. Bot.*, **17**, 689–742, 1903.

HÉRAIL, J.: Recherches sur l'anatomie comparée de la tige des dicotylédones, *Ann. Sci. Nat. Bot.*, 7 sér., **2**, 203–314, 1885.

JEFFREY, E. C.: The morphology of the central cylinder in the angiosperms, *Trans. Can. Inst.*, **6**, 599–636, 1899.

———: The structure and development of the stem in the pteridophyta and gymnosperms, *Phil. Trans. Roy. Soc. London*, **195B**, 119–146, 1903.

———: "The Anatomy of Woody Plants," Chicago, 1917.

——— and R. E. TORREY: Physiological and morphological correlations in herbaceous angiosperms, *Bot. Gaz.*, **71**, 1–31, 1921.

LAMOUNETTE, M.: Recherches sur l'origine morphologique du liber interne, Thesis, Fac. Sci., Paris, 1891.

PFEIFFER, H.: Das abnorme Dickenwachstum, *in* Linsbauer, K.: "Handbuch der Pflanzenantomie," IX, 1926.

RÖSELER, P.: Das Dickenwachsthum und die Entwickelungsgeschichte der secundären Gefässbündel bei den baumartigen Lilien, *Jahrb. Wiss. Bot.*, **20**, 292–348, 1889.

SARGANT, E.: A new type of transition from stem to root in the vascular system of seedlings, *Ann. Bot.*, **14**, 633–638, 1900.

SCHENCK, H.: Beiträge zur Biologie und Anatomie der Lianen. *In* SCHIMPER, A. F. W.: "Botanische Mittheilungen aus den Tropen," **4**, Jena. 1892.

SCHOUTE, J. C.: Die Stammesbildung der Monocotylen, *Flora*, **92**, 32–48, 1903.

SCOTT, D. H., and G. BREBNER: On the anatomy and histogeny of *Strychnos*, *Ann. Bot.*, **3**, 275–304, 1889.

———: On the internal phloem in the root and stem of dicotyledons, *Ann. Bot.*, **5**, 259–300, 1891.

SCHWENDENER, S.: Das mechanische Princip im Bau der Monocotyledonen, 1874.

SINNOTT, E. W.: The anatomy of the node as an aid in the classification of angiosperms, *Amer. Jour. Bot.*, **1,** 303–322, 1914.

─── and I. W. BAILEY: Investigations on the phylogeny of the angiosperms, No. 4. The origin and dispersal of herbaceous angiosperms, *Ann. Bot.*, **28,** 547–600, 1914.

SKUTCH, A. F.: Anatomy of the axis of the banana, *Bot. Gaz.*, **93,** 233–258, 1932.

STRASBURGER, E.: Ueber den Bau und die Verrichtungen der Leitungsbahnen in den Pflanzen, Histologische Beiträge III, Jena, 1891.

VAN TIEGHEM, P.: Sur les tubes criblés extralibériens et les vaisseaux extraligneux, *Jour. Bot.*, **5,** 117–128, 1891.

─── and H. DOULIOT: Sur la polystélie, *Ann. Sci. Nat. Bot.*, 7 sér., **3,** 275–322, 1886.

WEISS, J. E.: Das markständige Gefässbündelsystem einiger Dikotyledonen in seiner Beziehung zu den Blattspuren, *Bot. Centralbl.*, **15,** 280–295, 318–327, 358–367, 390–397, 401–415, 1883.

WORSDELL, W. C.: The origin and meaning of medullary (intraxylary) phloem in the stems of dicotyledons, II. Compositae, *Ann. Bot.*, **33,** 421–458, 1919.

CHAPTER XII

THE LEAF

The leaf is a specialized organ in which the function of photosynthesis is centered. In a sense, this is the most important of plant functions, since all other functions depend upon it or contribute to it either directly or indirectly. The structures having to do with absorption and conduction are important chiefly because they supply the leaf with materials used in photosynthesis and remove the products formed in this process. The complex mechanical systems of the larger plants for the most part support large leaf areas in such relation to light that photosynthesis can go on advantageously. Upon the products of photosynthesis, all forms of life, with the exception of some specialized bacteria, depend directly or indirectly for their food supply. The leaf is, therefore, an exceedingly important organ.

Morphology of the Leaf.—The morphology of the leaf varies with different major plant groups. In some of the more primitive groups it is apparently a lateral expansion of the axis in which epidermal, cortical, and stelar tissues take part. The stelar tissues—xylem and phloem, together with accompanying sclerenchyma—form the skeleton upon which the green tissues, which are cortical, are supported. The epidermis is continuous with the epidermis of the stem. The leaf in other groups—ferns and seed plants—is probably a reduced branch system with its members fused in various degrees. In many monocotyledons and some dicotyledons, the leaf is undoubtedly a phyllode, that is, a flattened, expanded petiole. In some other plants, for example *Phyllocladus*, the function of the leaf is carried on by fused branchlets, leaf-like in appearance, but in which the position and orientation of the vascular tissues within the flattened part indicate its branch nature. (Leaf type is further discussed on p. 333 and in Chap. XIV.)

Leaves consist typically of three parts: the expanded portion, or blade, where most of the green tissue is located; the petiole, which supports the blade on the axis and functions also in conduction; the stipules, small, paired lobes at the base. Stipules in many plants are wanting or are soon lost by abscission, but in some plants are persistent structures which may form an appreciable part of the photosynthetic system, as in *Pisum*. Anatomically, stipules are appendages at the base of the leaf, with a vascular supply derived exclusively from the foliar traces.

Leaves are classified as *parallel veined* when the main bundles traverse the leaf without anastomosis, and *net veined* when the main branches of the vascular system form a network. Pinnate and palmate, closed and open venation, and other familiar morphological classifications are also based upon the arrangement of the vascular bundles.

Whatever the arrangement of the larger vascular bundles, the ultimate divisions of the conducting strands of angiosperm leaves completely or partly encircle minute areas of the photosynthetic tissue with which they are in close contact (Fig. 144*B, C*). These divisions of green tissue have been called *vein islets*, and in a way represent more or less well-defined photosynthetic units. The vascular bundles surrounding the vein islets are in intimate contact physiologically with the photosynthetic tissue

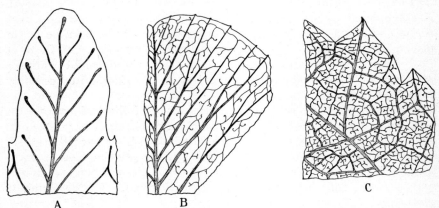

Fig. 144.—Vascular skeleton of leaves. *A*, pinnule of *Aspidium*, no definite vein islets exist; *B*, part of leaflet of *Trifolium*, vein islets rather indefinite; *C*, part of leaf of *Malus pumila*; small, definite islets, a frequent type.

since they lack the sclerenchyma normally present about the larger bundles. The size and shape of the vein islets vary with different types of venation and with different species, and in some plants, especially in the ferns and grasses, definite islets do not exist (Fig. 144*A*). As the leaf matures, the islets increase in size, their growth being a part of the general growth throughout the leaf. In full-grown leaves of any one species, the size of vein islets is apparently fairly constant, regardless of the size of the leaf or the age of the individual plant.

Arrangement of Leaves.—Leaf arrangement on the stem, or phyllotaxy, and variations in size and shape are doubtless in part correlated with the exposure of the photosynthetic surface to light. These variations, although of interest chiefly to taxonomic morphology, are naturally closely bound up with variations in the structure of the plant. Different types of phyllotaxy are related to different arrangements of the leaf traces in the primary vascular cylinder of the plant, and different shapes and sizes of leaf blade are associated with variations in the arrangement and amount of sclerenchyma and of vascular tissues.

Ontogeny of the Leaf.—Leaves originate in the promeristem of the growing point of the stem. The leaf primordium is first evident externally as a rounded or wedge-like protuberance on the side of the promeristem. This is initiated by anticlinal and periclinal divisions in the outer layers of the apical meristem just below the apex. At the tip of the protuberance is a group of meristematic cells that constitute the apical meristem of the leaf. The behavior of this meristem varies with the form and structure of the developing leaf. (See also p. 334.)

In the formation of a simple dicotyledonous leaf this terminal meristem builds up a short, finger-like structure flattened on the adaxial side. If the leaf is stipulate, stipular meristems arise early as swellings at the base of this structure. Lateral ridges next appear except at the base (Fig. 145A). These are the marginal meristems that build the leaf blade. As development of the blade begins, the base of the finger-like structure remains unexpanded and forms an intercalary meristem that builds the petiole. The body of the fingerlike meristematic structure becomes the midrib of the leaf. In pinnately compound leaves the first-formed apical meristem lays down the central axis or rachis of the leaf and the lateral leaflets originate from primordia which arise laterally on this structure and function as terminal meristems in forming the axis of each leaflet. Upon these leaflet axes, marginal meristems form the blades of the leaflets, as in simple leaves.

The leaf blade develops according to a more or less definite pattern which varies only in minor details. The epidermis is formed from the superficial layer of the marginal meristem by anticlinal divisions. Subepidermal cells on both sides of the leaf are formed by anticlinal divisions of subepidermal cells in the marginal meristem. These subepidermal layers remain fairly distinct in the early stages of development, keeping pace with leaf expansion by continuing anticlinal divisions. In later development, the adaxial layer gives rise to the upper layer of palisade cells and the abaxial layer produces the lower layer of spongy parenchyma and even some of the adjacent, more deep-seated layers of mesophyll by periclinal division (Fig. 147).

The central tissues of the mesophyll are usually formed by the periclinal division of subepidermal initials of the marginal meristem. In the inner daughter cells formed in this way, division in all planes gives rise to the central mesophyll, consisting of the inner part of the spongy parenchyma, the vascular bundles, and sometimes to inner layers of

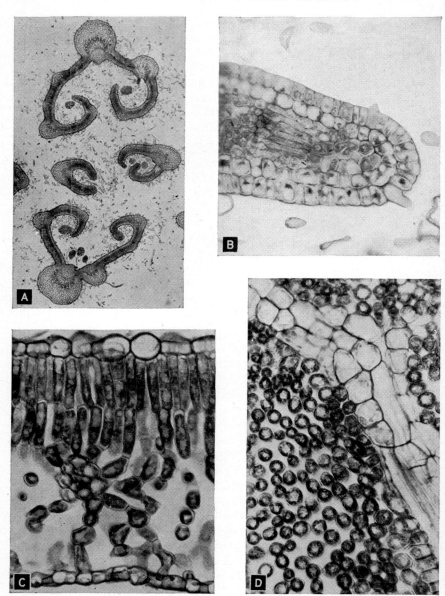

Fig. 145.—Development and structure of the leaf of *Malus pumila*. *A*, transverse section through vegetative shoot above growing point showing successive stages in development of the lamina: first and second leaf (left and right center respectively) show marginal meristems developing along ventral flanks of midrib; in third and fourth leaf (lower and upper respectively), the midrib and primary veins are differentiating; the small round structures are glands. *B*, transverse section of marginal meristem showing epidermal and subepidermal layer formed by anticlinal divisions alone, and central mesophyll by both

palisade tissue (Fig. 147). Periclinal divisions in the central-mesophyll area take place rapidly near the marginal meristem (Fig. 145B) so that close to this meristem the leaf is quickly built up to its maximum number of layers in thickness.

Ontogeny of the Petiole and Stipules.—The development of the petiole takes place simultaneously with the expansion of the leaf blade. Cell division occurs throughout the petiolar meristem followed by rapid elongation.

The ontogeny of stipules varies with the relation of these structures to the leaf base. In the majority of dicotyledons, the stipules are attached to the leaf base. In such plants, the young leaf in early stages is three-pronged, because the stipules develop rapidly from apical and marginal meristems. In dormant buds and near the terminal meristem of young shoots, the stipules may be as large as or larger than the leaf petiole as seen in transverse section. Most temperate-zone trees shed the stipules before or soon after the leaves are mature. Sheathing stipules, stipular thorns, and stipules attached laterally to the petiole have a somewhat different ontogeny.

Duration of Leaf Development.—Apical and marginal growth in leaves is of comparatively short duration in most plants but in fern leaves the apical growing point persists for some time, the tip continuing to develop after the base is mature. In some genera, apical growth continues even for more than a year and very long leaves are formed. In plants other than ferns, apical and marginal growth ceases early and general growth throughout the young leaf continues. The outline and fundamental structure of most leaves are developed while the leaf is yet minute. In the winter buds of many temperate-zone trees, for example, *Liriodendron,* small leaves are present, in shape resembling full-grown leaves, with the main vascular structure outlined, and a considerable part of the cells already present. Later growth consists in large measure of rapid increase in cell size and maturation of the mesophyll cells and vascular tissues, which takes place during the period of leaf expansion. New cells may be formed generally throughout the leaf during this time. After the leaf attains full size, the larger vascular bundles usually increase in diameter by secondary growth. There is, naturally, great difference in the time taken for leaf development in different plants. In most woody plants of the temperate zone, complete expansion and maturity are

---

anticlinal and periclinal divisions: left, differentiation forming a vein; right, a hair developing from epidermal cell. *C,* transverse section of small vein showing bundle sheath surrounding vascular elements and the filaments of palisade and spongy mesophyll cells bending around and making contact with the cells of the bundle sheath. *D,* section parallel to surface of leaf through palisade mesophyll and upper part of vein showing intercellular spaces in palisade mesophyll. (*After MacDaniels and Cowart.*)

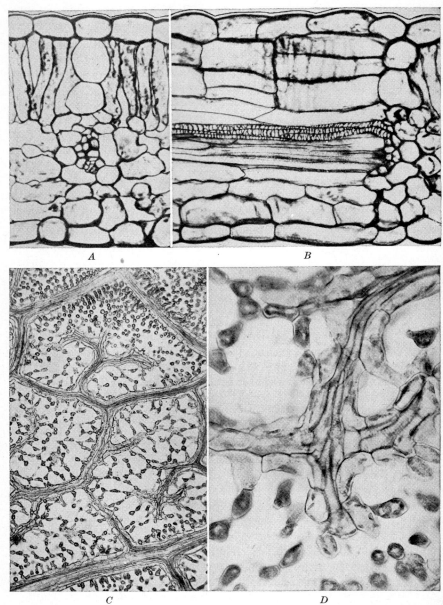

Fig. 146.—Vascular structure of leaf. *A*, transverse section of a minor vein in *Vitis vulpina* showing border parenchyma and vein extensions to epidermal layer; *B*, longitudinal section of vein in *Vitis vulpina* similar to *A* and transverse section of part of connected vein showing cells of vein extension and epidermal cells elongated in direction of vein; *C*, section parallel to surface of leaf of *Malus pumila* through spongy mesophyll showing ramification of vascular bundles, a vein islet (top), and bundle ends; *D*, bundle end in lower right corner of *C* at higher magnification showing the reduction in vascular elements. (*A, B*, after *Wylie*; *C, D*, after *MacDaniels and Cowart*.)

attained rapidly, the actual time depending upon seasonal temperatures. In many tropical plants and in herbaceous forms with very large leaves, development goes on over a longer period. Intercalary leaf meristems are common in leaves of the linear type, occurring in the grasses generally, and in such genera as *Iris, Allium,* and *Pinus.* However, such meristems also persist for only a comparatively short time, with the exception perhaps of those of the anomalous gymnosperm, *Welwitschia,* where they are apparently indeterminate in time of activity.

The Development of Vascular Tissue in the Leaf.—The primary vascular tissues of the leaf blade and petiole form a system continuous with the leaf trace with which they are joined. All parts of this system differentiate from procambium, though the time of maturity differs in the different parts of the leaf. Ordinarily, the first vascular tissue (both

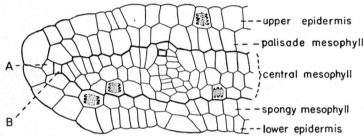

Fig. 147.—Origin of internal layers of leaf. Margin of young lamina of leaf of *Nicotiana tabacum* in transverse section showing anticlinal division of cells in epidermal and subepidermal layers and periclinal and anticlinal divisions in the central mesophyll. *A*, subepidermal initial from which *B* has arisen. (*After Avery, modified.*)

xylem and phloem) of the leaf to mature is the part of the leaf trace near the junction of leaf and stem. Here the tissue often differentiates soon after the formation of the leaf primordium near the growing tip.

Recent evidence indicates that the phloem tissues, particularly the sieve tubes, differentiate acropetally in the procambium strands from the functioning phloem below. No xylem is formed in the developing leaf trace until the progressive differentiation of sieve tube elements reaches the base of the leaf primordium. Xylem differentiation begins at this point, and extends both acropetally and basipetally. The order of maturing of the sieve tube elements may be acropetal from the mature vascular tissue below or may be discontinuous in the protophloem strand. As the leaf primordium elongates and expands, the vascular bundles are extended distally from the trace.

Within the developing leaf blade, the intricate network of veins has its origin in procambium strands that arise from the division of the meristematic cells in the central mesophyll of the young leaf (Fig. 147). Such division may take place close to the marginal meristem or elsewhere in the leaf where cell division is going on. Differentiation of vascular elements may begin in more or less isolated strands and develop in both directions to join with other parts of the vascular system.

In the formation of the conducting strands of the leaf, the same types of cells are developed, and the order is usually the same as in the vascular strands of the stems of the same plant. There is commonly, however, a larger proportion of the extensible protoxylem elements, especially near

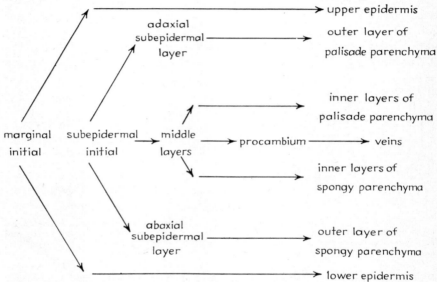

FIG. 148.—Schematic representation of tissue differentiation in the leaf. (*After Foster, modified.*)

the bundle ends (Fig. 146B,D). In leaves with secondary growth, this growth takes place soon after other parts of the leaf attain nearly full size.

Orientation of Vascular Tissue in the Leaf.—In the leaf traces of angiosperms and gymnosperms, before they leave the stele, the phloem is toward the outside of the stem. As the traces pass out of the stele and enter the petiole and blade, the xylem and phloem maintain their relative position, so that commonly in the petiole and usually in the blade, the phloem is on the lower or *dorsal side* of the leaf and the xylem on the upper or *ventral side* (Figs. 146A,B; 154B). Ventral and dorsal are used in the sense that the ventral side is the side next the axis and the dorsal side that away from the axis.

Although the orientation of the xylem and phloem is fairly constant for typical leaf blades, many variations occur in the petiole because of the different methods of fusion, division, or twisting of the leaf traces in their course through the petiole. In many plants, traces that enter the petiole separately remain distinct and pass to the blade without change of structure or orientation. In others, the traces fuse in the petiole to form a single strand of various cross-sectional shape, a hollow cylinder, or a group of more or less complete stele-like cylinders (Fig. 153A,B); in still others, the traces divide into several or many strands which become arranged and oriented in any one of many ways. Many of the bundles

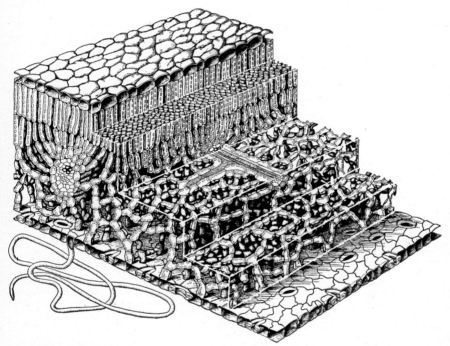

Fig. 149.—Structure of apple leaf showing relation of cells in the various tissues.

so formed may be amphicribral within the petiole. On passing out of the petiole into the blade, the vascular bundles usually again assume collateral structure with dorsal phloem; sometimes, however, the petiolar arrangement is maintained in the larger veins.

In many ferns the traces to a single leaf make up a considerable portion of the stele below the point of attachment. These large traces are in most ferns a part of an amphiphloic siphonostele, and as soon as freed from the stele become amphicribral bundles by the union of the external and internal phloem at the sides of the trace. The external and the internal endodermis also unite about the traces. These bundles pass up the petiole and eventually divide to form the smaller bundles of the compound leaves. Dichotomous branching of the veins is common in the smaller divisions of the leaves of many genera. Ultimately, the minute divisions come into close contact with the mesophyll (Fig. 144A), much as in angiosperm leaves. In small-leaved pteridophytes, for example, *Lycopodium* and *Equisetum*, the vascular supply consists of a single bundle which passes unbranched to the leaf tip.

The vascular supply of stipules is derived from the lateral leaf traces after the leaf traces have left the stele (Fig. 70K). Where the stipules are not attached to the petiole and where they sheath the stem, the stipular traces depart from the leaf traces in the cortex. Where the stipules

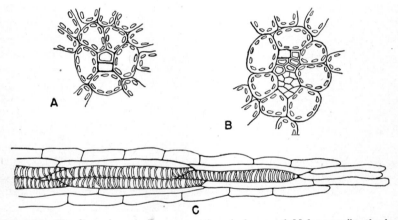

Fig. 150.—Bundle ends. *A, B*, cross sections in leaves of *Malus pumila*, the bundle sheathed by chlorophyll-bearing parenchyma: *A*, small bundle made up of one tracheid and one parenchyma cell; *B*, somewhat larger bundle with phloem. *C*, longitudinal section of *Freesia* petal, the tracheids surrounded by parenchyma.

are attached to the petiole and are an integral part of the leaf, the stipular traces may not depart from the leaf traces until after the latter have entered the petiole. Except in plants with foliaceous, persistent stipules, the stipular traces are usually small vascular bundles; the xylem consists of relatively few conducting elements and the phloem chiefly of parenchyma.

Elements of Xylem and Phloem in Leaves.—With the exception of some monocotyledons both primary and secondary vascular tissues of the petioles and larger veins of leaves resemble the stelar tissues of the same plant in the type of elements present in both xylem and phloem, for example, if scalariform vessels are characteristic of the stelar tissues, they occur in the leaves also. The vessels, sieve tubes, and parenchyma cells of the petioles and larger veins of dicotyledonous leaves are usually smaller than those of the stem and the bundles of these parts have relatively less secondary tissue as compared with primary tissue. The amount of secondary tissue in leaves varies with the species.

As the vascular bundles branch and become progressively smaller, the secondary tissues are reduced in amount until in veins of medium and small size, there is none. At the same time the size of the vascular elements is greatly reduced. In the leaf of a woody dicotyledon like the apple, the smaller veins consist of several close-spiral or reticulate elements accompanied by about the same number of small sieve tubes, companion cells, and parenchyma cells enclosed in the bundle sheath (Fig. 150B). With further reduction the xylem may consist of only a single spiral element (Fig. 150A) and the phloem of a single elongate parenchyma cell. In the transition from phloem consisting of sieve tubes, companion cells, and parenchyma cells to phloem made up of a single parenchyma cell, the sieve tube before its disappearance is reduced in diameter to that or less than that of the companion cell. In the smallest veins, the phloem mother cell fails to divide. The resulting parenchyma cell, which resembles a companion cell in form, has been called a *transition cell*.

Bundle Ends.—The ultimate divisions of the vascular bundles in leaves terminate in what are known as "bundle ends." In leaves that have more or less definite vein islets, the bundle ends bend into the mesophyll of the islet and end abruptly near its center. Some bundle ends may be somewhat enlarged at the tip or branched in various ways (Fig. 146C,D). In leaves with parallel veins, for example, the leaves of grasses, or in leaves without definite vein islets, the bundle ends may be merely short spurs from the veins extending into the mesophyll. The bundle ends and the small veins forming the vein islets supply the mesophyll with water and nutrients and absorb and remove the products of photosynthesis. Because of the distribution of the vein islets and bundle ends the distance through which water and materials in solution travel through the mesophyll is always about the same. In the apple leaf, for example, the distance between bundle ends or between bundle ends and veins bordering the vein islets is about 88 micra.

The structure of bundle ends varies somewhat with the species. Probably in mesophytic dicotyledons the bundle end commonly consists of a single spiral or reticulate xylem element accompanied by a single specialized parenchyma cell, the two elements surrounded by the parenchymatous bundle sheath. Such a bundle end in the leaf of the apple is shown in Figs. 150A, 146D. The bundle ends of hydathodes and some glands, especially the glands of insectivorous plants (Chap. XIV),

328 AN INTRODUCTION TO PLANT ANATOMY

which function in digestion, have more elaborate structure than typical bundle ends.

The Bundle Sheath.—With but few exceptions, notably in some aquatic plants and delicate ferns, the vascular bundles of leaves are enclosed in a structure known as the bundle sheath. The elements forming this sheath are parenchyma cells elongated parallel with the vein axis (Figs. 146C, 150B) and joined laterally in such a way as to surround the vascular tissue with a tight sheath resembling an endodermis but lacking

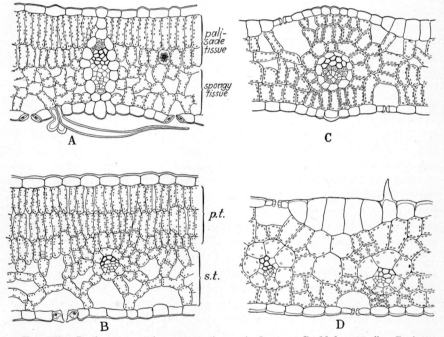

Fig. 151.—Leaf structure, in cross section. A, Quercus; B, Malus pumila; C, Avena sativa; D, Zea mays. A, B, with palisade and spongy parenchyma; C, D, without differentiated mesophyll.

Casparian strips. In many species, for example, Malus pumila, these cells contain chloroplasts and resemble the adjacent mesophyll cells except that they are more regular and elongated in shape and contain relatively fewer plastids than the mesophyll cells. In other species, for example, Carya Pecan, the bundle sheath cells are more sharply differentiated from the mesophyll in that they take a different stain and do not contain chloroplasts.

The cells of the bundle sheath are in intimate contact on their inner faces with the conducting elements of the vascular bundle or adjacent parenchyma and on the outer face with the palisade and spongy mesophyll cells which commonly are joined to form filaments ending at the bundle sheath (Fig. 145C). In some veins, flanges or spurs from

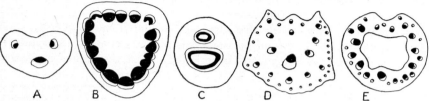

Fig. 152.—Vascular supply of petioles. Transverse sections through distal end of petiole showing *A, Asarum canadense*, orientation and number of bundles unchanged; *B, Spiraea Lindleyana*, bundles forming a ring; *C, Corylus Avellana*, bundles forming two stele-like cylinders; *D, Sagittaria sagittaefolia* and *E, Angelica*, division of bundles into numerous bundles. (*After Petit*.)

the bundle sheaths extend upward and downward to the upper and lower epidermis (Fig. 151A).

Physiologically considered, the bundle sheath forms a layer of living cells surrounding the vascular bundles through which water and solutes

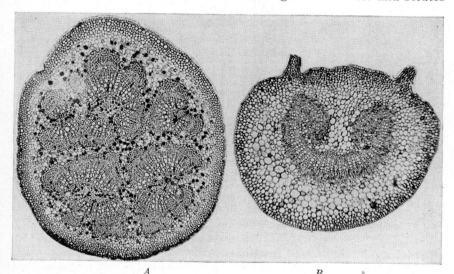

Fig. 153.—Structure of petioles. *A, Populus grandidentata*, vascular tissue arranged in cylinders, simulating steles; *B, Cephalanthus occidentalis*, vascular tissues in a crescent with ends recurved.

must pass to reach the tracheids, vessels, and sieve tubes. The fact that often the sheath is connected with the epidermis suggests that it may conduct materials to the epidermis also. In bundles with well-developed fibers above and below the vascular tissue, the bundle sheath has contact with the mesophyll on the sides of the bundle only (Fig. 157B).

The morphological nature of the bundle sheath is uncertain. The nature of the cells and their contents suggest that it is a part of the mesophyll which it resembles.

Mechanical Support in the Leaf.—Considered mechanically, the dicotyledonous leaf is a complex network of strands which supports a relatively large blade or photosynthetic surface at the end of a slender

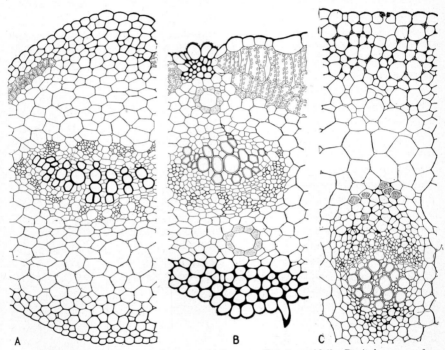

Fig. 154.—Collenchyma in leaves. A, *Ipomoea Batatas*, midrib; B, *Apium graveolens*, vein of leaflet; C, *Rheum Rhaponticum*, petiole. (*Courtesy of E. F. and E. Artschwager.*)

petiole. In this support, collenchyma, sclerenchyma, and the woody xylem play an important part. In early stages of development, collenchyma is of major importance in supporting the developing petiole and leaf blade. Collenchyma is most abundant in the outer cortical layers of the petiole and the larger veins (Fig. 154A,B). It is also commonly associated with veins of intermediate size where it lies above the vascular tissue. The turgor of the parenchyma itself is also important in support during the developmental stages of the leaf.

In dicotyledonous leaves fibers are usually associated with the vascular tissues of the petiole, where they occur as bundle caps adjacent to the phloem. Groups of fibers are found also on both the ventral and dorsal sides of many of the larger veins. In this position, often in combination with collenchyma, these fibers above and below the vascular bundles build up vertical plates of supporting tissue extending through the leaf and connecting the upper and lower epidermis (Fig. 157B). This so-called "I-beam" structure is considered most efficient in strengthening the leaf, and much has been written emphasizing the theory that this arrangement of mechanical tissues provides maximum strength with a minimum amount of supporting tissue.

The veins with their sclerenchyma make a network of plates through the leaf dividing the mesophyll into more or less definite compartments. Within these compartments, the upper epidermis supports the palisade tissue which is in a very real sense suspended from it. The upper epidermis is in turn supported by the plates of mechanical tissues extending through the leaf. As the larger bundles divide to form successively smaller and smaller vascular strands, the amount of sclerenchyma becomes much reduced, until in the smaller bundles mechanical elements are lacking. Some leaves are further strengthened or stiffened by large branching sclereids, scattered through the mesophyll, or by lignified epidermal and subepidermal cells.

Considered morphologically, the supporting elements of leaves and petioles may be formed from any of the tissues. Commonly, however, the fibrous strands or sheaths that are closely associated with the vascular bundles are either phloem fibers or pericyclic fibers. Occasionally, they are derived from xylem or from cortical tissues.

The Mesophyll.—The photosynthetic tissue between the upper and lower epidermis consists typically of thin-walled parenchyma known as *mesophyll*. This tissue usually forms the larger part of the substance of the leaf. The cells of the mesophyll show great variation in shape and arrangement, but, in general, they are grouped in two classes: the *palisade parenchyma*, or *palisade cells*, and the *spongy parenchyma*, or *spongy mesophyll* (Figs. 145C, 151A,B).

In the palisade parenchyma the cells are elongate and more or less cylindrical and arranged in one or more rather regular, relatively compact layers near the ventral, or upper side of the leaf with the long axis of the cells perpendicular to the leaf surface. The cells, which in transverse sections of leaves appear to be closely packed together, are really usually separate from each other or at least exposed to air spaces over a part of their surface (Figs. 145D, 149). In many species, these elongate cells are joined end to end to form filaments which connect with the upper epidermis at one end and with the bundle sheath at the other (Fig. 145C).

In leaves that stand more or less vertically or in a drooping position, palisade parenchyma may occur on both sides. A frequent modification of the columnar palisade cell is the cone-shaped type, which lies with its larger end against the epidermis. In many leaves, especially those of plants growing in water or in shaded situations, there is no well-developed palisade parenchyma; this is also the condition in the leaves of most gymnosperms (Fig. 182C), of grasses (Fig. 183C), and of other specialized types. The number of palisade layers and the density of the cell structure in those layers depend largely, either directly or indirectly, upon light intensity. There may, therefore, be great variation in the proportion and arrangement of the palisade parenchyma in the same species growing under different conditions. The mesophyll structure of leaves from different parts of the same plant also shows considerable variation depending on amount of light and other factors.

In the spongy parenchyma, the cells lack regularity in shape and are arranged loosely, so that a large part of their surface is exposed to the gases in the intercellular spaces. Some cells are very irregular in shape, with radiating arms connecting with the arms of similar cells, thus making an irregular network of green tissue (Fig. 146C). In some species, the spongy mesophyll cells form a network of filaments that arch over the stomatal regions and connect directly with the bundle sheaths (Figs. 145C, 149).

The lens-shaped chloroplasts in mesophyll cells are usually located next to the cell wall with the sides of the disk parallel with the wall. In elongate palisade cells the plastids may be arranged in several vertical rows. Chloroplasts can change their shape with changes in light intensity and other factors. With light of low intensity striking the leaf surface from above, the plastids may become more nearly spherical and thus expose a relatively large surface to the light. With intense light the plastids may become flattened so that light strikes only their thin edges. Other variations in chloroplast orientation have been reported.

The internal structural conditions in the leaf are advantageous to the function of photosynthesis. Briefly, these are, in part, the exposure of a large number of chloroplasts to sunlight, the exposure of a large cell-membrane surface to the intercellular spaces where interchange of gases takes place, and such an arrangement of cells in relation to each other and to the vascular bundles that the products of photosynthesis can be rapidly removed and the cells supplied with water and mineral nutrients.

Epidermis of Leaves.—The epidermis envelops the tissues of the leaf with a protective layer broken only by the stomata and the hydathodes. The general nature of this layer, the shape and size of the cells, and the structure and mechanism of stomata have been discussed in Chap. V.

The structure and the occurrence of hairs are also discussed in the earlier chapter. Cutinization and cuticularization have been treated in Chap. II. The epidermis of leaves is, naturally, of the greatest importance because of the nature of the soft mesophyll tissues within, which become seriously desiccated almost immediately upon injury to the protective layer. The modification of the epidermis which occur in plants growing in to different environments, especially in xerophytes, are discussed in Chap. XIV. In addition to protection from desiccation, the epidermis supports the palisade tissues and probably conducts solutes between the mesophyll cells and the veins.

Distribution of Stomata on Leaves.—In dicotyledonous leaves of the broadly expanded type, stomata normally occur in largest numbers on the dorsal surface. Some species have comparatively few on the ventral surface and some lack them on the upper surface. Plants with leaves that stand in a nearly vertical position may have stomata about equally distributed on both surfaces. Many of the monocotyledons have leaves of this type. Floating leaves of aquatic plants have stomata on the upper, exposed surface only, and submersed leaves lack stomata. In general, the distribution of the stomata varies with the environment in which the plants are living; those living under more severe conditions of water loss have stomata in the more protected positions.

In the majority of leaves, stomata are arranged apparently without regular pattern of orientation. They are spaced more or less equidistantly from each other and are only rarely found upon the veins. In some plants, for example, the grasses, stomata are in regular rows, all oriented in the same way. Xerophytes and some other plants often have stomata in groups or rows or sunken in furrows or otherwise protected (Fig. 180*C,D*). The number of stomata per unit area of epidermis varies greatly. Some plants, mostly xerophytes, have as few as 10 to 15 per square millimeter. At the other extreme, species of *Spiraea* have been reported with as high as 1300 stomata per square millimeter. Between these extremes, all numbers are found, the common number being about 400 per square millimeter, as in the apple leaf under ordinary mesophytic conditions. Apparently, the number of stomata per unit area bears little definite relation to habitat; other factors are of greater importance.

The Monocotyledonous Leaf.—Much of the description in the foregoing pages of this chapter applies primarily to the leaves of dicotyledonous plants. The monocotyledons as a group show greater diversity of specialized leaf types. The leaves of this group are not made up of stipules, petiole, and blade but in most species are phyllodes, petioles that have become flattened and function as leaf blades. In general, monocotyledonous leaves are parallel veined. Many apparent exceptions occur but these often can be shown by their venation to have been derived phylogenetically from parallel-veined types. Thus in the broad-bladed leaves of the skunk cabbage (*Symplocarpus*) and other Araceae, the many, scattered, separate vascular bundles traverse what is apparently the petiole and diverge directly into the blade-like structure at narrow angles and continue, more or less nearly parallel with each other, to the leaf margin. In the petiole of a dicotyledonous leaf of comparable size and shape, that of *Arctium*, the relatively few bundles, are arranged in a V, and give off branches which diverge from the midrib at a relatively wide and definite angle.

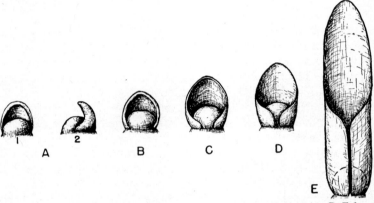

Fig. 155.—Stages in the ontogeny of a monocotyledonous leaf. *A1*, *B–E* face view; *A2* side view. *A*, showing emergence of protuberance on the side of the apical meristem; *B–D*, showing extension of protuberance laterally around apical meristem to form collar or sheath; *E*, showing enclosure of apical meristem and elongation of "blade" and sheath by intercalary meristems. (*Based on Trécul.*)

The monocotyledonous leaf has its origin in the primary meristem of the axis as a lateral protuberance resembling the meristem which forms the dicotyledonous leaf (Fig. 155*A*). Very soon, however, this leaf primordium extends laterally to form a ring or collar of meristematic tissue (Fig. 155*B–D*), which functions as an intercalary meristem building up the leaf from below. The apical leaf meristem is therefore relatively unimportant as compared with its counterpart in the dicotyledonous leaf in which it builds the foundation of the leaf blade.

In the development of the leaf, the median part of the meristematic collar grows much faster than the other parts and forms the "petiole" and blade-like part. This ring-shaped meristem naturally gives rise to a sheathing leaf base which in greater or less degree encloses the growing point of the stem (Fig. 155*D,E*). Examples of unusual leaf types of the monocotyledons are the tubular leaves characteristic of some species of *Allium* and *Narcissus*. The flattened leaves of *Iris*, V-shaped at the base, show bundles on the one side inverted with relation to those on the other (Fig. 156). The inversion of bundles may have come about by the longitudinal folding and fusion of the halves of a leaf, as in *Iris*, or by the flattening of a tubular leaf.

Of special interest are the leaves of the palms. Here are found the largest angiosperm leaves; some species have an overall leaf length of 25 ft. The compound leaves, both pinnate and palmate, are apparently the result of a splitting of the leaf blades as they expand. The leaves of

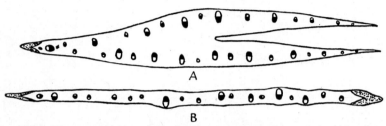

Fig. 156.—Monocotyledonous leaf, *Iris;* transverse sections of sheathing base (*A*) and upper part (*B*). *A*, showing folded leaf; *B*, showing sides of leaf fused and some of the vascular bundles inverted. (*After Arber.*)

the banana and other Musaceae are split by the wind to form what are essentially pinnately compound leaves.

The Grass Leaf.—The grass leaf merits special treatment because of the botanical and economic importance of the grass family and its complex leaf structure. The leaves of the grasses are alike in general structure, with variations resulting from adaptation to xerophytic or other ecological conditions. Fundamentally, the grass-leaf blade consists of a skeleton of parallel vascular bundles between which the mesophyll is supported. The bundles are of nearly the same size throughout their length and, except for occasional small connecting strands, are separate from one another. Frequently, the leaf blade has a large median bundle associated with a pronounced midrib projecting on the dorsal side. The other bundles may be of two or three sizes differing chiefly only in the amount of mechanical and conducting tissue present. The smaller bundles are spaced alternately between the larger. The entire leaf consists of a sheathing base and a linear blade; the former surrounds the culm for some distance and merges in a more or less prominent joint with the leaf blade, which is set at an angle with the culm. In many species the sheathing base extends above the joint in a short membranous structure called the ligule.

Sclerenchyma in the Grass Leaf.—Sclerenchyma in grass leaves is usually associated with the vascular bundles. In the larger bundles there are often two strands of fibers, one above and one below the vascular tissues. These join with the xylem and phloem to form a plate of hard tissue extending through the leaf (Fig. 157B). In smaller veins the strand of fibers may be on the dorsal side of the leaf only and joined either with the vascular tissue or free from it (Fig. 157B). In some species, strands of fibers occur beneath the epidermis on both sides of the bundle but free from it (Fig. 157D). Small bundles may lack sclerenchyma

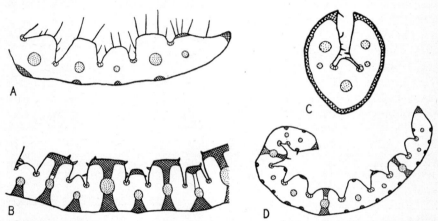

Fig. 157.—Types of sclerenchyma distribution in the grass leaf. *A, Festuca duriuscula*, showing sclerenchyma strand on the margin of the leaf and strands on the dorsal side free from the vascular bundles; *B, Elymus arenarius*, showing strands of sclerenchyma joined above and below with vascular bundles forming I-beam plates of tissue through the leaf; *C, Festuca ovina*, showing sclerenchyma over the dorsal surface only; *D, Agrostis canina*, showing strands of sclerenchyma on margins and on both sides of the leaf but free from the vascular bundles except for occasional I-beam plates. Vascular bundles, stippled; sclerenchyma, crosshatched. (*After Lewton-Brain.*)

altogether. There is thus great variation in the amount and position of the fibrous strands. In addition to the sclerenchyma associated with the bundles, nearly all grasses have a strand of fibers along each edge of the leaf (Fig. 157A,D). In leaves of some xerophytes, hard tissues may extend over the entire dorsal side (Fig. 157C).

Bundle Sheaths in the Grass Leaf.—Among the characteristic features of grass leaves are the specialized tissues immediately surrounding the vascular bundles. These occur commonly as two layers of cells, each one cell thick, completely surrounding the bundle next to the vascular tissues. The inner sheath is more or less thick-walled and lignified, resembling an endodermis in general appearance. Whether this is an endodermis in the strictly morphological sense is doubtful.

THE LEAF

The thickening of the cell walls of this layer is frequently heavier on the inner tangential and radial than upon the outer tangential walls, although in some plants it extends entirely around the cells. The inner walls are pitted and presumably allow transfusion between the vascular tissue and this layer. Various functions have been ascribed to this layer, but protection against crushing, especially of the softer phloem cells, seems most probable. In bundles with well-developed sclerenchyma, this lignified sheath abuts directly upon the fibrous strands, thus adding rigidity to the whole structure by completing the so-called "girder structure." In some species, the inner sheath is absent or present in the larger bundles only.

The outer sheath, often known as the *mestome sheath*, is made up of thin-walled parenchyma cells, which appear mostly isodiametric in transverse sections of the leaf, but much elongated in longitudinal section. These cells form about the bundle a parenchymatous girdle, to which has been ascribed the function of conducting the soluble food products from the photosynthetic tissues to the conducting tissues. In the larger bundles of some species the cells of this girdle lack chlorophyll, and are hence readily recognized. In the smaller bundles the cells may contain chlorophyll (Fig. 151C), but usually in less amount than other photosynthetic cells. In some species the cells of this sheath extend to the epidermis on one or both sides of the bundle.

Mesophyll of the Grass Leaf.—In the leaves of grasses, as in other leaves, the photosynthetic tissue fills the space between the vascular bundles more or less solidly. Usually there is no well-developed palisade layer of elongate cells such as is characteristic of dicotyledonous leaves generally (Fig. 151D). Sometimes, however, there is a weakly developed palisade layer next to the epidermis on one or both sides of the leaf. The cells of such a palisade layer are nearly isodiametric and differ from the spongy mesophyll cells chiefly in their more compact arrangement. The cells in the spongy mesophyll are also frequently irregular in shape, and arranged in branching conduction systems, as in analogous tissues of other types of leaves. In a few xerophytic grasses and some species of *Cyperus* the green tissues themselves are limited to a sheathing girdle about the vascular bundle.

Epidermis of the Grass Leaf.—Although variations in the structure of the epidermis of the grasses often occur, especially in plants adapted to growth in extremes in environmental conditions, the general features are fairly constant. The epidermal cells are elongate-rectangular in shape (Fig. 79I) and are nearly square in transverse section. They are arranged in regular rows extending lengthwise of the leaf. Over the vascular bundles the cells are often smaller and thick-walled, to some extent resembling sclerenchyma. Great variation occurs in the extent of cutinization. In surface view the cells do not show marked irregularity of outline as does the epidermis of many dicotyledons.

A type of cell common to the leaves of nearly all grasses is the so-called *motor cell*, or *bulliform cell*, which, apparently, constitutes the mechanism that functions in the rolling of grass leaves in dry weather (Fig. 151D). These cells have much more depth than the ordinary epidermal cells and are arranged in rows extending throughout the length of the leaf upon its upper surface, frequently lying at the bottom of well-defined grooves (Fig. 157B,D). The cells are thin-walled and lack chlorophyll, and with decrease in their turgor the leaf rolls upward and inward. Some species possess only one or two rows of this type of cells, whereas others have many. Few species lack them. At the edges of the groups of motor cells, as seen in cross section of the leaf, there are cell types transitional between typical motor cells and the ordinary epidermal cells.

Arrangement of the Stomata in the Grass Leaf.—The type of stoma found in the leaves of the grasses is surprisingly uniform, considering the diversity of habitat in which grasses are found. The stoma of the grass type is described in Chap. V (Fig. 79I,J,K). The stomata are elongate with their long axes parallel with that of the leaf and the epidermal cells surrounding them. They are usually arranged in rows alternating with rows of epidermal cells. Frequently several rows of stomata are spaced close together between wider bands of epidermal cells. Although in a considerable number of species stomata occur upon both surfaces of the leaf in approximately equal numbers, in the majority of species these openings are more numerous in the upper epidermis and in some species, chiefly xerophytic, may be wanting upon the lower side.

Various types of hairs are found in the epidermis of many species. Short, stiff projections, which give the surface of the leaf a harsh texture, are particularly common. In some forms these teeth give the leaf margin an effective cutting edge, as, for example, in species of *Leërsia*.

Persistence of Leaves.—A large proportion of the gymnosperms, many tropical and some broad-leaved, temperate-zone angiosperms, retain their leaves for more than one season. In evergreen plants possessing secondary growth, as in the gymnosperms and some dicotyledons, the persistence of leaves for more than one year involves the lengthening of the leaf trace as successive annual layers of xylem are added by the cambium. This lengthening is accomplished by the activity of a special meristematic layer in the trace itself (Chap. VI). Ordinarily, the needle leaves of the gymnosperms and the broad leaves of angiosperms do not persist more than three to five years, after which time they are cut off by abscission layers in the same manner as are deciduous leaves. In a few genera, evergreen leaves may persist for many years as, for example, on the trunks of *Araucaria*. In most ferns, some of the palms, and other monocotyledons the leaves persist for a considerable number of years, the older leaves gradually ceasing to function. In many of these plants, particularly some palms, the leaves are not cut off by abscission layers but cling to the stem until they disintegrate by weathering. The leaves of many annual and perennial herbs are also "withering persistent." In the majority of temperate-zone plants, both woody and herbaceous, the leaves function for only a single season at the end of which they wither or are abscised. A discussion of leaf abscission is to be found in Chap. IX.

References

ARBER, A.: The phyllode theory of the monocotyledonous leaf, with special reference to anatomical evidence, *Ann. Bot.*, **32,** 465–501, 1918.

AVERY, G. S.: Structure and development of the tobacco leaf, *Amer. Jour. Bot.*, **20,** 565–592, 1933.

BOWER, F. O.: On the comparative morphology of the leaf in the vascular cryptogams and gymnosperms, *Phil. Trans. Roy. Soc. London*, **175,** 565–615, 1884.

CLEMENTS, E. S.: The relation of leaf structure to physical factors, *Trans. Amer. Micro. Soc.*, **26,** 19–102, 1904.

COLOMB, G.: Recherches sur les stipules, *Ann. Sci. Nat. Bot.*, 7 sér., **6,** 1–76, 1887.

COPELAND, E. B.: The mechanism of stomata, *Ann. Bot.*, **16,** 327–346, 1902.

CROSS, G. L.: The origin and development of the foliage leaves and stipules of *Morus alba*, *Bull. Torrey Bot. Club*, **64,** 145–163, 1937.

DE CANDOLLE, C.: Anatomie comparée des feuilles chez quelques familles de dicotylédones, *Mém. Soc. Phys. et d'Hist. Nat. Genève*, **26,** 427–480, 1879.

DUFOUR, L.: Influence de la lumière sur la forme et la structure des feuilles, *Ann. Sci. Nat. Bot.*, 7 sér., **5,** 311–413, 1887.

EICHLER, A. W.: Zur Entwickelungsgeschichte des Blattes mit besonderer Berücksichtigung der Nebenblatt-Bildungen, Marburg, 1861.

FLOT, L.: Recherches sur la naissance des feuilles et sur l'origine foliaire de la tige, *Rév. Gén. Bot.*, **17,** 449 passim 535, 1905. **18,** 26 passim 508, 1906. **19,** 29 passim 192, 1907.

FOSTER, A. S.: Leaf differentiation in angiosperms, *Bot. Rev.*, **2,** 349–372, 1936.

GERRESHEIM, E.: Ueber den anatomischen Bau und die damit zusammenhängende Wirkungsweise der Wasserbahnen in Fiederblättern der Dicotyledonen, *Bibl. Bot.*, **81,** 1–66, 1913.

HABERLANDT, G.: Vergleichende Anatomie des assimilatorischen Gewebesystems der Pflanzen, *Jahrb. Wiss. Bot.*, **13,** 74-188, 1882.

LEWTON-BRAIN, L.: On the anatomy of the leaves of British grasses, *Trans. Linn. Soc. Bot. London*, 2 ser., **6,** 315–359, 1903.

MACDANIELS, L. H., and F. F. COWART: The development and structure of the apple leaf, *Cornell U. Agr. Exp. Sta. Mem.*, **258,** 1–29, 1944.

PÉE-LABY, E.: Étude anatomique de la feuille des graminées de la France, *Ann. Sci. Nat. Bot.*, 8 sér., **8,** 227–346, 1898.

PETIT, L.: Nouvelles recherches sur le pétiole des phanérogames, *Actes Soc. Linn. Bordeaux*, **43,** 11–60, 1889.

Rippel, A.: Anatomische und physiologische Untersuchungen über die Wasserbahnen der Dicotylen-Laubblätter mit besonderer Berücksichtigung der handnervigen Blätter, *Bibl. Bot.*, **82**, 1–74, 1913.

Sinnott, E. W., and I. W. Bailey: Investigations on the phylogeny of the angiosperms, 3. Nodal anatomy and the morphology of stipules, *Amer. Jour. Bot.*, **1**, 441–453, 1914.

Smith, G. H.: Anatomy of the embryonic leaf, *Amer. Jour. Bot.*, **21**, 194–209, 1934.

Strasburger, E.: Ein Beitrag zur Entwicklungsgeschichte der Spaltöffnungen, *Jahrbuch. Wiss. Bot. Pringsheim.*, **5**, 297–342, 1866/1867.

Trécul, A.: Mémoire sur la formation des feuilles, *Ann. Sci. Nat. Bot.*, 3 sér., **20**, 235–314, 1853.

Tyler, A. A.: The nature and origin of stipules, *Ann. N.Y. Acad. Sci.*, **10**, 1–49, 1897.

Wylie, R. B.: The role of the epidermis in foliar organization and its relations to the minor venation, *Amer. Jour. Bot.*, **30**, 273–280, 1943.

CHAPTER XIII

THE FLOWER—THE FRUIT—THE SEED

THE FLOWER

Anatomically, the flower is a determinate stem with crowded appendages, with internodes much shortened or obliterated. The appendages are of leaf rank but differ from those of the vegetative stem in function and appearance, and the uppermost ones are often so placed as to appear terminal on the axis. But the flower is a typical stem differing in no fundamental way structurally or ontogenetically from the normal stem with leaves.

Ontogeny of the Flower.—The flower develops from an apical meristem that differs in only minor ways (Chap. III) from that of a leafy stem. The early stages of floral-appendage development are closely like those of leaves. All organs develop laterally on the apical meristem, initiated by periclinal divisions below the surface layer. The sequence of development of the whorls of organs varies but is commonly acropetal. All organs have, at least for a time, an apical meristem, and those that are laterally expanded have marginal initials. Procambium develops acropetally in all organs as does the phloem. The xylem matures either acropetally or both acropetally and basipetally from one or more points of beginning near the base of the organ. In carpels, especially in syncarpous ovaries, connection of the early-formed xylem with the xylem of the receptacle may be long delayed, even until the fruit is partly developed. The carpel remains an immature organ until the fruit is mature. In various parts it retains meristematic tissues of different types. The other organs are usually mature in the flower. Where fusion is present, the fusion may be ontogenetic, as it is frequently between carpel margins, or it may be phylogenetic, the fused organs arising from a common meristem, as in typical inferior ovaries. The latter type of union has been called "congenital fusion." In adjacent organs that arise as independent organs from independent apical meristems, for example, the members of a gamopetalous corolla, the initiating regions may shift from the apex to the base and merge, and the organs then continue growth as a single structure with only the vascular tissue showing its compound nature. Adnate organs, in any number, may arise independently but later continue growth from a single meristematic region as a fused but superficially single structure. Such apparently simple structures may show evidence of their compound nature in the presence of independent vascular bundles for every organ within the common structure, or the vascular skeletons of the organs may also be so intimately fused as to show little or no evidence of their multiple morphological nature. The structure of mature floral parts is discussed in later pages and under Fruit.

The Vascular Skeleton of the Flower

The Floral Stele.—In structure, the pedicel is a typical stem with a ring of vascular bundles or an unbroken cylinder of vascular tissue. As the stele enters the receptacle, it conforms in shape to the shape of that part of the flower, commonly expanding and then constricting in its upper part. From the receptacular stele depart the vascular traces to the various floral organs (Fig. 158), traces which in origin, structure, and behavior are similar to those of leaves. Gaps accompany the exit of these traces, and the gaps and the crowding of the organs break up the cylinder into a network of strands (Fig. 158). Often the vascular cylinder is further dissected, as in the stems of herbaceous plants with reduced vascular cylinders. The receptacular stele may be simple but is usually highly complex.

The traces to sepals, petals, stamens, and carpels are given off successively from the stele in whorls or in spirals, according to the manner of arrangement of these organs in the flower. (Many organs that to external inspection are apparently whorled, such as the petals of some Ranunculaceae and the stamens of some Rosaceae, can be seen from anatomical evidence to be arranged in flat spirals.) The distal bundles of the receptacular stele commonly become the traces of the uppermost carpels (Fig. 158C,D), but in many floral types stelar bundles continue beyond the point where the last traces are given off and gradually fade out in the top of the receptacle (Fig. 158A,B). Such bundles are vestigial and usually consist of procambium or phloem only.

Traces of the Floral Appendages.—The traces to the various kinds of organs differ in number and, except in sepals, bear little or no relation in this respect to the leaf traces of the plant. The number—which is always odd—ranges from one to many. In all floral organs, constancy in trace number throughout angiosperms is high.

Sepals.—Sepals commonly resemble bracts anatomically. With few exceptions, they have a trace supply similar to that of the leaves of the plant.

Petals.—Typically, petals have a single trace, regardless of size, form, nature, or time of persistence. Three to several traces are present in some families.

Stamens.—A single trace is usually the vascular supply of stamens, without regard to size, shape, character of base, or period of persistence. In a few Ranalian families and rarely elsewhere, as in some members of the Lauraceae and Musaceae, three traces are present. (The three-trace

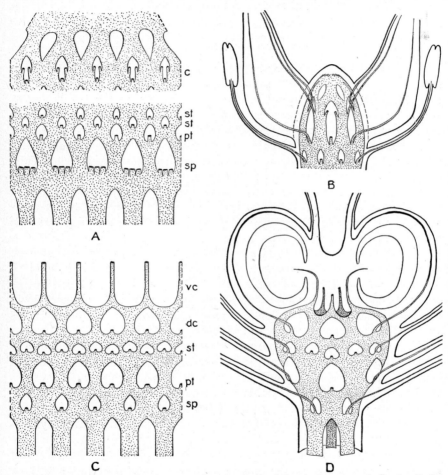

Fig. 158.—Diagrams showing vascular structure of simple flowers. *A, B, Aquilegia; C, D, Pyrola.* Description in text. *c*, carpel trace; *dc*, dorsal carpellary trace; *pt*, petal trace; *sp*, sepal trace; *st*, stamen trace; *vc*, ventral carpellary trace.

condition is apparently primitive—as is the similar condition in leaves and carpels—and one trace represents reduction from the primitive number.)

Carpels.—The carpel is generally considered a leaf-like organ folded upward (ventrally) along its midrib, with its margins more or less fused, and with ovules borne near the margins. This concept is borne out by its anatomy. The carpel has one, three, five, or several traces. The three-trace carpel is most common and the five-trace carpel is frequent, those with higher numbers infrequent. The one-trace carpel has clearly been derived by reduction from the three-trace. The median trace, which leaves the stele before (below) the other carpel traces, is known as the *dorsal trace* because it becomes the dorsal (midrib) bundle of the folded organ. (The term "trace" is applied in the flower, as it is in the leafy stem, to the strands of vascular tissue that depart from the stele as supply for appendages. The continuations of the traces within the appendages are called "bundles.") The two outermost traces are called *ventral* or *marginal traces* because they become the bundles that run along the ventral edge of the carpel—along or near the margins of the organ, if it were unfolded. Within the carpel, these bundles as compared with the dorsal bundle and the bundles of all other organs are inverted (Fig. 162). The upward and inward folding of the sides of the carpel brings about the inversion of these ventral bundles. In relation to the surfaces of the organ the bundles are normally oriented. Where there are more than three traces, the additional traces lie between the dorsal and ventral traces and are known as *lateral traces*. The orientation of the bundles that are the continuations of these traces depends upon the form of the carpel and the position of the bundle; many are half-inverted (Fig. 162*e, f*). Any or all of the bundles may be branched—and other branches commonly develop during fruit formation—but only occasional species have branching ventral bundles, and branches of these bundles running toward the margin are rare.

The vascular supply of carpels is the most complex in floral organs because of the folding of the carpels and the fusion of the margins; because the carpels constitute the uppermost organs and derive their traces at the top of the determinate receptacular stele where the vascular bundles are reduced and fade out; and because the carpels, owing to their close proximity to one another, tend to become fused and form a syncarpous ovary.

Ovules.—Ovule "traces" are derived, except in a few families and genera, only from the ventral bundles or from placental branches of these bundles (Fig. 163). The trace normally is a single small bundle that leads to the base of the ovule or enters it, passing as far as the chalazal region. It does not enter the nucellar region but in some genera sends branches into the integument. Where an ovule represents the surviving member of a group in which reduction has occurred, it may have "captured" the trace supply of two or more ovules, or even of two or more placentae. The ovules of angiosperms commonly have no vascular tissue but many, especially those of more primitive groups, have strong integumentary supplies.

Because of the large number of traces departing close together to the crowded floral organs and the gaps and other breaks in the receptacular stele, the vascular skeleton of most flowers is highly complex. In many genera it has become superficially simpler—though actually still more complex—by the fusion of traces that lie near together.

Anatomy of Simple Flowers

Examples of anatomically (and primitively) simple flowers provide a basis for the understanding of more complex types.

Aquilegia.—Figures 158A,B, and 159 show the vascular structure of the flower. In Fig. 158A the vascular cylinder is shown as split down one side and spread out; in Fig. 158B the top of the flower is shown as seen from one side. Figures 158A and B show the structure of the receptacular stele and the position and number of the traces of the appendages and their gaps. (Only two of the several rows of stamens are shown.) The stele continues in vestigial form beyond the level of departure of the uppermost appendages. Figure 159 shows the structure of the flower as shown in a series of diagrams of cross sections at successive levels from the pedicel to the top of the flower.

The pedicel has five bundles (Figs. 158A, 159a) which unite at the base of the flower, forming an unbroken cylinder. The sepal traces are then given off (Fig. 159b), three to each sepal, the three together leaving one gap. The traces to the petals, one to each, are next given off (Fig. 159c) just below the level at which the sepal gaps close. Then the stamen traces pass off, one to each organ, in several whorls (Fig. 159d–e). Above the uppermost whorl of stamens the vascular cylinder becomes complete again. Then the dorsal traces of the carpels are given off (Fig. 159f), followed by the pairs of ventral traces to each carpel. These ventral traces are derived from the sides of the gaps formed by the dorsal traces (Fig. 159g), and their gaps therefore merge with those of the dorsal traces. The ventral traces immediately twist, passing out inverted into the carpel. Figure 159h shows the condition in the receptacle just above the outward passage of the ventral carpel traces. The base of the loculus of each carpel is seen with the dorsal trace at the outside and the pair of ventrals on the inside. Internal to the ventrals the remainder of the vascular cylinder is seen as five masses of less strongly developed tissue (shown by dotted lines), Fig. 159h, chiefly phloem, with some weak xylem. This tissue (Fig. 159i, center), becomes smaller and weaker toward the rounded top of the receptacle and fades out (Fig. 158A,B). (This fading out of vascular tissue is similar to that often seen in the vascular cylinder of determinate leafy stems.) Figure 159*i* shows the carpels nearly free from the receptacle, with their three-bundle supply. Figure 159*j* (above the receptacle) shows cross sections of the typical follicles with their dorsal bundles, and their pairs of inverted ventral bundles.

Pyrola.—Figure 158*C,D*, shows the vascular structure of the flower. *C* shows the vascular cylinder split down one side and spread out; *D* shows a side view of the vascular cylinder with departure of traces to members of

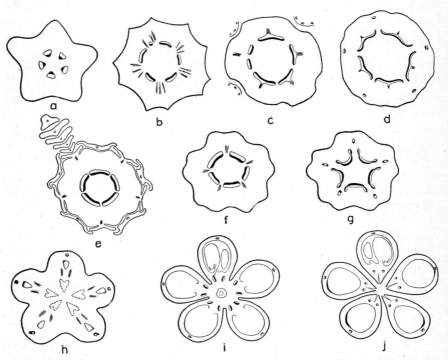

Fig. 159.—Diagrams representing vascular structure of flower of *Aquilegia* as seen in series of transverse sections from pedicel to level above top of receptacle. Detailed description in text.

all whorls of floral appendages. The receptacular stele of this flower has no vestigial part beyond the ventral traces to the carpels. The structure of the pedicel is similar to that of *Aquilegia*. In contrast with *Aquilegia* each sepal has but one trace. Petal and stamen traces are also similar to those of *Aquilegia*, although the number of stamens is less and the stamens are set in two close whorls. Above the departure of the dorsal carpellary traces the stelar bundles unite in twos to form five bundles, each representing the two ventral traces of a carpel. Into the formation of these traces goes all of the remaining vascular tissue.

Anatomy of More Complex Flowers

Fusion in the Floral Skeleton.—In the specialization of the flower, fusion and reduction of appendages have been the outstanding changes. Stages of *cohesion*—where members of the same whorl are fused to one another—and of adnation—where members of a whorl are fused to members of a whorl above or below—are found in many families. Evidence of reduction is to be seen in vestigial structures and through comparative study of related forms. In the evolutionary history of fusion, organs are

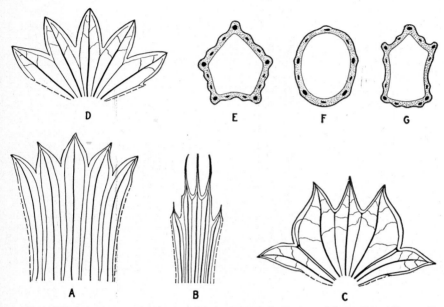

Fig. 160.—Anatomy of calyces, showing stages in fusion of lateral bundles under cohesion. *A, Nepeta veronica; B, Blephilia hirsuta; C, Trichostema dichotomum; D, Ajuga reptans. E, F, G* (cross sections): *E, Monarda didyma; F, Physostegia virginiana; G, Salvia patens.* Detailed description in text.

at first united externally, and internal structure is not modified. The line of fusion is histologically obscure or absent; the vascular tissues are unaffected. Ultimately the fusion becomes more intimate and, wherever the vascular bundles lie close to one another, fusion may involve them also. Fusion may be between the traces alone or between the traces and the bundles within the appendages, but it seldom extends throughout the length of the bundles. Often there is ontogenetic and histological evidence of the fusion.

Fusion under Cohesion.—The effect of cohesion on the vascular skeleton is chiefly upon lateral traces and bundles that are brought close together by gamosepaly, gamopetaly, syncarpy, and the marginal fusion of the upfolded margins of the carpel. It can be simply shown by examples. The gamosepalous calyces of the mints show all stages of fusion of the vascular tissue. In *Nepeta* (Fig. 160A) each sepal has a midrib and two lateral bundles; there is no fusion. The calyx of *Ajuga* (Fig. 160D) shows the pairs of adjacent laterals fused nearly to the sinuses. The zygomorphic calyx of *Blephilia* (Fig. 160B) shows two pairs of laterals fused and three pairs free; that of *Trichostema* (Fig. 160C) a similar condition, with the free pairs fused slightly at the sinuses. A taxonomic character, number of calyx ribs, sometimes used in generic distinction in the Labiatae, is based on the extent of this fusion of lateral

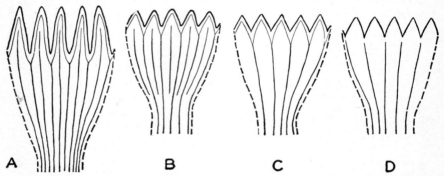

Fig. 161.—Vascular system of gamopetalous corollas showing fusion of lateral petal-bundles and loss of bundles in reduction. *A, Helianthus divaricatus; B, Senecio Fremontii; C, Chrysanthemum Leucanthemum; D, Xanthium orientale.* In all examples fusion of lateral bundles is complete nearly to the sinuses; in *Chrysanthemum*, the free ends of the lateral bundles have been lost. Stages in loss of median petal bundles are shown: *A*, bundles unreduced; *B*, basal parts lost; *C, D*, bundles absent. (*After Koch.*)

bundles. Examples are seen in *Monarda* (Fig. 160E), with 15 bundles, no fusion; *Physostegia* (Fig. 160F), with 10 bundles, all pairs of laterals fused; and *Salvia* (Fig. 160G), with 13 bundles, two pairs of laterals fused. In petals (Fig. 161) and carpels (Fig. 164) similar fusion is common.

The folding of the carpel presents the simplest type of fusion under cohesion in the gynoecium (Fig. 162). The two ventrals are brought close together, forming the pair of bundles characteristic of the ventral side of the primitive carpel, the follicle (Fig. 162a,b). In earliest phylogenetic stages the carpel margins merely touch or are lightly fused (Fig. 162a). (The carpel is even not completely closed in some genera.) With closer fusion, the epidermal layers on the surface of contact are lost and the two bundles form a double bundle (Fig. 162b,c); with complete fusion they become a single bundle that shows no histological or ontogenetic evidence of its doubleness (Fig. 162d).

Fusion in the vascular tissue of a carpel may be present in the ventral bundles from an origin as one trace throughout their length, or may exist only in part of the carpel (Fig. 163); where the ventral bundles arise as separate traces, they may unite at any point in their course. A carpel with two bundles, a dorsal and a ventral that consists of two bundles, may have one or two traces. Further, in much reduced carpels, especially in achenes, all three carpel traces may arise as one and separate only at the base of the loculus (Fig. 168f,j). In any carpel, fusion may also occur

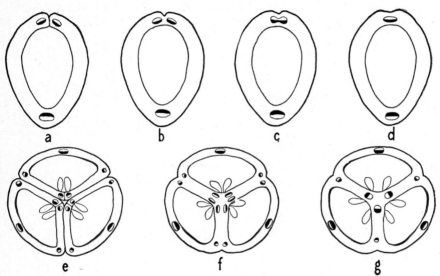

FIG. 162.—Diagrammatic cross sections of typical ovaries showing structural effects of cohesion, especially upon the vascular bundles: a–d, stages in the fusion of the ventral bundles of follicles; e–g, stages in the development of syncarpy, the fusion of carpels and of their ventral and lateral bundles. a, margins of carpel appressed, but unfused, epidermal layers persisting; b, margins fused, epidermal layers lost; c, the two ventral bundles slightly fused; d, the ventral bundles completely fused, without histological evidence of the fusion; e, the carpels cohering, but without structural fusion; f, the lateral carpel-walls fused, the ventral and lateral bundles of adjacent carpels approximated in pairs but not fused; g, the pairs of lateral and ventral bundles of adjacent carpels intimately fused.

among the bundles in the distal part of the ovary or in the style, when the traces and bundles are elsewhere free.

Syncarpy involves fusion changes similar to those in free carpels (Fig. 162e–g). The lines separating the carpels and their margins are lost (Fig. 162f). The ventral bundles, inverted, form a ring of bundles in the center. They usually lie in pairs, each pair consisting of the ventrals of the same carpel, or, more often, of bundles from each of two adjacent carpels. The members of a pair tend to fuse and all degrees of fusion are found. In the center of a syncarpous ovary with three carpels, there may be a ring of six or of three ventral bundles (Fig. 162)—if of three, each bundle is morphologically double and represents either the two ventral traces of one carpel or one from one carpel plus one from the adjacent carpel. Where syncarpy arose while the carpels were still open, fusing margin to margin and enclosing a common loculus (Fig. 164)—or where the septa have been reduced or lost, and placentation is parietal—the ventrals of adjacent carpels lie in pairs, or fused, in the outer wall of the ovary (Fig. 164*A*). Carpel limits in such ovaries run through the placentas, which are morphologically double structures, each half belonging to a different carpel.

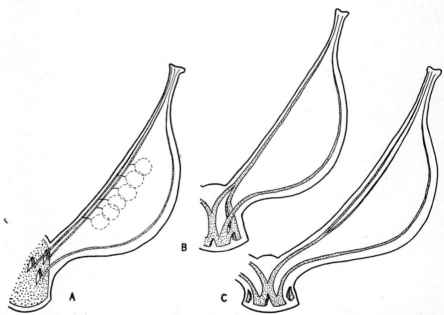

Fig. 163.—Vascular supply of the carpel. *A*, three traces from three gaps, the bundles within the carpels unfused to the stigma; *B*, three traces from a dissected stele, the ventrals soon fused, only two bundles in the carpel; *C*, the ventral traces arising fused but becoming free in base of ovary. In *B* and *C*, the top of the floral stele is used up in the formation of the ventral traces.

Fusion under Adnation.—Whenever members of one whorl become fused to those of other whorls—whether the whorls consist of like or of different organs—their vascular bundles tend to fuse, as under cohesion. Fusion takes place between bundles that are near together either radially or tangentially and may involve any number of bundles belonging to two or more whorls. Stages in a simple example—that of the epipetalous stamen—are shown in Fig. 165*F–J* and described in the legend. These stages can be seen in many families, the Crassulaceae, for example. Fusion involving several whorls, with union extending varying distances from the trace origin, is shown for Rosaceous flowers (Fig. 167*M–P*).

THE FLOWER—THE FRUIT—THE SEED

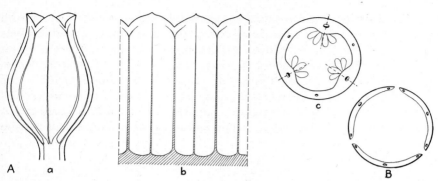

FIG. 164.—Diagrams illustrating cohesion between carpels. *A* based on *Reseda odorata:* *a*, lateral view showing origin of traces and course of bundles in ovary, the ventral bundles of adjacent carpels fused except at their tips; *b*, ovary wall split longitudinally and spread out; *c*, cross section through center of ovary, dotted lines indicating carpel limits. The ventral bundles have fused in pairs and the fused bundles and placentae lie on the line of carpellary fusion. *B*, illustrating theoretical position of carpels before fusion to form syncarpous ovary of the type shown in *A*.

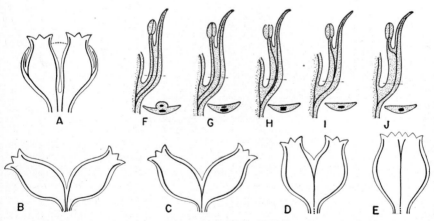

FIG. 165.—Diagrams showing fusion of vascular bundles under adnation. *F–J*, stages in fusion of stamen to petal, as seen in longitudinal and cross sections: *F*, the organs only lightly fused, externally still clearly two structures; *G*, fusion more complete, double nature not evident externally in the lower part, but in vascular structure clear throughout; *I,J*, the vascular bundles fused for various distances from the base, without histological evidence of double nature. *A*, adnation of bracts to flowers, *Lonicera caerulea*, the bracts adnate to the lower part of the inferior ovaries and the median bundle of the bract fused to one of the bundles in the ovary wall. *B–E*, adnation of flower to flower: *B*, *Lonicera canadensis;* *C*, *L. tatarica;* *D*, *L. fragrantissima;* *E*, *L. oblongifolia*. The four species show increasing fusion of ovaries, with similar increasing fusion of the vascular supply of the two flowers. (*A–E* after Wilkinson.)

The Inferior Ovary.—It is apparent that an inferior ovary can be formed by the adnation of the sepals, petals, and stamens to the carpels or by the sinking of the gynoecium in a hollowed receptacle, with fusion of the receptacle walls about the carpels. From anatomical evidence, it is clear that the inferior ovary in nearly all families is made up of appendages fused to the carpels. In the ovary walls of some plants, *Alstroemeria*, *Hedera*, for example, the traces of all adnate organs run free to the base of the ovary in the positions they would have if there were

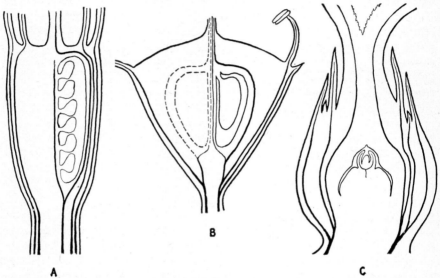

Fig. 166.—Diagrams of vascular structure in inferior ovaries. *A, Alstroemeria,* and *B, Hedera helix,* with bundles of the calyx, corolla, and stamens with no fusion with each other or with carpellary bundles in their course through the ovary wall; *C, Juglans nigra* (flower with adnate bracts and bractlets as well as sepals), bundles of the sepals, bractlets, and bracts fused with carpellary bundles for various distances from their origin. (Ventral carpellary bundles not in plane of section.) (*A*, after *Van Tieghem*; *C*, after *Manning*.)

no adnation (Fig. 166*A,B*). There is no fusion. In most species, however, with inferior ovaries, there is fusion within the ovary wall for various distances, as in *Juglans* (Fig. 166*C*) and some of the Ericaceae (Fig. 167*A–F*). The "ovary wall" in such forms, as is shown by the position and course of the bundles, clearly consists of the true ovary wall plus the tissues of the adnate organs.

The sinking of the gynoecium in the receptacle necessarily involves invagination of the top of the stele. An "ovary wall" so formed contains two cylinders of stelar bundles, the inner inverted. Within the cavity, the distal part of the receptacular stele, from which the carpel traces are derived, turns downward and the morphologically uppermost carpels are the lowest. *Darbya* (Fig. 167*L*) shows this well for the Santalaceae, the only family in which the inferior ovary has been shown to be of receptacular nature. In a few genera the tip of the receptacle is sunken and the carpels line the cavity, as in *Rosa* (Fig. 167*K*) and *Calycanthus*, but the sides of the receptacular cup are not fused to the carpels and the ovaries remain superior. The fleshy fruit of the rose is, in its lower part, receptacle; in its upper part, it consists of fused appendages, as clearly shown by the course of the vascular bundles (Fig. 167*K*). Stages in the development of the peculiar rose-type of fruit are shown by related genera (Fig. 167*I–K*). It is commonly believed that the rose hip and the apple represent fleshy receptacles of similar nature. But the rose and the apple tree belong to widely separated groups within the rose family, as shown by taxonomic and cytological, as well as anatomical evidence, and their fruits are not homologous. The ovary and fruit of the apple tree and closely related genera are entirely appendicular—the traces to all the floral parts arise from a receptacle of normal form and there are no inverted stelar bundles. Adnation is extensive (Fig. 167*P*): in one set of radii, sepal median trace, stamen trace, and dorsal-carpellary trace are united; in radii alternate with these, the lateral traces of adjacent sepals, petal trace, and three stamen traces are united. Fusion of the traces of the various organs extends for different distances, that of the dorsal carpellary with other traces only a short distance. The ventral carpellary traces alone show no fusion with other traces. Histologically the limits of the carpels can be seen in some varieties of apple. Morphologically, the flesh of the apple consists of tissues of all parts of the flower, with the receptacle forming only an insignificant bit at the base. Genera related to the apple show stages in the evolutionary development of the ovary (Fig. 167*M–P*).

The inferior ovary in most families—orchids, irises, evening primroses, for example—shows extreme fusion of the bundles of all appendages so that a cross section shows only a few bundles in the outer ovary wall (Fig. 167*G,H*).

Adnation of Flowers with Other Flowers and Other Organs.—Two or more flowers borne close together on a peduncle may become anatomically fused, as in *Mitchella*, *Maclura*, some species of *Lonicera* and *Cornus*. Such fusion usually involves only the basal parts, especially the ovaries. Various species of *Lonicera* show all stages in the fusion of the ovaries of two flowers (Fig. 165*B–E*); *Mitchella* shows complete fusion of the ovaries of two, rarely three and four flowers. In these genera, fusion is so complete in the most intimately fused species that some of the bundles of one flower are fused to those of the other. One bundle within the double ovary may supply perianth parts of two flowers.

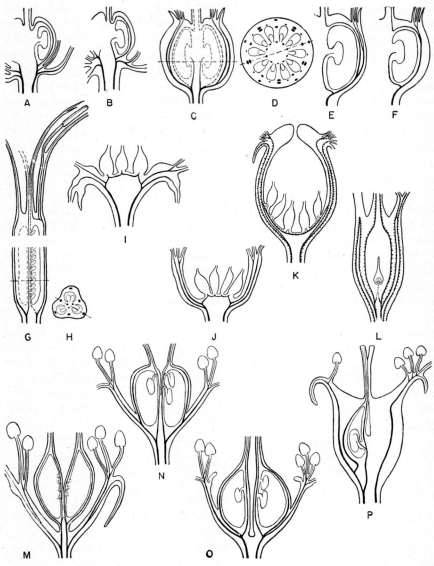

Fig. 167.—Vascular anatomy of flowers with relation to adnation and the inferior ovary. *A–F*, flowers of the Ericaceae showing stages in increasing adnation of appendages: *A*, *Pyrola secunda*, ovary superior, no adnation; *B*, *Andromeda glaucophylla*, ovary superior, stamens fused to petals at base, with fusion of the vascular bundles of the two organs nearly to point of separation; *C*, *D*, *Gaylussacia frondosa*, ovary inferior, calyx, corolla, and stamens adnate to carpels but vascular supply of these organs free in their course through region of adnation, except close to base, where the bundles that lie in some radii—those of sepals

(*For remainder of legend see page 355*)

Bracts may be fused to the flowers that they subtend, as in *Juglans*, and *Lonicera* (Figs. 165*A*, 166*C*). Fusion of all the parts of an inflorescence—flowers, bracts, and axis—is illustrated by the pineapple (*Ananas comosus*). The many flowers, each subtended by a bract, are closely packed in spirals. The inferior ovary of each flower is completely sheathed by bracts, the anterior half by its own bract, the remainder by parts of three other bracts. Fusion between the ovary and the four surrounding bracts is histologically complete to the top of the ovary. This fusion extends throughout the inflorescence, uniting all ovaries, all bracts, and the axis into one structure. The bundles of the flower are wholly free from each other and from those of the bracts, but bundles of adjacent bracts are often fused and supply two bracts.

Where vegetative parts are similarly fused, for example, bract to peduncle (*Tilia*); axillary peduncles to the adjacent stem (*Sparganium, Streptopus*)—giving supra-axillary inflorescences—the vascular tissue may also be fused and structurally complex.

Placental Vascular Supply.—Because in many flowers the ventral bundles give rise directly to ovular traces, these bundles have been, incorrectly, called the "placental supply." In follicles, achenes, and some other types of ovaries, the placenta is merely a position, and there is no placental supply as such. Where the placenta is a definite, often fleshy enlargement of an area along the margin of the carpel, it is supplied by branches from the ventrals, which branching further, give rise to the ovule traces, as in the Ericaceae and Cucurbitaceae. The ovules on a placenta in a syncarpous ovary obviously may receive their vascular supply from different carpels. Where the number of ovules has been reduced from an earlier larger number, the surviving ovules may have strong vascular traces; in extreme reduction—some types of basal and

---

and outer stamens and those that lie in alternate radii (those of petals and inner stamens)—fused; *D*, cross section through center of ovary showing position and freedom of bundles of organs adnate to carpels (dorsal carpellary bundles not fused with other bundles because lying in different radii); *E*, *F*, *Vaccinium corymbosum* and *V. macrocarpon*, showing fusion, progressively greater than in *C*, of petal and stamen bundles with dorsal carpellary bundles. *G*, *H*, *Iris versicolor*, typical inferior ovary, the bundles of the adnate organs in each radius fused throughout their length as far as the top of the ovary. *I*, *J*, *K*, *Dalibarda repens*, *Rubus triflorus*, *Rosa setigera*, respectively: *I*, *J*, showing flattening of receptacular tip; *K*, showing inversion of tip. *K*, with stelar bundles inverted at about one-third distance to top; the lower third, with its inverted stelar bundles is receptacular; the upper two-thirds, "calyx tube"—sepals united with petals and stamens—similar to and homologous with that in *I* and *J* but in-arched, and covering the sunken carpels. *L*, *Darbya*, ovary inferior, formed by inversion of top of receptacle as evidenced by inverted vascular bundles. *M–P*, series of related Rosaceous genera, showing stages in adnation of "calyx tube" to carpels: *M*, *Physocarpus opulifolius*, the carpels superior, the sepals, petals, and stamens, fused, but the "calyx tube" free from the carpels; *N*, *Sorbus sorbifolia*, the "calyx tube" fused to the base of the carpels; *O*, *Spiraea van Houtei*, the "calyx tube" adnate to carpels halfway to the top of the ovary; *P*, *Malus pumila*, the "calyx tube" adnate to the ovary to the top, bundles in some of its radii fused with the dorsal carpellary bundles. (*I–K*, *M–O*, after Jackson; *L*, after Smith.)

free-central placentation—one or two ovules may have the vascular supply of more than one carpel (Juglandaceae; Polygonaceae). (Anatomical structure demonstrates that no angiosperm ovules are cauline.)

Vestigial Vascular Tissue.—Evidence of the structure of ancestral floral types is frequently found in the vascular skeleton of modern flowers. The presence of vestiges of the vascular stele at the tip of the receptacle has been discussed (Figs. 158, 159). Remnants of the vascular supply of lost organs may be present in the tissues of the receptacle when all external evidence of the organs has disappeared. Only rarely does the vascular supply of a vestigial organ disappear while external remnants are still present. The vascular vestiges of lost appendages are usually stubs of traces buried in the cortex of the receptacle. The lost organs so represented may be entire whorls, especially of petals and stamens, or individual members of persisting whorls. Traces to lost petals are common in apetalous flowers: *Aristolochia; Rhamnus* (apetalous forms); *Salix; Quercus* (some species); to lost stamens, for example, in Scrophulariaceae, Labiatae, Cucurbitaceae, *Lysimachia, Chionanthus, Lychnis*, to lost carpels in Caprifoliaceae, Ericaceae, Rutaceae, Valerianaceae. In unisexual flowers there are commonly external remnants of lost organs; in those in which such vestiges have disappeared, vascular traces may be present within the receptacle: Fagaceae; Caryophyllaceae; Urticaceae.

Traces to lost ovules are among the commonest vestigial structures in the flower. They are frequent in some families, especially the Ranunculaceae, Umbelliferae, and Rosaceae (Fig. 168*B,E,N*).

Reduction of Vascular Supply within an Organ.—As reduction occurs in an organ in evolutionary specialization—for example, as a follicle with several ovules becomes an achene with one ovule; as a corolla becomes greatly reduced in size in the crowding together of small flowers in a head, as in the Compositae—the vascular supply of the organ is greatly reduced. In carpels either the dorsal or the ventral bundles may be shortened, the latter extending only to the ovules. In extreme reduction the entire carpellary supply consists of a small bundle running directly from the receptacular stele to the ovule. Stages and variety in this loss are shown in achenes (Fig. 168). In the stamen, the bundle, which commonly extends to the anther, may enter only the base of the filament. In the petal, either median or lateral bundles may be shortened or lost. In the gamopetalous corolla, the median bundle is more often lost (Fig. 161*B,C,D*) and the fused laterals persist; the latter may be reduced in length (Fig. 161*D*).

The Sepal and the Petal.—Both the sepal and the petal are commonly leaf-like in form and general appearance. Structurally also the former is like the leaf except when it is petaloid. The petal is leaf-like in grosser

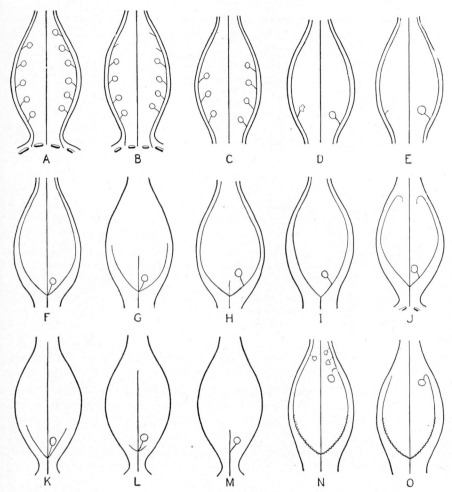

Fig. 168.—Diagrams of carpel structure showing reduction of ovules and fusion and reduction of vascular supply. *A–E*, follicles: *A, Helleborus viridis*, typical follicle, with many ovules and three traces; *B, Trollius laxus*, ovules reduced in number, traces to lost ovules persisting; *C, Aquilegia canadensis*, upper ovules and their traces lost; *D, Hydrastis canadensis*, all ovules but lower two lost, one of these abortive; *E, Waldsteinia fragarioides*, all ovules but one lost, trace to one other persisting. *F–O*, achenes: *F–M* with basal ovule surviving; *N, O*, with an upper ovule surviving. All with dorsal and ventral traces united at the base, in some as far as the ovule, which then appears to be attached to the dorsal (*F, G, K, L, M*). *F, Geum rivale*, with distal parts of all bundles present; *G, Duchesnea indica*, dorsal and ventral bundles greatly shortened; *H, Fragaria vesca*, with dorsal bundle greatly shortened; *I, Agrimonia striata*, with dorsal bundle lost beyond its union with the ventral bundles; *J, Ranunculus Ficaria*, the ventral bundles recurving, not entering the style; *K, R. Flammula*, the ventral bundles greatly shortened; *L, R. Cymbalaria*, only vestiges of the ventral bundles persisting, the dorsal bundle greatly shortened; *M, R. aquatilis*, the ventral bundles lost beyond the ovule, the dorsal bundle continuing hardly beyond it; *N, Potentilla recta*, one ovule surviving, others vestigial, the ventral bundles unreduced; *O, P. canadensis*, one ovule surviving, the ventrals lost beyond the ovule. (*After Chute, modified in part.*)

358 AN INTRODUCTION TO PLANT ANATOMY

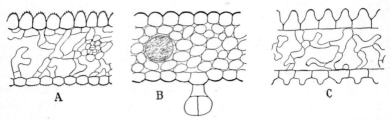

Fig. 169.—Cross sections of petals. A, *Amelanchier laevis;* B, *Lysimachia Nummularia,* showing glandular hair and secretory chamber; C, *Pinguicula vulgaris.*

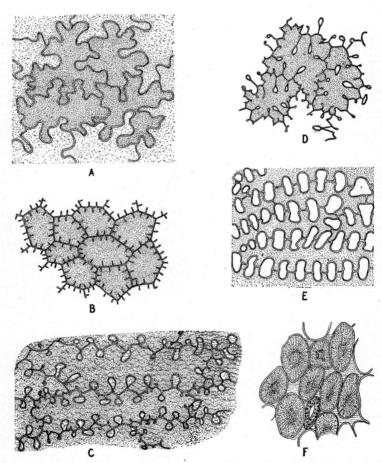

Fig. 170.—Epidermis of corolla. A, *Calceolaria;* B, *Pelargonium;* C, *Clarkia;* D, *Anchusa;* E, *Linum;* F, *Erythrina.* (After *Hiller.*)

structure but histologically differs in many ways from the typical leaf (Fig. 169). The vascular system is commonly reduced in amount and in supporting tissues. The mesophyll is simple in structure: there is usually no palisade layer and the spongy tissue consists of few cell layers. Chromoplasts or colored cell sap, or both, are present. The epidermis is less simple than that of leaves. Its cells are commonly weak-walled and of complex form, often undulate, stellate, or irregularly lobed in outline and dovetailed with one another. With this shape and arrangement they doubtless form a layer mechanically stronger than one consisting of cells of simpler form. The cells of one or both surfaces are usually papillose. Stomata are fewer than those on leaves and frequently nonfunctional or absent. The guard cells lack chloroplasts except when the mesophyll also has these plastids. Secretory areas and glandular hairs are common. Intercellular spaces, usually lacking in the epidermis of leaves, are frequent. They lie in the loops or lobes of the cell wall (Fig. 170C,D,E) but are always covered by cuticle. The cuticle varies greatly in thickness being very thin on delicate, ephemeral petals.

The extreme types of petals show many variations from, and exceptions to, the typical structure described above. The less highly specialized type of petal has a strong vascular supply, a weak palisade layer, chloroplasts, a nonpapillose epidermis with many stomata and often a hypodermis, and supporting tissue about the larger veins. The most highly specialized type has the vascular tissue weak, with veinlets and even most of the main bundles lost, and wholly without accompanying supporting tissue. The mesophyll consists of one to three poorly defined layers of widely spaced cells which are lacking along the margins and in distal parts, so that part of the petal consists of epidermal layers only. The epidermis is papillose, with chromoplasts or colored cell sap, and without stomata or glands.

The Stamen and the Carpel.—Stamens and carpels consist chiefly of unspecialized parenchyma. Their traces continue in the organs as weak or strong, often amphicribral bundles, for varying distances, and with varying amounts of branching. In the stamen, the trace or traces commonly run unbranched to the anther, but in delicate stamens may fade out anywhere in the filament. In some stout stamens, as in those of *Magnolia*, small lateral branches are present. The traces of the carpel may pass as unbranched bundles to the style or stigma, or may branch in varying degrees, some showing leaf-like venation in the branching. Within the stigma the bundles frequently branch. In reduced carpels, all branching is lost, and the main bundles are shortened by loss of the distal portions. For example, the ventral bundles may not extend beyond the placentae and all bundles may fail to enter the style. In extreme reduction, as in small achenes, the vascular supply may be reduced to little more than a basal remnant (Fig. 168M). Vestigial bundles are frequently found. These are the weak or abortive remnants of traces or bundles representing the supply to organs lost in evolutionary modification. The most common vestigial bundles are the traces to lost petals, stamens, and carpels, represented by stubs arising from the receptacular stele as normal traces but ending blindly in the cortex of the receptacle, and ovule traces that end blindly in the placenta or the margin of the carpel (Fig. 168B,E). The epidermis is of typical structure, the cells usually with straight walls and thin cuticle. Stomata occur on stamens only when these structures are expanded and leaf-like. In carpels they occur freely on the outer surface and are even found in the epidermis lining the ovarian cavity.

THE FRUIT

Morphology of Fruits.—A fruit is a developed and ripened ovary or ovaries together, often, with adjacent floral organs and other plant parts. The fruit develops directly from the flower and its structure is therefore fundamentally that of the floral parts and other structures from which it has arisen. Fruits range in complexity of morphological structure from a single carpel, such as a legume, to a fruit like the pineapple, which is an entire inflorescence including the ovaries, floral parts, bracts, and the inflorescence axis all fused into a single succulent mass.

Generally, the morphological nature of a fruit can be determined from the mature fruit alone, particularly if anatomical characters are considered. The outlines of at least some of the organs—carpels, receptacle, sepals, bracts—may be visible, and the fundamental vascular supply persists. Although these organs can often be identified, they are usually swollen and distorted. The terms used in floral description are not applicable and are supplanted by special terms. The body of the fruit developed from the ovary wall, which surrounds and encloses the seed, is known as the *pericarp*. When this is not homogenous histologically, and distinct outer, inner, and median parts are evident, these parts are known as *exocarp, endocarp*, and *mesocarp*, respectively. In many fruits the pericarp is more or less nearly homogeneous and not separable into parts; in other fruits only exocarp and endocarp are evident. Strictly speaking, the term "pericarp" refers only to the modified ovary wall, but in general usage it is applied to the outer tissues of fruits regardless of their morphological nature.

Ontogeny of Fruits.—Since the fruit is derived directly from the flower, the ontogeny of the fruit can be considered to begin with the formation of the floral parts that contribute to it. This is particularly true of the ovary which has its basic structure determined in the flower. In some genera, for example, in *Lycopersicon* (tomato) and *Vaccinium*, cell division in the ovary ceases when the flower opens, and subsequent development of the fruit consists of the enlargement and specialization of the existing cells of the ovary, but in most fruits, the flowering period is followed by cell division in the ovary and accessory fruit parts. Fertilization and the growth of the embryo are normally necessary to set in motion the mechanism of fruit development, and in this sense the ontogeny of the fruit begins with fertilization.

Fruits normally go through a cycle of development that is constant for any one species, but differs widely among species. This consists of an early stage of cell division with the laying down of basic form and structure; a period of cell enlargement and tissue differentiation in which final form is reached; and a period of ripening. These stages may overlap somewhat and are usually related to stages in the growth of the endosperm, seed coats, and embryo.

In the ontogeny of fruits, various organs or parts of organs may be lost. For example, the flower of the coconut (*Cocos nucifera*) has a three-carpelled, three-celled ovary, but in the formation of the mature nut only one carpel develops. Remains of the abortive carpels form a part of the endocarp of the fruit.

The shape of fruits is determined by the place and rate of cell division, by the plane of cell division, and by the direction in which enlarging cells elongate. Some fruits, for example, the date, have a basal meristem which contributes to the formation of a much elongated fruit. Elongate fruit form in some of the cucurbit fruits is related to the restriction of the divisions of meristematic cells largely to the plane perpendicular to the fruit axis and to the elongation of the new cells chiefly in the direction of this axis.

The Structure of Fruits

Vascular Skeleton of Fruits.—A fruit possesses as its basic vascular supply that of the floral parts from which it has been derived. This vascular supply is strengthened and extended as the fruit develops; the diameter of the bundles is increased either by primary or secondary growth, or by both these methods of increase, and the bundles are extended by terminal primary growth and by the addition of branches that extend through the developing tissues. But the main vascular system of a fruit, though often obscured by branches, remains fundamentally the same as in the flower. In the apple flower, for example, there are ten principal bundles in addition to those that supply the carpels. These bundles are the vascular supply of the sepals, petals, and stamens.

In the fruit, these appear in relatively the same position as in the flower but much increased in size and with the addition of a highly developed, anastomosing branch system extending to all parts of the fleshy outer layers. The smaller branches extend throughout the fleshy tissues as the smaller bundles of a leaf spread throughout the mesophyll. The ultimate branches, like those in a leaf, are very slender and consist of only a few elements which are chiefly protoxylem cells accompanied by a very few elongate parenchyma cells. In the pericarp of fleshy fruits, these finer bundles are more numerous than in the pericarp of dry fruits.

Epidermis of Fruits.—The cells of the epidermis on both the inner and the outer surfaces of the carpels resemble the epidermal cells of stems

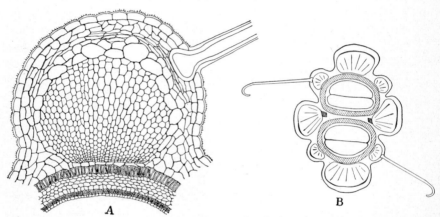

Fig. 171.—Dry fruit of *Circaea latifolia*. *A*, detail of portion of the cross section (*B*) of entire fruit, showing the pericarp with its various layers, the ridge of corky secondary tissue, and the large, rigid hairs.

and leaves in form, in wall structure, and in the presence of cuticle and stomata. The inner epidermis is more delicate than the outer, but usually has a thin cuticle and may have stomata. The cells of the epidermis are most commonly isodiametric, polygonal cells, as in the Rosaceae and Liliaceae. In the Ranunculaceae, some Scrophulariaceae, and other families, the cells are sinuous in outline, lobed and dovetailed with one another, as in the epidermis of many petals and some leaves. In berries and some drupes, and probably in other fleshy fruits, for example, in *Rubus* and *Vitis*, the epidermal cells are polygonal and very small with rather thin walls. In many other fruits, the epidermal cells are thick-walled with a heavy cuticle and in early stages of development may contain tannin. The number of stomata in the epidermis of fruits varies from many as in *Pyrus* to none as in *Vitis* and *Vaccinium*.

Periderm in Fruits.—In a few fruits of the "russet" type, such as russet apples and pears and *Achras* and *Calocarpum*, a periderm layer arises in or close beneath the epidermis and is apparent on the surface of ripe fruit as a rough, brownish layer. In the form of lenticels, restricted periderm occurs in apples and pears where it forms the "dots" on the fruit. Corky ridges and nodules, which represent restricted periderm, are common on dry fruits, as in *Circaea* (Fig. 171).

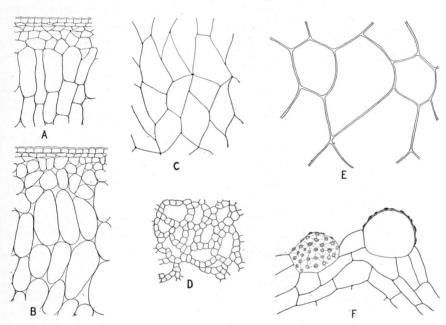

Fig. 172.—Fleshy pericarp. *A*, *B*, ripening and ripe outer pericarp of *Prunus Cerasus* (var. Montmorency); *C*, ripe, fleshy pericarp of *P. Cerasus* (var. Morello); *D*, "dry," fleshy pericarp of *Zizyphus* (jujube); *E*, fleshy pericarp of *Citrullus*; *F*, outer pericarp and epidermis of fruit of *Floerkia*; the walls of some epidermal cells with prominent centripetal thickenings.

Fleshy Fruits.—Fleshy fruits consist in some considerable part of succulent parenchyma, whereas dry fruits consist of sclerenchyma and nonsucculent parenchyma in varying amounts. In fleshy fruits, the pericarp (here referring to the tissues enclosing the seed cavities regardless of their morphological nature) may be succulent throughout, or the outer layers may be succulent and the inner layers dry and stony. The fleshy layer may be homogeneous, as in the apple, peach, and cherry (Fig. 172*A*,*B*), or the pericarp may consist of mingled succulent parenchyma and sclereids in varying proportions, as in the pear, quince, and black walnut. The fleshy tissues are usually thin-walled parenchyma, more or less turgid with fluids. Some cells may contain tannin, as in *Diospyros*, or raphides, as in some species of *Vitis*. Mucilage-bearing cells are found in some fruits as in the rind of *Musa*. Lysigenous glands are characteristic of the rinds of *Citrus* and laticiferous ducts are abundant in *Carica*.

Sclereids in the fleshy layers frequently occur in groups or clusters and give the flesh a gritty or sandy texture. From these aggregates as a center, elongate parenchyma cells may radiate, as in *Pyrus serotina*. The individual sclereids may be isodiametric or very irregular in outline.

The pulp of citrus fruits is of different nature from that of most fruits. It consists of numerous multicellular emergences developed from the inner

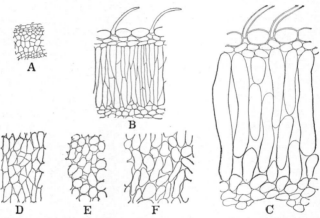

Fig. 173.—Fleshy pericarp. *A, B, C,* successive stages in the development of the exocarp of *Rubus strigosus: A*, green fruit soon after fertilization; *B*, half-ripe fruit; *C*, ripe fruit, all sections radial and drawn to same scale. *D, E, F,* fleshy receptacle of *Fragaria: D*, half-ripe fruit, radial section; *E, F,* ripe fruit, tangential and radial sections, respectively.

surface layers of the carpels. The cells of these projecting structures become turgid with fluid and the emergences fill the carpel cavities.

Whether the fleshy pericarp is strictly homogeneous in texture or contains more than one cell type, there are usually several more or less distinct layers. Only the outer and inner epidermis and the fleshy layers are distinct in some berries such as fruits of the Solanaceae (tomato, ground cherry). More often a hypodermis, several layers of cells thick, underlies the outer epidermis. This may consist of simple parenchyma, as in *Rubus* (Fig. 173*B*); collenchyma, as in *Prunus cerasus* and *Persea;* sclereids, as in *Pyrus communis;* or other cell types and combinations of types. The size and shape of the cells in the different layers varies greatly. Usually the cells of the epidermis and hypodermis are small compared with those of the middle fleshy layer.

The ripening process in fleshy fruits involves changes in structure and in chemical composition. Chemically, stored starch is changed to sugar, or in some species to fat; tannin is reduced in amount or disappears; acids are usually reduced in amount and various esters are produced. Histologically, ripening may involve increase in cell size and change in cell shape, the cells becoming turgid with fluid and the walls often excessively thin and delicate; it may also involve, to a greater or less extent, the separation of the cells from one another. In some plants the cells are already freed in part while the fruit is still immature (Fig. 173*D,E*) with resulting increase in intercellular spaces. These air spaces in the apple fruit may make up as much as twenty-five per cent of the volume of the fruit at the "firm ripe" stage. The dissolution of the middle lamella and outer wall layers (which in some plants, as in *Crataegus*, are very thick) sets the cells free. The fluid filling the cells of the flesh is said to come in part from the formation of more soluble substances, from organic acids, and from the breaking down of various substances such as starch, pectic acid, and cellulosic materials. Probably in almost all fruits, fluids formed within the fruit or brought in during ripening, are retained within the cells of the flesh, even when the fruit is very ripe. A small amount of free fluid coming from the dissolved middle lamellae may be present between the cells, but the apparently free fluid found on the opening of soft fruits is probably mostly intracellular fluid set free by the rupture of the tenuous walls of turgid cells. Very slight pressure is sufficient to break such cells, possibly even the changes in tissue tension brought about by the breaking of the epidermis.

Fruits fall roughly into two groups on the basis or rate of increase in size in the final stages of ripening. In one group, including the cherry, blueberry, and raspberry, growth is extremely rapid at the end of development, with great enlargement in a few days or even hours. This rapid growth consists mostly of the radial enlargement of cells and of the formation of intercellular spaces between separated cells. In partly grown fruit, the cells of the developing flesh, which were at first isodiametric, have often become tangentially elongated; when ripe, these same cells become isodiametric again or even, as in most fruits with very rapid growth at the last, radially elongated (Fig. 173*A,B,C*).

In the other group, the fruits reach full size before they are ripe and may even shrink slightly through loss of water during the ripening process. This is true of apples, pears, plums, and many other fruits that are considered mature when they are harvested but may require days or even months before they are ripe in the sense that they are ready to eat. During the interval between firm-ripe and edible-ripe stages, changes take place in the middle lamella and cell wall which result in easier separation of the cells. The fruit industry is based on the practice of harvesting fruits when they are mature but still firm and transporting them to markets before they ripen and become soft.

The ripening process eventually leads to the breakdown and disintegration of fruits. In the apple and pear the cells are freed from one another and the fruit becomes mealy or mushy. In some of the soft fruits the cell walls disintegrate releasing the juices. Early stages of breakdown are followed in most fruits by fungous or bacterial decay which destroys the fruit.

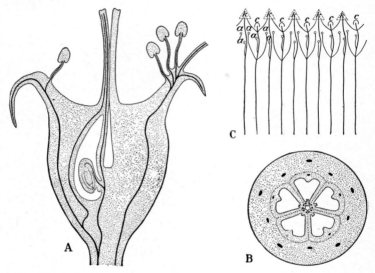

Fig. 174.—The vascular system of the flower of *Malus pumila*. *A*, vertical section; *B*, median cross section; *C*, vascular system, exclusive of carpellary supply, spread out in one plane—*k*, *c*, *a*, bundles to sepals, petals, and stamens respectively. (*C, after Kraus and Ralston.*)

Variety in texture of fleshy fruits is related to structure. Thus the crispness of an apple depends upon the turgor of rather firm cells. In the peach and high quality pears, large thin-walled cells that break easily and release their contents give soft flesh and abundant juice. In some fruits that have buttery or custard-like consistency, as for example, in *Asimina* and *Diospyros*, the cells of the flesh disintegrate when the fruit becomes edible. Aggregates of sclereids next to the epidermis or scattered in the pulp give the granular texture characteristic of some pears. Masses of sclerenchyma in the outer layers give a hard rind as in species of *Cucurbita*. Heavy cuticle combined with collenchyma gives toughness to the outer layers of some types of *Persea* fruits. The flesh of some fruits, such as those of *Gaultheria* and *Zizyphus*, has a rather tough, dry consistency resulting from the presence of many air chambers in the flesh and low water content (Fig. 172D).

Stony Layer of Fruits.—In many fruits, particularly drupes, the fleshy outer layers enclose a hard or stony layer. In simple drupes like the plum and cherry, the fleshy and stony layers make up the ovary wall or pericarp. In other fruits, floral organs other than the ovary may enter into the formation of the pericarp. Because the stony layer is usually next to the seed cavities, it is in most drupes derived from the carpels. The hard layer usually consists of sclereids and fibers, but it may be cartilaginous, as in the apple and pear, or tough and leathery, as in the mango. The outer surface of this layer may be smooth, as in species of *Prunus*, or may extend in prongs of hard tissue into the soft pulp, as in *Spondias*, or in tough, hair-like processes, as in *Mangifera*.

The stony layer develops during the early stages of enlargement of the fruit and usually attains full size and becomes lignified before the embryo starts rapid growth. In *Juglans, Carya,* and *Prunus,* embryo enlargement is delayed for weeks until the stony layers are fully developed. Obviously, after lignification of this tissue, no further change in size of this part of the fruit is possible. The stony layer commonly consists of the inner epidermis of the pericarp and the inner pericarp. Usually, it is itself differentiated into two or more layers of sclerenchyma differing from each other in cell shape and orientation. In *Prunus cerasus*, the inner layer next to the seed coats has its cells elongated at right angles to the fruit axis; in the outer layer, the cells are elongated parallel with this axis. The type and arrangement of cells and the number of layers in the stony endocarp of different species apparently vary considerably.

Dry Fruits.—Dry fruits are of essentially the same structure as fleshy fruits during early stages of development in that all meristematic tissues consist of soft parenchyma. The basic structural pattern and particularly the vascular system are the same in both. The fruits of the almond are essentially identical with those of the peach through all early stages of growth, but in the final ripening period the exocarp of the almond becomes dry instead of fleshy as in the peach and splits along the suture to free the nut which is the stony endocarp and its enclosed seed. The difference between these fruits lies chiefly in the changes which occur during the later stages of cell differentiation and particularly in the final ripening period.

In dry fruits, a large proportion of the cells mature as sclerenchyma of various kinds. In the ripening process the parenchyma cells lose their protoplasts and become dry; their walls often become lignified or suberized.

The three characteristic layers of the pericarp are more commonly present in dry fruits, where sclerenchyma and 'dry' parenchyma are abundant, than in fleshy fruits. These three layers vary in nature and in extent in different fruits, being either sclerenchymatous or parenchymatous, and ranging from one layer to many layers in thickness. The parenchyma cells may be closely appressed to one another, or loosely arranged with prominent intercellular spaces. The sclerenchyma cells are fibers of many types and sclereids. Often an exocarp is highly complex, consisting of successive layers of different types of sclerenchyma, sometimes including layers of parenchyma of different cell shapes, types, and orientation. Such a structural unit is apparently very strong mechanically. An essentially four-layered pericarp is formed in many fruits, such as the achenes of the Compositae, where the mesocarp consists of two parts.

In many dry fruits, the thin-walled parenchyma collapses to a greater or less extent as the fruit dries in maturing. When the fruit is a firm, thin-walled capsule, as in the Caryophyllaceae, the pericarp is made up largely of closely packed, firm-walled parenchyma that does not collapse on drying.

Dry fruits show a great variety of "accessory structures" such as corky ridges, spines, hooks, hairs, and various kinds of "ornamentation." Hairs of the internal epidermis of the fruit of *Ceiba* constitute the kapok of commerce. In winged fruits, for example those of *Acer* and *Ailanthus*, the part of the pericarp constituting the wing is usually of firm but light structure. In these fruits, the supporting cells are chiefly those of vascular bundles with their fibrous sheaths. Between these bundles, the tissue is usually loose, with intercellular spaces.

The Placenta.—In many fruits, especially those of the fleshy type, such as tomato and watermelon, enlarged placentae form a considerable, even a large, part of the fruit. Even in some dry fruits, such as the capsule of *Epigaea*, the trailing arbutus, the placenta is large, fleshy, and berry-like. (The placenta is, of course, a part of the carpel and not usually considered a part of the pericarp.) Fleshy placentae consist of masses of soft, thin-walled parenchyma and vascular bundles. The parenchyma cells are large and, as in the watermelon (Fig. 172E), can often be seen with the naked eye.

Accessory Fruit Parts.—Fleshy fruits that consist in part of structures other than the carpel have the same general histological structure as simple fleshy fruits. The accessory structures are morphologically diverse but some flower parts are found frequently. In many fruits, such as those of *Pyrus*, *Vaccinium*, and the Cucurbitaceae, developed from an inferior ovary, the "calyx tube" is adherent to the carpels and forms a part of the fruit. (The "calyx tube" consists morphologically, of the fused bases of calyx, corolla, and stamens.) A soft, fleshy receptacle makes up a large part of some fruits, for example the blackberry and the strawberry. The pulp of the strawberry is said to be developed in large part by a cortical, phellogen-like meristem, which forms many cells centripetally. Inside this, there is the vascular cylinder and the enlarged pith. In multiple fruits arising from an entire inflorescence, such as those of *Ananas, Artocarpus,* and *Morus,* the receptacle forms a considerable part of the fruit and becomes fleshy together with the fused calyx lobes and the bracts subtending the flowers. The fruit of *Ficus* is an inflorescence with its flower-bearing branches fused around a central cavity; the flesh of the fruit consists mostly of the cortical tissues of the branches.

Dehiscence of Fruits.—The place and manner of dehiscence in dry fruits differ widely but are constant and characteristic for a species. In many plants there seems to be no elaborate histological structure associated with dehiscence; the opening takes place along "sutures," which are lines of structural weakness. These lines of dehiscence are often those of marginal contact of carpels where there is incomplete fusion. In other plants a line of separation is formed by the development of rows of special cells weakly held together. Separation occurs later between these rows by the tearing apart of the tissues during changes in tension due to drying. The development and specialization of these "opening cells" accompanies that of the cells of the rest of the fruit and does not follow at a much later period as does the development of separation layers in leaves and stems. Dehiscence is apparently not commonly the result of the separation of individual cells, as is abscission, or of the disintegration of tissue. In some semifleshy fruits that are dehiscent, for example, those of *Carya,* the middle lamellae along preformed suture lines appear to dissolve. Along these lines, the fruit opens as it dries. The stresses brought about by unequal growth or by unequal drying out of the pericarp may cause sudden and even explosive rupture of the fruit along lines of weakness.

THE SEED

The enlarged and matured ovule with its enclosed embryo constitutes the seed. Before fertilization, the ovule consists morphologically of the embryo sac, nucellus, and one or two integuments. Histologically, it is simple, made up of parenchyma, with the exception sometimes of a few procambium or vascular cells. After fertilization the endosperm, and later, the embryo develop rapidly within the enlarging ovule, and the integuments enlarge and their tissues become complex. The nucellus

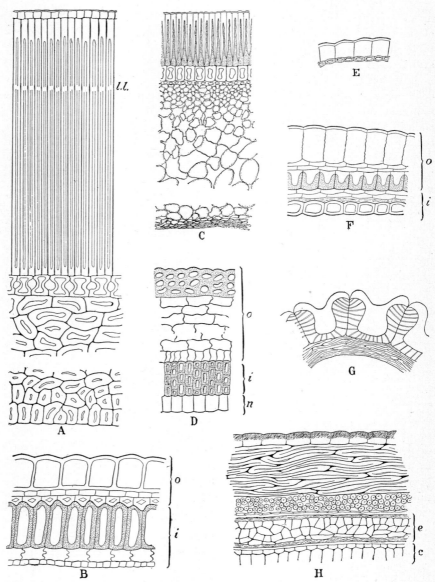

Fig. 175.—Seed coats. *A, Gymnocladus dioica* (only one-fifth of the stony inner layer shown); *B, Viola tricolor; C, Phaseolus multiflorus* (only one-third of the soft inner layer shown); *D, Magnolia macrophylla*, only part of the fleshy layer shown; *E, Plantago lanceolata; F, Lepidium sativum: G, Vaccinium corymbosum*, the epidermal cells very large, with heavy inner, and delicate outer walls; *H, Malus pumila. c*, cotyledon; *e*, endosperm; *i*, inner integument; *l. l*, "light line"; *n*, nucellus; *o*, outer integument. (*B, D, E, F, after* Brandza.)

may persist and increase in volume becoming a prominent part of the seed, as in the beet, but it usually fails to increase and is soon destroyed by other tissues. In many plants the endosperm forms a considerable part of the mature seed, but frequently it is partially or wholly absorbed by the growing embryo. (Many seeds described as "exalbuminous" have small amounts of endosperm.) In general, the histological simplicity of the ovule is continued in the endosperm and embryo, but the histological structure of the seed coats is complex.

Seed Coats. *Morphology.*—Throughout the different groups of seed plants the ovule has either one or two integuments. In the dicotyledons the majority of the Gamopetalae and Apetalae and some Polypetalae have but one integument, as do part of the monocotyledons; other angiosperms have two integuments. Sometimes all of the integument, or integuments, enters into the composition of the seed coat (Fig. 175*B*), but more often the seed coats are developed from a part only of the integuments and the other parts are absorbed as the seed develops. When such absorption takes place, it is the innermost or sometimes the median layers of an integument that are removed. Further, whether the integuments develop as a whole or in part, the nucellus may contribute to the seed coats, becoming distinguishable only with difficulty from the adjacent inner layers of the integument (Fig. 175*D*). In probably the majority of seeds, however, the nucellus is wholly absorbed and is not represented in the seed. The reduction of the tegumentary tissues goes very far in some seeds, chiefly in indehiscent fruits where only the two or three outermost layers—sometimes only the outer epidermis—of the outer integument persist in the ripe seed, as in the Umbelliferae. In some of the Compositae, extreme reduction has occurred: in the mature achene all of the integuments except a thin layer of crushed disorganized tissue (Fig. 177*B*) and also the inner layers of the pericarp have disappeared so that the fruit is an achene in which the seed and the fruit tissues are intimately associated and not readily separable. In the seeds of some monocotyledons, for example *Zea*, the integuments are entirely absorbed.

Where the ovule has two integuments, various conditions exist in the seed. Both integuments may be present in the seed coat—the inner, represented by all its cell layers, the outer, by all or by only the outermost two or three layers. In these seeds, the inner integument forms the important part of the seed coat; its outer part forms a protective layer (Fig. 175*B*)—a condition found in the Malvaceae, Tiliaceae, Violaceae, Hypericaceae, and other families.

Again, where both integuments are present in the seed, the outer is strongly developed with protective layers, and the inner, though several-layered, is rather unspecialized (Fig. 175*F*). Of such structure are the seed coats of the Cruciferae, Berberidaceae, Papaveraceae, and certain lilies, irises, and aroids. In the Onagraceae, Lythraceae, Aristolochiaceae, and other families, both integuments form lignified protective layers and at least the outer layers of the nucellus contribute to the seed coats. In the Magnoliaceae, the inner integument becomes the protective seed coat with a part of the nucellus attached to it; the outer integument often becomes fleshy (Fig. 175D). Complete absorption of the inner integu-

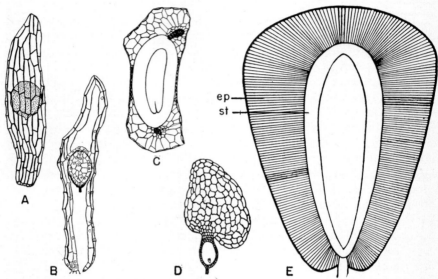

Fig. 176.—Types of seeds. *A, B, Cypripedium parviflorum:* A, seed showing transparent seed coat enclosing embryo (stippled) suspended in pocket of air; B, longitudinal section of seed showing seed coat consisting of funiculus and outer integument, shriveled inner integument forming envelope around embryo, suspended embryo a mass of undifferentiated cells. *C, Clethra alnifolia,* longitudinal section showing embryo more highly differentiated than in B, and integument several cells thick but not differentiated into layers; *D, Pterospora andromedea,* seed showing wing developed from integument; *E, Punica granatum,* seed showing outer integument consisting of a stony layer (*st*) and a fleshy epidermal layer (*ep*) of greatly elongated cells; the inner integument is membranous and is not shown. (*A, B,* after Carlson; *C, D,* from Netolitsky; *C,* after Peltrisot; *D,* after Baillon; *E,* after Tung.)

ment and of the nucellus occurs in the Ranunculaceae, Leguminosae, and in certain lilies and amaryllises.

In ovules with one integument, only rarely does the entire structure become seed coat. A larger or smaller inner or median part is usually absorbed and the outer layers, together with the inner epidermis, form the seed coat, as in the Polemoniaceae, Plantaginaceae (Fig. 175E), Balsaminaceae, and other families. So various and obscure is the morphological nature of seed coats that in most plants this can be determined only by ontogenetic studies.

Histological Structure.—Seed coats show variation in histological complexity quite unrelated to their morphological nature. In some dry, indehiscent fruits, seed coats are lacking; the embryo is surrounded by

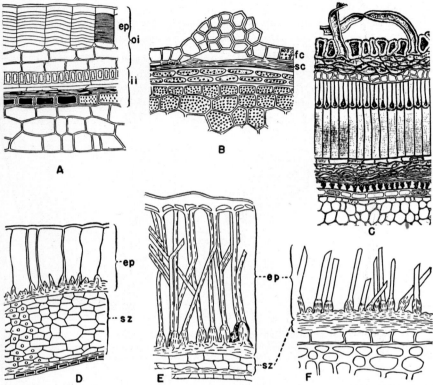

Fig. 177.—Seed coats in transverse section. *A, Linum usitatissimum*, showing stratified layers of mucilage in epidermal cells (*ep*), outer integument (*oi*) and inner integument (*ii*); *B, Lactuca sativa*, showing fruit coat (ovary wall) (*fc*), and crushed, disorganized seed coat (*sc*); *C, Gossypium*, showing epidermal hairs (cotton "fibers"). *D–F, Lycopersicon esculentum*, showing stages in development of seed coat: *D*, showing thickening of inner tangential walls of epidermal cells and beginning of thickening in bands on radial walls, numerous cells in subepidermal zone; *E*, showing epidermal cells much elongated, thickening of bands on radial walls greatly increased and extending over parts of outer tangential walls, subepidermal zone mostly absorbed; *F*, showing epidermal "hairs"—the thickened bands remaining after disintegration of outer tangential walls and thin parts of radial walls, the absorption and disorganization of the cells of the subepidermal zone resulting in a membranous layer; the seed coat consisting, therefore, of the epidermal "hairs," a noncellular membranous layer and cellular inner layer one cell in thickness. (*A*, after *Tschirch and Oesterle; B,* after *Kondo; C,* after *Winton; D–F* after *Souèges*.)

ovarian tissues. Extreme simplicity is found in the seeds of orchids in which an almost undifferentiated embryo is enclosed in a flimsy, transparent sheath of thin-walled cells (Fig. 176*A,B*). Great complexity is characteristic of the seed coats of certain types of seeds: very hard seeds, seeds with "ornamented" coats (Fig. 178A–E), and seeds with fleshy coats (Figs. 175D, 177D).

The seed coats of many species, for example, those of some Leguminosae and *Canna*, are highly impervious to water and gases. Such seed coats have a thick cuticle that is so impervious to water that germination is prevented unless the cuticle is ruptured. The scarification of seeds and treatment with sulfuric acid to increase the percentage of germination are based in large part upon the necessity of rupture or destruction of the cuticle to permit the entrance of water and oxygen.

In many hard seeds there is a characteristic protective layer beneath the epidermis made up of close-packed, radially placed, columnar

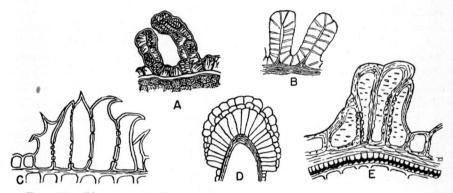

Fig. 178.—"Ornamentation" on seeds. A, *Gentiana stylophora*; B, *Wrightia Barteri*; C, *Dicliptera resupinata*; D, *Ruellia squarrosa*; E, *Delphinium*. (All from Netolitsky; A, after Guerin; B, after Wahl; C, D, after Schaffnit; E, after Mitlacher.)

cells (Fig. 175A). This layer is often called the palisade layer, and its peculiar cells are sometimes referred to as "Malpighian cells" (because first described by Malpighi). This palisade layer somewhat resembles the palisade layer of leaves but it is a sclerenchymatous layer without intercellular spaces. The walls are commonly unevenly thickened (Fig. 175C,F,G); they may be of cellulose or be heavily cutinized or lignified. They doubtless give protection against mechanical injury and against changes in water content in the seed. The palisade layer nearly always shows a "light line," a band-like region running transversely to the long axis of the cells—and hence tangentially in the seed (Fig. 175A)—where light refraction is different from that in the rest of the cells. In some species this "light line" is formed by the deposition of wax globules in the cells. Within the heavy-walled protective layer, or rarely external to it, are other layers of various thicknesses, and of many kinds of cells.

Among these are sclereids of many types; fibers; parenchyma cells of the greatest variety, both in size and shape, in content, and in nature of the wall. The successive layers are typically made up of cells of very different kinds: often the cells of adjacent layers, though similar, are differently oriented, lying with long axes at right angles to the surface in one layer and parallel with it in another, or with the long axes parallel with the long axis of the seed in one layer, and at right angles to it in the other (Fig. 175H). Cells with peculiarly and irregularly thickened walls (Fig. 175F,G), and also those of remarkable shape and unusual relation to adjacent cells are common.

The outer layers of the seed coats of some seeds are fleshy. In genera such as *Magnolia*, *Caulophyllum*, and *Ginkgo*, that have seeds exposed at maturity, the seed coats are histologically similar to the outer layers of fleshy fruits. In species in which the seeds are not exposed, the histological structure is less complex. For example, in *Punica* (pomegranate), the edible fleshy layer is derived from the epidermis, the cells of which elongate radially to many times their original diameter and become turgid (Fig. 176E). In the tomato (*Lycopersicon*), the fleshy pulp surrounding the seed is in part derived from the epidermis of the seed coat. The epidermal cells as they develop elongate radially to several times their original diameter, the inner tangential wall thickens greatly, and parallel rod-like thickenings are deposited on the thin radial walls. The enlarged cells become fleshy and turgid. When the seed is mature, the outer and the radial walls break down between the rods which then appear as hair-like projections of the seed coat (Fig. 177D,E). In the dry seed, the hair-like structures are free except at the base (Fig. 177F). The mature fleshy seed is surrounded by a fleshy, aril-like upgrowth of the placenta.

Mucilaginous seed coats are found in some genera. In the seeds of *Linum*, the radially elongate epidermal cells are filled with mucilaginous material deposited in stratified layers (Fig. 177A). When the dry seeds are moistened, the contents of the cells swell, rupturing the outer walls and cuticle.

The surface of seed coats shows structural markings of great diversity. These are characteristic and constant for a species and furnish means of identification in seed analysis. Ridges, rods, folds, hooks, and other structures are usually modifications of the epidermal cells, but subepidermal layers of the seed coat may also take part in their formation. Expansions of the seed coat consisting of thin-walled, air-filled cells form wings and other buoyant structures (Fig. 176C,D). Of special interest are the epidermal hairs ("fibers") of the seed coats of *Gossypium* (Fig. 177C).

Vascular Bundles of Seeds.—The bundles of seeds are confined to the raphe and the outer integument. In general the integuments of small seeds lack vascular tissue, but the presence or absence of bundles is to some extent a family character. Strands in the raphe represent the ovule supply in an adnate funiculus. Where the seed is large and the seed coat of complex structure a network of bundles may be present.

Embryo and Endosperm.—The embryos of different families show great diversity in stage of differentiation in the mature seed. In the orchid seed the embryo consists of relatively few undifferentiated meristematic cells with no indication of radicle, plumule, and cotyledons (Fig. 176*A,B*). In others, mostly large seeds, for example, those of *Persea* (avocado), the well-developed embryo has radicle, massive cotyledons, and a well-developed plumule. Vascular bundles are present in the procambium stage. Between these extremes are many intergrading forms.

In thick cotyledons that do not expand on germination the tissues are mature, consisting of rounded or polyhedral cells packed with starch or aleurone. Intercellular spaces are present. In such cotyledons the vascular system is much simplified and stomata are lacking. In cotyledons that expand on germination, the tissues are immature, but the mesophyll may be differentiated into spongy and palisade tissue. Stomata may be present in the epidermis. The vascular system in all embryos is usually in the procambium stage; only rarely are protoxylem and protophloem elements mature, as in *Castanea* and *Aesculus*.

The endosperm consists always of polyhedral parenchyma cells more or less isodiametric in shape. The walls are largely cellulose and commonly thin, but in some plants, such as the date (*Phoenix*) and the persimmon (*Diospyros*), (Fig. 21*E,B*), they are greatly thickened. The additional cellulose is reserve food. This very hard endosperm is commonly known as "horny" endosperm. In the large seeds of some palms, for example, *Phytelephas*, endosperm of this type finds commercial use in the manufacture of buttons. The storage materials in typical endosperm are starch, aleurone, and oils; starch and aleurone are said not to occur in the same cell.

References

THE FLOWER

ARBER, A.: Floral anatomy and its morphological interpretation, *New Phyt.*, **32**, 231–242, 1933.

BECHTEL, A. R.: The floral anatomy of the Urticales, *Amer. Jour. Bot.*, **8**, 386–410, 1921.

BONNE, G.: "Recherches sur le pédicelle et la fleur des Rosacées," Paris, 1928.

BROOKS, R. M.: Comparative histogenesis of vegetative and floral apices in *Amygdalus communis*, with special reference to the carpel, *Hilgardia*, **13**, 249–306, 1940.

CHESTER, G. D.: Bau und Function der Spaltöffnungen auf Blumenblättern und Antheren, *Ber. Deut. Bot. Ges.*, **15**, 420–431, 1897.

CHUTE, H. M.: The morphology and anatomy of the achene, *Amer. Jour. Bot.*, **17**, 703–723, 1930.

DOUGLAS, G. E.: Studies in the vascular anatomy of the Primulaceae, *Amer. Jour. Bot.*, **23**, 199–212, 1936.

———: The inferior ovary, *Bot. Rev.*, **10**, 125–186, 1944.

EAMES, A. J.: The vascular anatomy of the flower with refutation of the theory of carpel polymorphism, *Amer. Jour. Bot.*, **18**, 147–188, 1931.

GRÉGOIRE, V.: La valeur morphologique des carpels dans les Angiospermes, *Bull. Acad. Roy. Belg.*, **17**, 1286–1302, 1931.

GUMPPENBERG, O. VON: Beiträge zur Entwicklungsgeschichte der Blumenblätter mit besonderer Berücksichtigung der Nervatur, *Bot. Arch.*, **7**, 448–491, 1924.

HANCY, A. J.: The vascular anatomy of certain ericaceous flowers, Thesis, Cornell University, 1916.

HENSLOW, G.: On the vascular systems of floral organs, and their importance in the interpretation of the morphology of flowers, *Jour. Linn. Soc. Bot. London*, **28**, 151–197, 1891.

———: "The Origin of Foral Structures through Insect and other Agencies," New York, 1888. (Vascular anatomy.)

HILLER, G. H.: Untersuchungen über die Epidermis der Blüthenblätter, *Jahrb. Wiss. Bot.*, **15**, 411–451, 1884.

JACKSON, G.: The morphology of the flowers of *Rosa* and certain closely related genera, *Amer. Jour. Bot.*, **21**, 453–466, 1934.

KOCH, M. F.: Studies in the anatomy and morphology of the Composite flower, I. The corolla, *Amer. Jour. Bot.*, **17**, 938–952, 1930.

KRAUS, E. J., and G. S. RALSTON: The pollination of the pomaceous fruits: III. Gross vascular anatomy of the apple, *Ore. Agr. Coll. Exp. Sta. Bull.*, **138**, 1916.

KUHN, G.: Beiträge zur Kenntnis der intraseminal Leitbündel bei den Angiospermen, *Bot. Jahrb.*, **61**, 325–379, 1928.

MCCOY, RALPH W.: Floral organo-genesis in *Frasera carolinensis*, *Amer. Jour. Bot.*, **27**, 600–609, 1940.

MANNING, W. E.: The morphology of the flowers of the Juglandaceae, II. The pistillate flowers and fruit, *Amer. Jour. Bot.*, **27**, 839–852, 1940.

MULLER, L.: Grundzüge einer vergleichenden Anatomie der Blumenblätter, *Nova Acta K. L-C., Deutsch. Akad. Naturforscher*, **59**, 1–356, 1893.

NEWMAN, I. V.: Studies in the Australian Acacias, VI. The meristematic activity of the floral apex of *Acacia longifolia* and *Acacia suaveolens* as a histogenic study of the ontogeny of the carpel, *Proc. Linn. Soc. N.S.W.*, **61**, 56–88, 1936.

SATINA, S., and A. F. BLAKESLEE: Periclinal chimaeras in *Datura stramonium* in relation to development of leaf and flower, *Amer. Jour. Bot.*, **28**, 862–871, 1941.

SMITH, F. H., and E. C. SMITH: Anatomy of the inferior ovary of *Darbya*, *Amer. Jour. Bot.*, **29**, 464–471, 1942.

——— and ———: Floral Anatomy of the Santalaceae and related forms, *Oregon State Monographs. Studies in Botany*, **5**, 1942.

TILLSON, A. H.: The floral anatomy of the Kalanchoideae, *Amer. Jour. Bot.*, **27**, 595–600, 1940.

TSCHECH, K.: Der Gewebebau grüner Kelchblätter, *Oesterreich. Bot. Zeitschr.*, **88**(3), 187–199, 1939.

VAN TIEGHEM, P.: Recherches sur la structure du pistil et sur l'anatomie comparée de la fleur, *Mém. Acad. Sci. Inst. Imp. France*, **21**, 1–262, 1875.

WILKINSON, A. M.: The floral anatomy and morphology of some species of *Cornus* and of the Caprifoliaceae, Thesis, Cornell University, 1945.

THE FRUIT

BARBER, K. G.: Comparative histology of fruits and seeds of certain species of Cucurbitaceae, *Bot. Gaz.*, **47**, 263–310, 1909.

DU SABLON, L.: Recherches sur la déhiscence des fruits a péricarpe sec, Thesis, Paris, 1884.

FARMER, J. B.: Contributions to the morphology and physiology of pulpy fruits, *Ann. Bot.*, **3**, 393–414, 1889.

GARCIN, A. G.: Recherches sur l'histogénèse des péricarpes, Thesis, Paris, 1890.

HOUGHTALING, H. B.: A developmental analysis of size and shape in tomato fruits, *Bull. Torr. Bot. Club*, **62**, 243–252, 1935.

JULIANO, J. B.: Origin, development and nature of the stony layer of the coconut, *Phil. Jour. Sci.*, **30**, 187–197, 1926.

KRAUS, G.: Ueber den Bau trockner Pericarpien, *Jahrb. Wiss. Bot.*, **5**, 83–126, 1866.

LAMPE, P.: Zur Kenntnis des Baues und der Entwickelung saftiger Früchte, *Zeitsch. Naturwiss.*, **59**, 295–323, 1886.

LONG, E. M.: Developmental anatomy of the fruit of the Deglet Noor date, *Bot. Gaz.*, **104**, 426–436, 1943.

MACDANIELS, L. H.: The morphology of the apple and other pome fruits, *Mem. Cornell Univ. Agr. Exp. Sta.*, 230, 1940.

OKIMOTO, M. C.: Morphology and anatomy of the pineapple inflorescence and fruit, Thesis, Cornell University, 1943.

SINNOTT, E. W.: A developmental analysis of the relation between cell size and fruit size in cucurbits, *Amer. Jour. Bot.*, **26**, 179–189, 1939.

TUKEY, H. B., and J. O. YOUNG: Histological study of the developing fruit of the sour cherry, *Bot. Gaz.*, **100**, 723–749, 1939.

———: Gross morphology and histology of developing fruit of the apple, *Bot. Gaz.*, **104**, 3–25, 1942.

WAHL, VON, C.: Vergleichende Untersuchungen über den anatomischen Bau der geflügelten Früchte und Samen, *Bibl. Bot.*, **40**, 1–25, 1897.

WINTON, A. L.: The anatomy of edible berries, *Conn. Agr. Exp. Sta. Rept.*, **26**, 288–325, 1902.

THE SEED

BRANDZA, M.: Développement des téguments de la graine, *Rév. Gén. Bot.*, **3**, 1–32, 105–117, 150–165, 229–240, 1891.

CARLSON, M. C.: Formation of the seed of *Cypripedium parviflorum* Salisb., *Bot. Gaz.*, **102**, 295–301, 1940.

GODFRIN, J.: Étude histologique sur les téguments seminaux des angiosperms, *Bull. Soc. Sci. Nancy*, **5**, 109–219, 1880.

———: Recherches sur l'anatomie comparée des cotylédons et de l'albumen, *Ann. Sci. Nat. Bot.*, 6 sér., **19**, 5–158, 1884.

GUIGNARD, L.: Recherches sur le développement de la graine et en particular du tégument séminal, *Jour. Bot.*, **7**, 1–14, 21–34, 57–66, 97–106, 141–153, 205–214, 241–250, 282–296, 303–311, 1893.

HAUSS, H.: Beiträge zur Kenntniss der Entwicklungsgeschichte von Flügeinrichtungen bei höheren Samen. *Bot. Arch.*, **20**, 74–108, 1927.

Jumelle, H.: Sur la constitution du fruit des Graminées, *Compt. Rend. Acad. Sci., Paris*, **107**, 285–287, 1888.

Kayser, G.: Beiträge zur Kenntnis der Entwickelungsgeschichte der Samen mit besonderer Berücksichtigung des histogenetischen Aufbaues der Samenschalen, *Jahrb. Wiss. Bot.*, **25**, 79–148, 1893.

Kondo, M.: Über die in der Landwirtschaft Japans gebrauchten Samen, *Ber. Ohara Inst. f. Landw. Forsch.*, **1**, 399–450, 1919.

Netolitsky, F.: Anatomie der Angiospermen Samen, *In* K. Linsbauer: "Handbuch der Pflanzenanatomie," X, 1926.

Pammel, L. H.: Anatomical characters of the seed of Leguminosae, chiefly genera of Gray's Manual, *Trans. Acad. Sci. St. Louis*, **9**, 91–273, 1899.

Schnarf, K.: Anatomie der Gymnospermen Samen, *In* K. Linsbauer: "Handbuch der Pflanzenanatomie," 1937.

Souèges, M. R.: Développement et structure du tégument séminal chez les Solanacées, *Ann. Sci. Nat. Bot.*, 9 sér., **6**, 1–124, 1907.

Tschirch, A., and O. Oesterle: "Anatomischer Atlas der Pharmakognosie und Nahrungsmittelkunde," Leipzig, 1900.

Tung, C.: Development and vascular anatomy of the flower of *Punica granatum* L., *Bull. Chinese Bot. Soc.*, **1**, 108–128, 1935.

Winton, A. L.: The anatomy of certain oil seeds with especial reference to the microscopic examination of cattle foods, *Conn. Agr. Exp. Sta. Rpt.*, **27**, 175–198, 1903.

CHAPTER XIV

ECOLOGICAL ANATOMY

In the foregoing chapters, emphasis has been placed upon normal plant structure as it develops in regions of average or optimum water supply. Such an environment is called *mesophytic* and is that which obtains generally in the important agricultural districts of the temperate zone and also in parts of the tropics, especially at median altitudes. The plants that live in this environment are known as *mesophytes*, and include the majority of the best known wild and cultivated plants of temperate regions.

There are also many plants living under extremes of available water supply. There is an extensive flora of vascular plants, known as *hydrophytes*, living on the surface of bodies of water or submersed at various depths down to those where diminished light intensity and other conditions become limiting factors for existence of plants of this type. Much greater in number of species and more diversified in structural complexity is the great group of plants that live in regions where the supply of available water is deficient, that is, in a so-called *xerophytic* environment. This group, known as *xerophytes*, includes species from many families in no way closely related phylogenetically, which, in this environment, have come to resemble each other more or less closely in vegetative structure. Between the extreme xerophytes and the hydrophytes, all gradations in form and all degrees of structural variation occur in plants whose natural habitat is intermediate between mesophytic and xerophytic on the one hand and mesophytic and hydrophytic on the other.

Types of Xerophytic Environment.—The structural modifications common to xerophytes occur under many different environmental conditions. The most common of these is that found in deserts or semiarid places where there is a deficiency of rainfall during either a large part or all of the year. Here are found many plants, not typically xerophytic in structure, that grow only during a short rainy season and pass the dry season as seeds, or as dormant bulbs, corms, or roots beneath the surface of the ground. In such locations, with the exception of extreme deserts where no higher plants can grow, there are also a considerable number of species which maintain stems and leaves, or only stems, above ground during the whole year. In addition to these habitats, which are characterized by actual lack of water, are those environments where water is apparently abundant, but for some reason is physiologically unavailable to the plant. Thus there is recognized a group of plants, known as *halophytes*, which grow in salt marshes or in certain types of alkaline soil that are only slightly toxic. These plants, although frequently standing in water, have elaborate structural modifications that prevent water loss, much like those found in plants of the desert flora. The water, though present, is of such high osmotic concentration that it cannot be readily absorbed by the plant; a somewhat similar condition is found in the peat bogs of the colder temperate and the subarctic regions where the water is not absorbed, not because its concentration is osmotically too high, but probably because of a toxic effect upon the roots of the plant which hinders their development or because of the very low temperature of the soil.

Another set of physiologically xerophytic conditions is that found in regions where there is no actual lack of water, but which are cold for a part of the year. Here low temperatures greatly cut down the rate of absorption by the roots and reduce conduction generally, so that even if transpiration is also reduced, there is a physiological shortage of water. This situation is more extreme where the soil freezes to a considerable depth. Plants that hold their leaves under such conditions—for example, the many species of needle-leaved gymnosperms which are generally distributed in the temperate and subarctic zones—show typical xerophytic structure.

Exposure to persistent winds of high velocity and to intense light and heat also produce xerophytic conditions. Many xerophytic situations are the result of a combination of the environmental factors described above; in a desert situation, lack of moisture, both in soil and atmosphere, intense light and heat, and winds of high velocity may be operative.

Xerophytes.—Xerophytes differ from mesophytes both physiologically and structurally. In some plants the normal mesophytic organs and tissues are not changed in structure, but have become physiologically more effective, for example, plants growing in dry situations, such as the olive tree, develop an extraordinarily large and deeply penetrating root system capable of efficient absorption. Other examples of this kind of modification are those salt-marsh plants that have a higher osmotic concentration within the plant than is found in common mesophytes; this concentration makes possible the absorption of water from the salt-marsh soil. With this high osmotic concentration is coupled the ability of these plant cells to resist the toxic action of the salt solution, which would undoubtedly be injurious to many other plants with normal osmotic concentration. Along with physiological, are also commonly found structural modifications that reduce the loss of absorbed water. Some of the specialized structures are highly complex and may involve nearly all of the tissues of the plant.

Lignification and Cutinization.—Heavy cuticularization and extreme cutinization of the epidermis and even of subepidermal cells are common in xerophytes. All gradations are found in the thickness of the cuticle from a thickness only slightly greater than normal, like that of plants

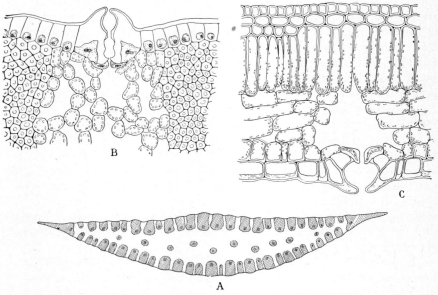

FIG. 179.—Structure of xerophytic leaves. *A*, diagram of cross section, *Dasylirion serratifolium*, sclerenchyma singly cross-hatched, vascular tissue doubly cross-hatched; *B*, detail of small marginal portion; *C*, cross section, *Cycas*, showing lignified hypodermis and partly lignified palisade tissue.

of semixerophytic habitats, to the elaborate thickenings of extreme xerophytes in which the cuticle may be as thick as, or thicker than, the diameter of the epidermal cells. In addition to increased thickness of the cuticle, the walls of the epidermal cells themselves are frequently cutinized, and sometimes also those of the underlying cells. Along with well-developed cutinized layers there are frequently found different degrees of lignification of the epidermal and subepidermal cells. In some organs, for example, the leaves of *Cycas*, lignification may extend even to parts of the palisade parenchyma cells (Fig. 179*C*). Similar to cuticularization is the formation of wax as a covering on the epidermis. Many plants secrete wax externally in small amounts, but certain genera, for example, *Copernicia* and *Ceroxylon*, the sources of carnauba wax, produce this substance in quantities sufficient to be commercially valuable. (Wax is sometimes produced by plants not growing under xerophytic conditions.)

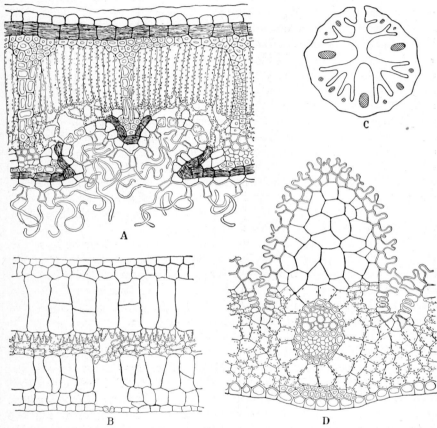

Fig. 180.—Structure of xerophytic leaves. *A*, a sclerophyllous type, *Banksia*; *B*, a malacophyllous type, *Begonia*; *C*, *D*, a rolling type, *Spartina*. *A*, an extreme type with very thick cuticle, outer hypodermal layer filled with mucilage and tannin, mesophyll pockets enclosed in sclerenchyma, stomata in hair-filled pockets, spongy parenchyma sparse; *B*, mesophyll thin, protected by layers of mucilage-containing cells, stomata unprotected; *C*, diagram of cross section of leaf in rolled condition; *D*, detail of portion of same; stomata in furrows nearly closed by interlocking epidermal cells.

In addition to a cutinized epidermis, many xerophytic plants possess one to several layers of cells immediately beneath the epidermis that form a *hypodermis* (Figs. 179*C*, 182*B,D*). Its cells are often much like the epidermal cells in structure and are sometimes derived in ontogeny from the young epidermis, but in most plants, the hypodermis of leaves is morphologically mesophyll and may be in the form of a layer of sclereids or a sheet of fibrous tissue. The hypodermis of stems belongs to the outer cortex. The hypodermis of stems and leaves may be cutinized to some extent or, more frequently, lignified. Gums and tannins are common in this layer (Fig. 180*A*).

Sclerenchyma.—In addition to the hypodermis, xerophytes generally have a larger proportion of sclerenchyma in their leaf structure than is found normally in mesophytes. This tissue, as masses either of fibers or sclereids, is usually arranged in more or less regular layers between the mesophyll of the leaf and the epidermis or hypodermis. In some plants, as in *Banksia* (Fig. 180*A*), there is a continuous, rather thin sheet of sclerenchyma between the hypodermis and the mesophyll. In other forms, for example, in *Dasylirion* (Fig. 179*A*,*B*), there are heavy, parallel strands of fibers below the epidermis. These strands cover the mesophyll except for small openings leading from the stomata to the interior of the leaf. Such sclerenchyma layers prevent water loss as well as aid in the support of the organ, and may also act as a partial screen against intense light. Xerophytes that have increased cutinization and sclerification of the leaves are commonly called *sclerophyllous*.

Hairs.—The modifications so far described prevent water loss from the plant by the formation of layers which are themselves more or less impervious to water. Moreover, the cutting down of the circulation of air over the leaf surface prevents rapid evaporation through the stomata. In many plants, especially those of alpine regions exposed to strong winds, a covering of matted epidermal hairs on the under side of the leaves, or wherever stomata are abundant prevents water loss. Hairs may also be abundant over the entire aerial part of the plant. Hairs form dead-air spaces next to the epidermis where air remains at a relative humidity approaching that on the inside of the leaf. Xerophytes that have a hairy covering on the leaves and stems are known as *trichophyllous*.

Rolling of Leaves.—The leaves of some xerophytes, of which the xerophytic grasses are the outstanding example, roll tightly under dry conditions. In such leaves the stomata are located on the upper or ventral surface only, so that when the edges of the leaf roll inward until the edges touch or overlap, the stomata are effectively shut away from the outside air. An extreme example of this kind is seen in *Spartina*, a salt-marsh grass, in which the tight upward folding of the leaf and also the sheltered location of the stomata in furrows (Fig. 180*C*,*D*) greatly reduce air movement over stomatal areas. As described in Chap. XII, in the leaves of many grasses special motor cells on the upper surface form a rolling device. In the xerophytic grasses, these cells are particularly well developed. The linear leaves of other plants, for example, species of *Dasylirion*, and those of some broad-leaved genera, such as *Rhododendron*, have leaves that fold or roll when the water supply is deficient.

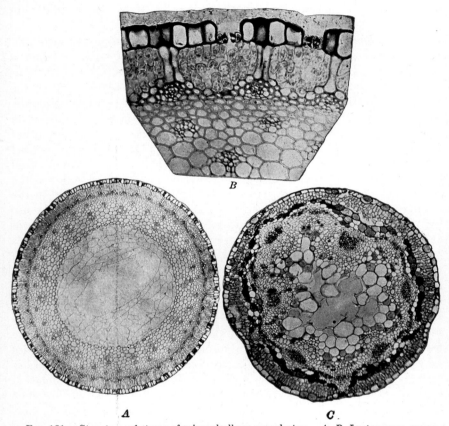

Fig. 181.—Structure of stems of microphyllous xerophytes. *A, B, Leptocarpus*, monocotyledon: *A*, cross section of stem; *B*, small portion of *A*, showing specialized cortical photosynthetic tissue, sunken stomata, lignified epidermis, the leaves, nonfunctioning scales, appearing in section outside epidermis in *B*; *C, Polygonella*, cross section of stem, showing similar condition in dicotyledon.

Stomatal Structure.—Stomatal openings are essential for the intake of carbon dioxide and oxygen and possibly for the passage inward and outward of other gases. When the stomata are open, water escapes even when water loss is harmful to the plant as a whole. For this reason the reduction of transpiration is of the utmost importance in xerophytes, and reduction in number of stomata—either by reduction of leaf surface or of stomatal number per unit area—and elaborate modification of the structure of the stomatal apparatus both reduce transpiration. Water loss in xerophytes is reduced not only by hairy coverings and by the rolling or folding of leaves, but also by the position of the stomata sunken below the level of the other epidermal cells. In these stomata the accessory cells may be of such shape and arrangement that they

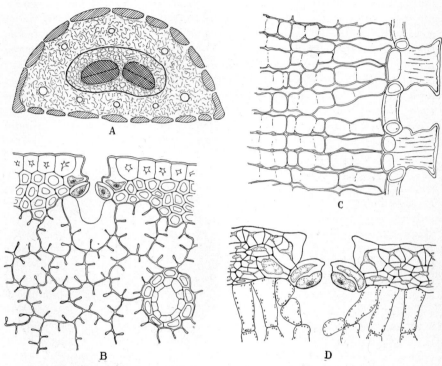

Fig. 182.—Structure of microphyllous xerophytes. *A, B, C,* leaf of *Pinus nigra,* diagram of cross section, detail of small part of same, and of small part of longitudinal section through stomatal furrow, respectively. *A,* the two vascular bundles, doubly cross-hatched, surrounded by parenchyma sheath and endodermis, the mesophyll with resin canals, protected by external sclerenchyma, singly cross-hatched, broken by stomatal furrows; *B, C,* the mesophyll uniform, dense, with infolded cell walls; *D,* cross section through stomatal region in stem, *Equisetum hyemale,* showing cortical photosynthetic tissue, sunken stoma, heavily silicified epidermis and hypodermis.

form one or more outer chambers connected by narrow openings with the stoma itself. In this way the opening between the guard cells is shut off from the outside air with its low humidity. Stomata of this type are common in extreme xerophytes, for example, in *Pinus* (Fig. 182*B,C*), *Equisetum* (Fig. 182*D*), *Cycas* (Fig. 179*C*), and *Leptocarpus* (Fig. 181*B*). The walls of the accessory cells of such stomata are ordinarily very thick and heavily lignified or cutinized, as are frequently also parts of the walls of the guard cells. Sunken stomata, as in *Casuarina* and *Banksia* (Fig. 180*A*), may have the further protection of hairs.

Reduced Leaf Surface.—*Microphyllous* plants form one of the larger groups of xerophytes. In this group the reduction of the leaf surface partly prevents water loss because the total exposed surface of the plant body is relatively small as compared with that of normal mesophytes. Abundant illustrations of this type of xerophyte occur in all groups of plants, for example, *Equisetum, Pinus, Casuarina, Asparagus,* the cacti, *Polygonella* (Fig. 181*C*). In these forms the leaves, if normal in function, are very small; often they are wanting in the mature plant or persist as small scales or bracts which for the most part do not function as leaves. In some genera, for example, *Equisetum, Leptocarpus,* and *Polygonella,* photosynthesis takes place in the stem where photosynthetic tissues are well developed. The reduction of leaf surface is commonly accompanied by increased sclerenchyma, sunken stomata, and water storage.

Needle Leaves of the Gymnosperms.—The needle leaves of the gymnosperms are an important microphyllous type. Many species of *Pinus, Picea,* and other coniferous genera are exposed to xerophytic conditions only during the winter when low soil and air temperatures prevent water absorption and conduction and do not prevent water loss from the leaves. In the leaves of these plants, in addition to reduced leaf surface there are heavy cutinization and sunken stomata. A characteristic needle leaf is the leaf of *Pinus* (Fig. 182*A–C*). In this leaf the vascular tissue consists of two collateral bundles each made up of about equal amounts of xylem and phloem, which are largely secondary in origin. Immediately surrounding the vascular tissues is a zone of so-called "transfusion" tissue (Chap. IV) limited by a well-defined endodermis. Some of the cells of the transfusion tissue, which is in the pericyclic position, show bordered pits even though the tissue has otherwise the appearance of parenchyma. Exterior to the endodermis is the photosynthetic tissue, which is made up of a special type of mesophyll cells. In these cells the inward projections of the wall greatly increase the wall surface along which the chloroplasts are distributed. The outer walls of the epidermal cells are cutinized, and the cutinization extends along the middle lamellae between and around the cells. The wall is thickened to such an extent that the lumen of the cells is nearly occluded. Beneath the epidermis is a well-developed hypodermis several layers thick, consisting of elongate sclerenchyma cells extending parallel with the long axis of the leaf. The stomata are somewhat sunken and arranged in definite longitudinal rows. Their structure as seen in transverse and longitudinal sections of the leaf is shown in Fig. 182*B, C*. The needle leaves of other gymnosperms, though differing from those of *Pinus* are for the most part similar in general features.

Fleshy Xerophytes.—A fourth large group of xerophytes, those possessing fleshy leaves or stems, are described as *malacophyllous*. In these plants, tissues that store water and mucilaginous substances are prominent. In leaves these tissues are located beneath the upper or the lower epidermis, upon both sides of the leaf, or, in extreme types, on both sides of the leaf and in the center. The storage cells are usually large and often thin-walled, as in *Begonia* (Fig. 180*B*). Frequently, the walls are reinforced in a manner that prevents collapse when turgor is reduced. Such storage tissue apparently may actually serve as a source of reserve water during drought, or may screen the underlying tissues from excessive light. In some species the thickened leaves are terete with the vascular bundles arranged in a stele-like cylinder. The mesophyll, consisting frequently of a large proportion of palisade tissue, is more compact than in mesophytes, and the amount of photosynthetic tissues is greater relative to the amount of leaf surface exposed. The leaf surface is, therefore, proportionately reduced. Water-storage tissue is abundant in many microphyllous xerophytes, especially the cacti and plants of similar habit, and some salt-marsh genera, such as *Salicornia*.

Epiphytes.—The structure of epiphytes varies with the environment. Many are xerophytic in their general structure, for example, *Tillandsia* and many orchids; the former has a hairy covering and the latter heavily cutinized epidermis and relatively small leaf surface. The root system of epiphytes may consist, in part, of holdfasts, which anchor the plant; in part, of absorbing roots, which are in contact with the substratum, and in some genera, of aerial roots. Some epiphytes live in habitats that are always moist and have no xerophytic structure.

Hydrophytes.—The number of species of vascular plants that grow submersed in water, or floating, is much smaller than the number of xerophytes and they show much less variety in structure than xerophytes; the uniformity of aquatic environments is in contrast with the variety of xerophytic environments. The factors affecting aquatic plants are chiefly those of temperature, osmotic concentration, and toxicity—the last two dependent upon the amount and the nature of the substances in solution. The structural characteristics in aquatic plants are mainly those of reduction of the protecting, supporting, and conducting tissues, and the presence of air chambers. (The effectiveness of xerophytic structure against drying out can be seen when a xerophyte, such as *Sedum*, and a hydrophyte, such as *Potamogeton*, are exposed to drying conditions; the former will remain alive for many days, whereas the latter will dry out in a very few hours.)

Epidermis in Hydrophytes.—The difference in the structure and function of the epidermis in hydrophytes as compared with that of plants in an aerial habitat is outstanding. In aquatics, the epidermis is not protective but absorbs gases and nutrients directly from the water. This layer in

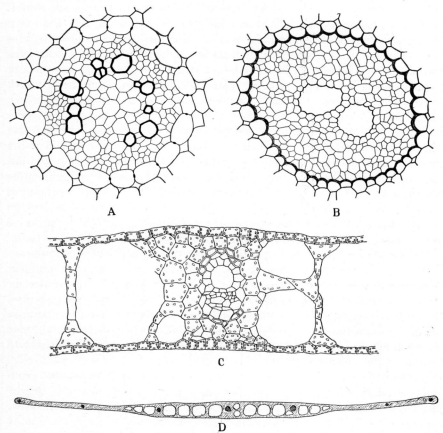

Fig. 183.—Structure of hydrophytes. *A, B*, cross section of stele of stems, *Elatine Alsinastrum* and *Potamogeton pectinatus*, respectively: *A*, xylem cylinder greatly reduced, phloem cylinder complete; *B*, xylem lacking, its position occupied by lacunae, phloem well developed and surrounding the lacunae, inner wall of endodermis much thickened. *C, D*, cross sections of submersed leaf of *Potamogeton epihydrus: D*, diagram, showing air chambers and bundles; *C*, detail of central portion of *D*, showing reduced bundle with the xylem lacking, sclerenchyma reduced, all other cells thin-walled, epidermis with chloroplasts, stomata lacking. (*A, B, after Schenck.*)

the typical hydrophyte has an extremely thin cuticle, and the thin cellulose walls permit ready absorption from the surrounding water. Commonly in aquatics, the epidermis contains chloroplasts and may thus form a considerable part of the photosynthetic tissue, especially where the leaves are very thin. Stomata are wanting in submersed hydrophytes (though sometimes vestigial); the gaseous interchange takes place directly through the cell walls. The floating leaves of aquatic plants have abundant stomata on the upper surface.

Dissected Leaves.—In many species of aquatic plants the submersed leaves are finely divided so that there is, proportionately, a much increased leaf surface in contact with the water. The slender terete segments of such leaves permit close contact between photosynthetic tissues and the water. Such genera as *Myriophyllum* and *Utricularia* are examples of aquatics with divided leaves of this type. Some species with divided, submersed leaves have floating or aerial leaves which are entire, toothed, or lobed.

Air Chambers.—Chambers and passages filled with gases are common in the leaves and stems of submersed plants. Water-storage cavities are structurally single cells, but air chambers are large, usually regular, intercellular spaces extending through the leaf and often for long distances through the stem. Good examples of this type of structure are found in the leaves of *Potamogeton* (Fig. 183*D,C*) and *Pontederia* (Fig. 184*A*). Such spaces are usually separated by partitions of photosynthetic tissue only one or two cells thick. The chambers provide a sort of internal atmosphere for the plant. In these spaces the oxygen given off in photosynthesis is apparently stored and used again in respiration, and the carbon dioxide from respiration is held and used in photosynthesis. The cross partitions of air passages, known as diaphragms, perhaps prevent flooding. The diaphragms are perforated with minute openings (Fig. 184*B*) through which gases but not water may supposedly pass. Air chambers also give buoyancy to the organs in which they occur. Another type of specialized tissue frequent in aquatic plants that gives buoyancy to the plant parts on which it occurs is *aerenchyma*. It is characteristic of species of *Decodon* and *Lythrum*, for example. Structurally, it is a very delicate tissue in which thin partitions enclose air spaces (Fig. 184*C*). Aerenchyma is phellem formed by a typical phellogen of epidermal or cortical origin. The phellem cells are small and thin walled, and a very delicate tissue is built up as the young phellem cells mature. At regular intervals individual cells of each layer elongate greatly in the radial direction while the other cells of this layer remain small. The elongation of these cells pushes outward all the cells of the last-formed layer and forms many elongate air chambers that lie parallel with the plant axis on which they are borne. The radially elongated cells form the radial walls of the chambers and the nonelongate cells which have become separated from the adjacent tangential rows of cells, the tangential walls. In a physiological sense, the term "aerenchyma" is applied to any tissue with many large intercellular spaces. Such tissues may be morphologically a part of the cortex or pith, and hence quite distinct from the typical aerenchyma described above which is of secondary origin.

Absence of Sclerenchyma.—Submersed plants usually have few or no sclerenchymatous tissues and cells; the water itself supports the plant

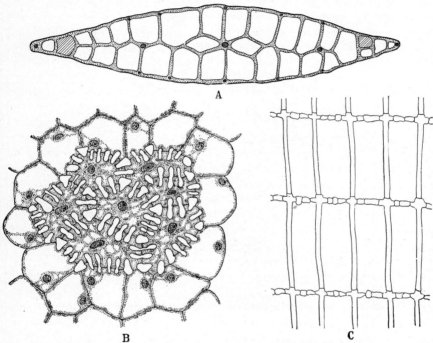

Fig. 184.—Structure of hydrophytes. *A, B, Pontederia cordata*: *A*, cross section of submersed leaf, showing air chambers and reduced vascular bundles (doubly cross-hatched), diaphragms in air chambers singly cross-hatched. *B*, diaphragm in detail showing perforations (unshaded). *C*, aerenchyma in *Decodon*, cross section, showing air spaces formed by elongation of certain cells of each phellem layer.

and protects it in part from injury. Density of tissues, thick walls, and in some plants collenchyma, give some rigidity, but submersed plants, especially the leaves and smaller stems, are usually flaccid and collapse when removed from the water. Sclerenchyma strands occasionally occur, especially along the margins of leaves; here they apparently increase tensile strength.

Reduction of Vascular and Absorbing Tissues.—An aquatic plant is, in reality, submersed in or floating upon a nutrient solution. Those structures that in land plants absorb mineral nutrients and water from the

soil and conduct these substances through the plant are greatly reduced
or absent in aquatic plants. The root system is often greatly reduced,
functioning chiefly as holdfasts or anchors, and a considerable part of the
absorption takes place through the leaves and stems. All degrees of
reduction of the root system are found. Even where reduction of roots is
not extensive, root hairs are usually lacking, and the roots probably do
not absorb water to any extent. In the vascular tissues, the xylem shows
the greatest reduction and in many species consists of only a few elements, even in the stele and main vascular bundles (Fig. 183*A*). Less
commonly in the stele and large bundles, and frequently in the small

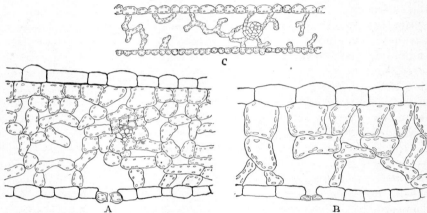

Fig. 185.—Structure of shade leaves. *A*, *Jeffersonia diphylla*; *B*, *Circaea alpina*; *C*, *Cryptogramma Stelleri*. Showing mesophyll loose, palisade layer weak or lacking, and epidermis weakly cutinized—*C*, extreme form, suggesting hydrophytic condition, with chloroplasts in epidermis.

bundles, xylem elements are lacking. In these plants there is usually a
well-defined xylem lacuna in the position of the xylem (Fig. 183*B*,*C*).
Such spaces resemble typical air-chambers. The phloem of aquatics,
though reduced in amount as compared with that of mesophytes, is in
most species fairly well developed as compared with the xylem. It
resembles the phloem of reduced herbaceous plants, in that the sieve tubes
are smaller than those of the woody plants. These reduced bundles are
illustrated in the leaf of *Potamogeton* (Fig. 183*C*,*D*). In aquatics an
endodermis is commonly present, though often weakly developed.

Shade Leaves.—The leaves of deeply shaded habitats have weakly
developed palisade layers. This type of leaf structure is found in many
mesophytic plants such as low-growing woodland plants, for example,
Jeffersonia (Fig. 185*A*). The more extreme plant types in such habitats,
such as *Cryptogramma* (Fig. 185*C*), a fern of moist, shaded cliffs, lack

palisade layers and may possess an epidermis that contains chloroplasts and becomes a part of the photosynthetic tissue. Hydathodes, where water is exuded, are frequently found in plants growing in a saturated atmosphere.

Parasites.—In parasitic, saprophytic, and insectivorous vascular plants in which the plant is dependent, wholly or in part, upon other organisms, there are prominent structural modifications. In parasitic vascular plants there are commonly no roots except in the seedling stage before connection with the host is made. Such plants as *Cuscuta* and *Conopholis* are familiar examples in which, early in the life of the seedling, the parasite establishes direct connection by means of its haustoria with the conducting system of the host. In the region of this connection, the xylem and phloem of the parasite join directly with the same tissues of the host, supplying the parasite with water, mineral nutrients, and elaborated plant foods. The xylem and phloem of the parasite come in direct contact with the xylem and phloem of the host through a dissolution of the outer tissues of the host by enzymes secreted by the parasite at the points of contact where haustoria are formed (Fig. 186*B*). New secondary vascular tissues are then formed by the parasite, and these continue the union with the newly formed tissues of the host.

In true parasites, all photosynthetic tissues are greatly reduced, exist as vestigial structures, or are wanting. In such plants as *Cuscuta* and *Arceuthobium*, the leaves are reduced to scales, and chlorophyll is usually absent throughout the plant. *Cuscuta* may form a large number of host connections. Other plants, for example, *Conopholis* and *Orobanche*, make only one host connection but penetrate deeply into the root of the host. In such species the parasite may form a flowering axis only. Besides these complete parasites, which depend wholly upon the host for food and water, there are many so-called "half-parasites," which, although connected directly with the vascular tissues of the host, apparently manufacture some part of their own food supply. The mistletoes, *Viscum* and *Phoradendron*, are familiar examples. Here water and mineral nutrients are obtained entirely from the host. Carbohydrates are probably, at least in part, formed by the parasite. The vascular tissues of parasites are usually much reduced. Sclerenchyma often supplies nearly all the support of the stems.

"Saprophytes."—The term *saprophytes*, as commonly applied to vascular plants, refers to a group of chlorophyll-lacking forms that have been said to secure their food supply from decaying organic matter, much as do many fungi. Many and probably all such plants are associated more or less closely with fungi in their underground parts. The method by which such plants obtain food materials is apparently bound up in some way, not well understood, with the physiological activities of these fungi. But clearly such plants are not strictly saprophytes—they are, to a certain extent, at least, symbiotic with or parasitic upon the lower forms, or are associated in nutrition with them. It is undesirable to call such plants "saprophytes" until their methods of nutrition are fully understood. Plants of this type are usually greatly reduced in structure—the leaves are scale-like and the stems are reduced largely to inflorescence axes. Internally, there is also great reduction and simplification. The xylem and the phloem are small in amount and the cells of

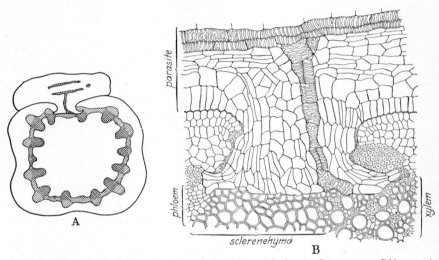

Fig. 186.—Haustorial connection of parasite with host—*Cuscuta* on *Bidens*. *A*, diagram, showing cross section of stem of host, oblique section of stem of parasite, and longitudinal section of haustorium penetrating host as far as the vascular tissues; *B*, detail of haustorium and surrounding tissues. The vascular tissues of parasite connected with those of host.

these tissues often abortive, and sclerenchyma is scarce. The roots may be well developed and abundant in proportion to the aerial parts, as in *Monotropa*, or may be lacking, with rhizomes taking their place, as in *Corallorrhiza*. Where roots are present, they are usually of peculiar structure. Roots of unusual form, together with the fungus hyphae associated with them, form *mycorrhizae*.

Two types of such roots are recognized: *ectotrophic*, those in which the mycelium forms a more or less superficial layer or coating about the roots, and *endotrophic*, those in which the hyphae lie within or between the root cells themselves. In ectotrophic forms the fungus hyphae form a tissue-like weft about the root, enclosing it like a glove about a finger. The mycelium is closely appressed to the root, and some of the hyphae penetrate between and around the epidermal and subepidermal cells. In some forms there is little penetration between the outer cells, in others the cells of the epidermis and of several outer layers of the cortex may be surrounded by hyphae. Rootlets invested by fungus hyphae in this way are of unusual structure—they have neither root cap nor root hairs, have limited growth in length, and are fleshy, with their conducting tissues greatly reduced. In endotrophic forms also, the roots may be enlarged and fleshy, but there is less reduction of vascular tissues. The hyphae are usually restricted to a special part of the cortex where a definite layer of infected cells may occur. In nongreen plants the interrelationship of the two plants is obscure.

Mycorrhizae occur on many green plants also; they are known in many genera of forest trees and in many members of the heath family. The relationship of the fungus and the green vascular plant is not understood but is possibly parasitism of basidiomycetous fungi upon the green plant. In some endotrophic forms, symbiosis may exist.

The above discussion is only a brief statement of some structural modifications of plants. The details of such modifications are almost infinite in number and variety. All can be structurally interpreted by comparison with mesophytic plants.

References

CLEMENTS, E. S.: The relation of leaf structure to physical factors, *Trans. Amer. Micro. Soc.*, **26**, 19–102, 1905.

CONSTANTIN, J.: Études sur les feuilles de plantes aquatiques, *Ann. Sci. Nat. Bot.*, 7 sér., **3**, 94–162, 1886.

DUFOUR, L.: Influence de la lumière sur la forme et la structure des feuilles, *Ann. Sci. Nat. Bot.*, 7 sér., **5**, 311–413, 1887.

HABERLANDT, G.: Anatomisch-physiologische Untersuchungen über das tropische Laubblatt, II. Ueber wassersecernirende und -absorbirende Organe, *Sitzungsb. Math. Naturwiss. Classe Kais. Akad. Wiss. Wien.*, **103**, 489–538, 1894. **104**, 55–116, 1895.

HANSON, H. C.: Leaf structure as related to environment, *Amer. Jour. Bot.*, **4**, 533–560, 1917.

HAYDEN, A.: The ecologic foliar anatomy of some plants of a prairie province in central Iowa, *Amer. Jour. Bot.*, **6**, 69–85, 1919.

LECLERC DU SABLON: Recherches sur les organes d'absorption des plantes parasites (Rhinanthées et Santalacées), *Ann. Sci. Nat. Bot*, 7 sér., **6**, 90–117, 1887.

McDOUGALL, W. B.: On the mycorhizas of forest trees, *Amer. Jour. Bot.*, **1**, 51–74, 1914.

SAUVAGEAU, C.: Sur les feuilles de quelques monocotylédones aquatiques, *Ann. Sci. Nat. Bot.*, 7 sér., **13**, 103–296, 1891.

SCHENCK, H.: Ueber das Aërenchym, in dem Kork homologes Gewebe bei Sumpfpflanzen, *Jahrb. Wiss. Bot.*, **20**, 526–574, 1889.

———: Vergleichende Anatomie der submersen Gewächse, *Bibl. Bot.*, **1**, 1–67, 1886.

Schimper, A. F. W.: "Plant Geography upon a Physiological Basis," Engl. trans., Oxford, 1903.
Starr, A. M.: Comparative anatomy of dune plants, *Bot. Gaz.*, **54,** 265–305, 1912.
Warming, E.: "Ecology of Plants," Engl. trans., Oxford, 1909. (Extensive bibliography.)
Woodhead, T. W.: Ecology of woodland plants in the neighborhood of Huddersfield, England, *Jour. Linn. Soc. Bot. London*, **37,** 333–406, 1906.

INDEX

(Figures in boldface indicate pages on which illustrations occur)

A

Abies, stem apex of, **71, 72**
  trabeculae of, **48**
  wood of, 215, 216, 225
Abscission, 266–275
  of floral parts, 270–271
  of leaves, 267–268, **269,** 270
  of phloem, 231
  of stems, 271, **272**
  zone of, 267, 275
Absorption, by root hairs, 277, 285
  through epidermis, 277
Abutilon, collenchyma in, **84**
Acacia, wood of, 225, 229
Acer, bark of, 258, 259
  cuticle of, **51,** 54
  fruit of, 368
  leaf abscission in, 268
  root-stem transition in, 296
  wood of, 214, 215, 225, 226
    grain of, 227
    parenchyma in, 216
    pith-ray flecks in, 228
    pit-pairs in, **43**
    rays in, 218
    vessel element in, **96**
Aceraceae, leaf traces in, 148
Achene, bundles in, 360
  pericarp in, 368
  reduction in, 356, **357**
  seed coat in, 371
Achras, latex, source of chicle, 118
  periderm in fruit of, 261, 363
Acorus, vascular bundles in, **306**
  vascular system in, **150**
Adlumia, stem of, 303
Adnation (*see* Fusion)
Aerenchyma, 390, **391**
Aeschynomene, wood of, 225

Aesculus, plasmodesmata in endosperm of, 35
  seed of, 376
Agathis, abscised branches of, 272
  heartwood in, 223
  kauri gum, 246
  leaf-trace cambium in, 185
  pit-pairs in tracheid of, **43**
  resin of, 118
Agave, fibers in, 88
Agrimonia, achene of, **357**
  phloem-ray cells of, 242
  stem of, 299
  vestigial interfascicular cambium in, 178
Agrostis, sclerenchyma in leaf of, **336**
Ailanthus, abscission of leaves of, 268
  abscission of stems of, 271
  fruits of, 368
  starch in, **21**
Air chambers, 8, 367, **389,** 390, **391**
Ajuga, fusion in calyx of, **347,** 348
Alburnum, 223
Aleurone grains,19, 22, **23,** 376
Alkaloids, 19
Allium, intercalary meristem in, 323
  leaf of, 335
Almond (*see Amygdalus*)
Alnus, aggregate xylem rays in, 217
  leaf trace and pith form of, **153,** 154
Aloe, bundle development in secondary growth of, **301**
  cambium in, 201
  periderm in, 260
  secondary growth in, 302
Alsinastrum, stem of, **389**
Alstroemeria, inferior ovary of, **352**
Amaranthaceae, accessory cambium in, 312
  medullary bundles in, 314
Amaryllis, seed coat of, 372

Amelanchier, petal of, **358**
Amentiferae, leaf traces in, 147
Amygdalus, fruit of, 367
Ananas, adnation in, 355
    fruit of, 360, 369
Anastomosis, 152, 301
Anchusa, epidermis of corolla of, **358**
Andromeda, ovary of, **353**
Anemarrhena, root-stem transition in, 296
Angelica, oil canal in, **115**
    petiole bundles in, **329**
Angiosperms, anomalous structure in, 315
    collateral bundles of, 139
    companion cells in, 109, 239
    endodermis in, 161
    fibers, gelatinous in, 49
        size of, 9
    leaf gaps in, 145
    leaf traces in, 147
    marginal ray cells in, 220
    origin of lateral roots in, **287**
    persistence of leaves in, 338
    phloem in, 104, 105, 232, 239
        internal, 304
        primary, in roots, 285
    pits in tracheids of, 92
    protoxylem in, 133
    root apex of, **77**
    stelar types in, 142
    stem apex of, **72**
    tyloses in, 222
    vascular bundle in, 307
        (*See also* Dicotyledons; Monocotyledons)
Annual rings (*see* Growth rings)
"Annual rings," of cork, 256, **257**
Anomalous structure, in stems, 309, **310,** 311, 312, **313, 314,** 315
Anther wall, 48
Anthocyanins, 18
Apical-cell theory, 67
Apical cells, 65–67, **75, 76**
    of axis, 124
    of floral apex, **74**
    of leaf, 319
    of root, **75, 76**
    of stems, **70, 71, 72, 73**
        (*See also* Initials; Meristem)
Apical meristems, **64,** 123
    initials of, 61, 65

Apical meristems, mature cells of protophloem in, 131
    in monocot leaf, **334**
    origin of cambium from, 63, 175
Apios, crystals in endodermis of, 158
    stem of, 303
Apium, collenchyma in, 87, 88, **330**
Aplectrum, stomata in, **168**
Apocynaceae, internal phloem in, 305
    intrusive growth in, 12
    nonarticulate latex ducts in, 119, 120
Aponogeton, chloroplasts in, **15**
Appendages, 1, 123, 126, 341
Apple (*see Malus pumila*)
Apposition, 34
Aquatic plants, endodermis in, 161
    stomata in, 333
    vessels lost in, 98
        (*See also* Hydrophytes)
Aquilegia, flower of, **343, 345, 346**
    follicle of, **357**
Araceae, roots in, 291
    seed coats of, 372
    venation in, 334
Aralia, leaf traces in, 148
Araucaria, absence of wood parenchyma in, 102
    compression wood in, 228
    extension of leaf trace in, 185
    leaf-trace cambium in, 185
    persistent leaves of, 339
Arceuthobium, 393
Arctium, venation in, 334
Arisaema, chromoplasts in fruit of, **16**
    protoxylem in fruit of, **135**
Aristolochia, anomalous stem of, 302, **310,** 311
    crushing of pith in, 156
    traces to lost petals in, 356
Aristolochiaceae, seed coats of, 372
Aroids (*see* Araceae)
Artemisia, stem of, **299**
Artocarpus, fruit of, 369
Asarum, collenchyma in, **84**
    petiole bundles in, **329**
    tyloses in, 222
Asclepiadaceae, internal phloem in, 305
    intrusive growth in, 12
    nonarticulate latex ducts in, 119
Asclepias, collenchyma in, **84**

INDEX

Asclepias, parenchyma in, **83**
  primary vascular system of, 154
  vascular bundles in, 308
Asimina, fruit of, 366
Asparagus, leaf of, 387
Aspidium, vascular skeleton of leaf of, **318**
Aster, collenchyma in, 85
  compression of outer tissues in, 182
Aubrietia, hairs on, **170**
Avena, hairs on, **170**
  leaf of, **328**
  stem apex of, 73
Avocado (*see Persea*)
Axis, aerial and subterranean, 1
  ontogeny of, 124, **125**
  parts of, 1, 123, 277, 293
  structure of, **3**

B

Balsaminaceae, seed coats in, 372
Banana (*see Musa*)
Banksia, leaf of, **383**, 384
  stomata of, 387
Bark, 258
  exfoliation of, 259, 260
  inner, 231
  "ring," 260
  scale, **259**
  uses of term, 254
Bars of Sanio, 49
Basswood (*see Tilia*)
Bast, of commerce, 246
  definition of, 111, 112, 231
  fibers, 112, 246
Bauhinia, anomalous stem in, **310**
Beet (*see Beta*)
Begonia, anomalous stem in, **310**
  leaf of, **383**, 388
  leaf-trace bundles in, 314
  root germs in, 290
Berberidaceae, derivation of woody stems of, 300
  medullary bundles in, 314
  seed coats of, 372
Berries, epidermal cells of, 362
Beta, anomalous structure in, 313, **314**
  collenchyma in, 87
  seed of, 371
Betula, abscission of leaves in, **269**, 270

Betula, periderm of, 255
  "annual rings" of cork in, **251**, 256
  "bark" of, 254, 258
  cork cells in, 250
  lenticels in, 265, 266
  phellem in, **251**
  wood of, clustered vessels in, 215
  curly grain in, 227
  diffuse-porous, 214
  heartwood in, 223
  pith-ray flecks in, 228
  vessel element in, **96**
Bidens, host to *Cuscuta*, **394**
Bignonia, anomalous stem in, **310**
Binucleate cells, 27
Birch "bark" (*see Betula*)
Black walnut (*see Juglans*)
Blackberry (*see Rubus*)
Blephilia, fusion in calyx of, **347**, 348
Blueberry (*see Vaccinium*)
Boraginaceae, medullary sheath of, 156
Bordered pit-pair (*see* Pit-pairs)
Branch, abscission of leafy, 267, **272–274**
  scar periderm of, **274**
  shedding of leafy, 275
Branch bases, burial of, 197, **198–200**
  stripping of phloem at, 197
Branch gap, **143**, 144, **145**, 146
  relation of secondary growth to, 186
  in woody stem, 297
Branch traces, **143–145, 149, 183**
  burial of, 185
Brazilian rubber tree (*see Hevea*)
Bryophyllum, rooting in, 290
Bryophytes, apical cell in, 66, 67
Bud scales, cork layers on, 261
  secretory cell in, **115**
Budding, cambium in, 201
Buds, meristems of adventitious, 63, 163
Bundle ends in leaf, **322, 326, 327**
Bundle sheath, **320, 326,** 328–330, 336
Burial, of branch bases, 197, **198–200,** 275
  of branch traces, 185
  of leaf-trace bases, **181, 184**
  of wound, 201
Burls, formation of, 201

C

Cactaceae, fleshy xerophytes, 388
  leaf of, 387

Cactaceae, leaf traces in, 314
  stem apex of, 73
Calceolaria, epidermis of corolla in, **358**
Callus, 200–201
  definitive, 108
  development in sieve elements, 107–109, 245
  formation of, 200–201
Calocarpum, periderm in fruit of, 261, 363
Calycanthaceae, cortical bundles in, 314
Calycanthus, flower of, 353
Calyptrogen, 75
Calyx, anatomy of, **347**
  hairs on, **170**
"Calyx tube," 368, 369
Cambium, 60–67, 175–202, **188, 189, 193, 195**
  accessory, 63, 312–314
  in budding and grafting, 201
  cylinder, **177**, 179
  duration of, 180
  effect on primary body of, 180–182
  fascicular, **177**, 178
  formation of xylem and phloem by, **186, 196**
  growth about wounds, 200–201
  initials of, 61, **99**, 176, 187–190, **192**, 194
    cell division in, **26**, 191
    primary pit-fields in, 28
  interfascicular, **177**, 178
  in monocots, 201, **202**, 302
  position in axis of, **3, 4**
  in roots, 119, 176, 179
  in stems, 164, 178–179
  storied, 190
  structure of, 187, **193**
  time of activity of, 195–197
"Cambium-miners," 228
Campanulaceae, internal phloem in, 305
Canal-like cavities, of septal nectaries, 117
  of simple pits, 39
Canals, 8
  oil, **115**
  in pericycle, 158
  resin, **115**
Canna, seed coat of, 374
  tyloses in, **222**

Cannabis, compression of outer tissues in, 182
  fibers in, **86**, 111, 157
  development of, 88
Caprifoliaceae, sieve-tubes in, 107
  traces to lost carpels in, 356
  vestured pits in, 48
Carica, druse in cell of, **20**
  latex, 118, 364
Caricaceae, articulate latex ducts in, 120
  lignification in wood of, 225
Carnivorous plants, 117
Carotenes, 18
Carotinoids, 18
Carotins, 18
Carpels, 359–360
  bundle fusion in, **350, 357**
  fusion of, **349, 351**
  reduction of bundles in, 356, **357**
  syncarpy, **349**
  traces of, 344, 359
  traces to lost, 356, 360
  vascular structure of, **350, 357**
Carpinus, aggregate rays in, 217
  periderm in, 255
Carrot (see *Daucus*)
Carya, bark of, 258, 259
  bundle sheath in, 328
  crystals in nutshell of, **20**
  fruit of, 367, 369
  phloem of, 232, **233, 235**, 237, 239–241
  root cuttings of, 290
  wood, cell arrangement in, 328
    crystals in, 20
    durability of, 225
    libriform fibers in, **94**
    parenchyma in, 102
Caryophyllaceae, fruit of, 368
  leaf traces in, 148
  vestigial traces in flower of, 356
Casparian dots, 158, **159**
Casparian strips, 48, 158, **159**
Castanea, curly grain in wood of, 227
  durability of wood of, 226
  embryo in seed of, 376
  inflorescence, abscission in, 275
  leaf abscission in, 268, **269**
  parenchyma in, **83**
  phellogen in, 253
  secondary phloem in, 241, 243

Castilloa, abscised branches in, 272
Casuarina, leaf-trace bundles in, 314
  stomata of, 387
Catalpa, abscission of leaves of, 268, **269**
  wood of, durability of, 222, 226
    ring-porous, 214
    tyloses in, 221
    vasicentric parenchyma in, 216
Caulophyllum, seed coats of, 375
Ceiba, fruit of, 368
Celastrus, stem of, 302
Celery (*see* Apium)
Cell, **7**
  albuminous, 110, 242
  annular, 133, 134
  arrangement of, **7**, 136, 214, 215
  cambiform, 240
  color in, 18
  definition of, 6
  development of, 9, **17**
  division, 9, 24, **25, 26,** 119, 191, **192**
  enlargement in fruit, 361, **363, 364,** 365
  inclusions, 13
  passage, 158
  plate, **24, 25,** 26
  sap, 7, 13, 17, **359**
  shape of, 8
  size of, 9
  transfusion, 158
    (*See also* Cell wall; Cells)
Cell wall, 6, **7,** 23
  calcium carbonate in, 50
  calcium oxalate in, 50
  centrifugal thickening of, 49
  checking in, **44**
  chemical nature of, 49
  of collenchyma, **84**
  "dentations" of, **45,** 48
  of fibers, 87
  formation of, **24–26, 29**
  gelatinous, 34, 49, 87, 93, **94,** 95, 215, 268
  gross structure of, 27–29, **30**
  method of building of, 34
  mineralization of, in grasses, 50
  minute structure of, 30–34, **31, 33, 34**
  origin of, 24
  of parenchyma, **83**
  in phloem fibers, 111
  of sclereids, 89

Cell wall, sculpture and modification of, 37, **38–41, 43–48**
  septa in fiber-tracheids, 95
  silica in, 50
  tyloses in, 220, 221
  in wood, 224
    (*See also* Primary cell wall; Secondary cell wall)
Cells, of abscission zone, 267–268
  cambium, 187–191, **194**
  endodermal, 163
  epidermal, 165, 166, 167
  grit, 88
  of lenticels, **263**
  meristematic, 61
  of periderm, 249, 250, **251,** 252
  permanent, 61, 82, 124
  phloem mother, 186, 188
  phloem ray, 242
  tyloses in, 220, **221**
  xylem mother, 186
    (*See also* Cell; Cell wall)
Cellulose, in cell wall, 30, **31**
  in endosperm, 376
  in fruit, 365
  orientation of fibrils in cell wall, **33**
  in wood, 224
Central cylinder, 1, 68, 123
  ontogeny of, 127
  of root, 281
Centrospermae, roots of, 291
Cephalanthus, petiole of, **329**
  secondary phloem in, **233,** 237, 241. 242
Ceroxylon, wax on leaves of, 383
Chaenomeles, fruit of, 363
Chenopodiaceae, accessory cambium in, 312
  medullary sheath in, 156
Chenopodium, anomalous stem structure in, **313**
  collenchyma in, 85
  pit-pairs in cells of, **44**
  primary vascular system of, **150**
Cherry (*see* Prunus)
Chestnut (*see* Castanea)
Chicle, latex, source of, 118
    (*See also* Achras)
Chimaeras, periclinal, plasmodesmata in, 37

Chionanthus, traces to lost stamens in, 356
Chlorophyll, 18
Chloroplasts, in *Aponogeton,* **15**
   in epidermis, **389, 392**
   in guard cells, 169
   of mesophyll cell, **7,** 332
   structure of, **15**
   in *Todea,* **15**
   type of plastid, 14, 15
   in *Zea,* **14**
Chondriosomes, 17
Chromoplasts, 14, 15, **16**
   in petal, 359
Chrysanthemum, fusion in corolla of, **348**
Cinnamon, 246
Circaea, fruit of, **362,** 363
   leaf of, **392**
Citrullus, fruit of, **363,** 368
   pit-pairs in, **44**
Citrus, lysigenous glands in, 364
   oil cavity in, 115
Citrus fruits, chloroplasts in seeds of, 17
   cuticle on, **51**
   oil cavity in, **115,** 118
   pulp of, 364
Clarkia, epidermis of corolla in, **358**
Clematis, dormant phloem in, 245
   interfascicular cambium in, 178
   parenchyma in, **83**
   pit-pairs in pith of, **44**
   vascular rays absent in, 208, 241
   woody stems of, **298, 302,** 303
Clethra, seed of, **372**
Closing membrane (*see* Pit membrane; Pit-pairs)
Club mosses, absence of leaf gaps in, 145
   exarch xylem in, 129
   pits in tracheids of, 92
   protostele in, 142
Coconut (*see* Cocos)
Cocos, fruit of, 361
   rhytidome-like layer in, 260
   sclerenchyma in, **86**
Cohesion (*see* Fusion)
Coleus, abscission of leaves in, 270
   tyloses in, 222
Collenchyma, cell wall in, 33
   in cortex, 163, **164**
   in fruit, 364, 366

Collenchyma, in hydrophytes, 391
   in leaves, **330**
   simple tissue, 82, **84,** 85
Color, autumn, 18
   in cells, 18
Columella, 279
Combretum, interxylary phloem in, 311
Commelinaceae, secondary root hairs in, 280
Companion cells, 104, **106,** 109–110, **238,** 239
   ontogeny of, 109, **195**
Compositae, articulate latex ducts in, 120
   flower of, reduction in, 356
   leaf traces in, 148
   lignification in phloem of, 246
   medullary sheath in, 156
   pericarp of, 368
   phloem in, internal, 305
      secondary, 239
   root hairs in, 166
   seed coats of, 371
   stem of, 298
Conduction, horizontal and vertical in phloem, 243, 244
   in roots, 277
   in secondary xylem, 204
   in trees, 209
   in xylem rays, 217
Conifers, bordered pit-pairs in tracheids of, 40
   compression wood in, 228
   perforations in pit membrane in, **41**
   trabeculae in tracheids of, 48
   tylosoids in, 221
   (*See also* Gymnosperms; listed genera of conifers)
Conjunctive tissue, 301, 312, **313**
Conopholis, 393
Convolvulaceae, internal phloem in, 305
Convolvulus, tyloses in, 222
Copernicia, wax on leaves of, 383
Corallorrhiza, 394
Cordyline, secondary growth in, 302
   storied cork in, 260, **261**
Coreopsis, hairs on leaf of, **170**
Cork, commercial, 250, **251,** 256, **257,** 258
   lenticels in, 266
   secondary tissue, 123
   (*See also* Phellem)

Cork cambium, 63, 65, 67
  (*See also* Phellogen)
Cork oak (*see Quercus suber*)
Cornus, adnation in, 353
  cuticle on stem of, **51,** 54
  phloem parenchyma in, **236,** 239, 240, 244
  ray cells in, 243
Corolla, epidermis of, **358**
  fusion in, **348**
  hairs on, **170**
  reduction in, 356
Corpus, 69, 70, 73
Cortex, 1, **3,** 68, **77,** 123, 163, **164,** 165
  bundles of, 314
  crystals in, **20**
  effect of cambial activity on, 181–182
  in fruit, 369
  latex cells in, **115**
  meristem in, 369
  of monocot stem, 301
  origin of aerenchyma in, 390
  parenchyma in, 83
  photosynthetic, **385, 386**
  primary, 163
  of root, 280, **281,** 283
  suberization of primary cells of, 260
  tannin in, 22
Corylus, petiole bundles in, **329**
Cotoneaster, root germs in, 290
Cotton "fibers," **373**
Cotyledons, plasmodesmata in, 35
Crassulaceae, fusion in flower of, 351
Crassulae, **40,** 49
Crataegus, fruit of, ripening in, 365
  scelerenchyma in, **86**
Crotch angles, distortion of tissues in, 190, 198, **200**
  phloem in, 197, **198**
  wood parenchyma in, **198**
  zone of weakness in, 199
Cruciferae, seed coats of, 372
  vestured pits in, 48
Crushing, of cortex, 182
  of endodermis, 163, 181
  of outer stem tissues, 300
  of pericycle, 182
  of phloem, **125,** 131, 132, 181, 182, 196, 232, **238,** 244, **285**
  of pith, 156

Cryptogramma, leaf of, **392**
Cryptomeria, stem apex of, 72
Cryptostegia, nonarticulate latex ducts in, 119
Crystal sand, 19
Crystalloids, 22
Crystals, 13, 19, **20, 83**
  in cell wall, 50
  in endodermis, 158
  in epidermis, 165
  in phloem parenchyma, 110, 113, 244, 246
  protein, 22
  in tyloses, 220
  in xylem, 103
Cucumis, hairs in, **170**
  stomata in, **168**
Cucurbita, cambiform cells in, 240
  development of secondary vascular tissue in, **196**
  fruit of, 366
  phloem of, companion cells in, 239
    internal, 304
    primary, 132
  primary skeleton in, 149
  root-stem transition in, 296
  tyloses in, 222
  vessel elements in, **100**
Cucurbitaceae, bundles in, bicollateral, 139, 305
  medullary, 314
  placental, 355
  fruits of, 361, 368
  traces to lost stamens in, 356
Curcuma, storied cork in, 260, **261**
Cuscuta, 393, **394**
Cuticle, **51,** 52–54, 260
  on fruits, 362, 366
  on petals, 359
  on seeds, 374
Cuticular pegs, **52,** 53
Cuticularization, 53, 382
Cutin, in cell wall, 32, 49
Cutinization, of cell wall, 34, 49, 53
  in herbaceous monocots, 260
  in roots of pteridophytes, 291
  in xerophytes, 382, 384, 387
Cycads, anomalous structure in, 315
  girdling traces in, 148
  stem apex of, **72**

Cycas, leaf of, **382**, 386
Cyperus, leaf of, 337
Cypripedium, seed of, 372
Cystoliths, 19, 50
Cytoplasm, **7**, 10, 13, **17, 24, 25**

D

Dalibarda, ovary of, **354**
Dandelion (*see* Compositae; *Taraxacum*)
Darbya, ovary of, 353, **354**
Dasylirion, leaf of, **51, 382**, 384, 385
Date (*see Phoenix*)
Daucus, chromoplasts in, 15, **16**
  lenticels in root of, 263
Deciduous plants, rupture of leaf trace in, **184**
Decodon, aerenchyma of, 8, 390, **391**
Defoliation, growth-ring formation after, 208
Dehiscence, 369
Delphinium, seed coat of, **374**
Dermatogen, **68, 76, 125**
Dianthera, bundle arrangement in, 152
Diaphragms, 390, **391**
Dicliptera, seed coat of, **374**
Dicotyledons, bundle arrangement in, 152, 307, **308**
  bundle ends in, 327
  endodermis in, 161
  leaf in, 317, 319, 330
  leaf abscission in, **269**
  medullary bundles in, 314
  primary type endodermis in, 159
  root apex of, **77, 78**
  secondary phloem in twigs of, 239
  seed coats of, 371
  stems of, **289, 299**
  venation in, 334
  xerophytic stem of, **385**
  (*See also* Angiosperms)
"Dictyostele," **141**, 142
Differentiation, in meristems, 60, 65, 67, 71
  in time and space, 124
  of tissues in fruits, 361, **363, 364**
  of tissues in leaf, **324**
  of xylem and phloem in leaf, 323
Digitalis, primary vascular cylinder in, 154
  stem of, **299**

Dionaea, digestive glands of, 117
Dioscoreae, secondary growth in, 302
Diospyros, fruit of, 364, 366
  fusiform cambium cells in, 190
  plasmodesmata in endosperm of, 35, **36**
  seed of, 376
  wood of, **213**, 218, **223**, 225
    cell arrangement in, 215
    pit-pairs in, **44**
Dipsacus, root-stem transition in, 295
Dirca, abscission of stems in, 274
  phloem of, fibers in, 111, 241
    secondary, 237
Distortion, of secondary vascular tissue, **195, 196, 308**
  of tissue in crotch angle, **198**
  of tracheid, **91**
Dracaena, cambial growth in, 201, **202**
  cuticle on stem of, **51**
  sclerenchyma in, **86**
  secondary growth in stem of, 302
  storied cork in, 260
Drosera, digestive glands of, 117
Drupes, epidermal cells of, 362
  stony layer of, 367
Druses, 19, **20**
  in phloem parenchyma and sclerenchyma, 246
Dry fruits, **362**, 367–368
  pericarp of, **362**
Duabanga, vestured pit-pairs in, **47**
Duchesnea, achene of, **357**
Dulichium, primary vascular system in, **150**
Duramen, 223

E

Ebony (*see Diospyros*)
Ecological anatomy, 380–396
  epiphytes, 388
  hydrophytes, 388, **389**, 390, **391**, 392
  parasites, 393, **394**, 395
  saprophytes, 394
  shade leaves, **392**
  xerophytes, 380–381, **382, 383**, 384, **385, 386**, 387, 388
Elatine, stem of, **389**

Elongation, of cells in xylem maturation, **195**
  of protophloem and protoxylem, 134
  in protoxylem, **135, 136**
  in roots, 285
Elymus, sclerenchyma in leaf of, **336**
Embryo, 361, **372**, 376
  chloroplasts in, 16
Emergence, 1, **2**
Endocarp, 360
  sclerenchyma in, **86**
Endodermis, 158, **159**, **160**, 161–163
  in axis, **3**, 123, **125**
  effect of cambial activity on, 181–182
  function as diffusion layer, 161–162
  in hydrophytes, 392
  inner, **159**, 160, 161, 304
  in leaf, **386**, 387
  origin of some meristems, 163
  outer, **159**
  primary, 158, 159
  in root, **77**, 277, **282**, **287**
  in stems, 157, **389**
Endosperm, 361, 369, **370**, 376
  exalbuminous, 371
  "horny," 376
  plasmodesmata in, 35, **36**
  plastids in, **16**, 17
Entada, interxylary phloem in, 311
Entomophilous plants, 117
Environment, 380
  aquatic, 380
  mesophytic, 380
  xerophytic, 380–381
Enzyme, activity of, causing gummosis, 228
  activity in parasites, 393, **394**
  protein-digesting, 117
  secretion, 117
Ephedra, primary vascular system of, **150**
Epidermis, 1, **3**, 68, 123, **164**, 165–171
  biseriate, multiseriate, 165
  of corolla, **358**, 359
  cuticle on, **51**
  effect of cambial activity on, 181
  of fruits, 362, **363**
  function of, 166
  of hydrophytes, **389**
  of leaves, 331–333, 337
  lignification of, 382, **385**

Epidermis, in monocots, 260
  origin of aerenchyma in, 390
  of root tip, **77**
  of seed coats, **373**
  of xerophytic leaves, **383**, 384, **386**
Epigaea, fruit of, 368
  hairs on corolla of, **170**
Epiphytes, 388
Equisetum, apical cells of, 65, 67, 76
  endodermis in stem of, **159**
  intercalary meristem in, 66
  leaf of, 387
  leaf gaps absent in, 145
  mineralization of cell wall in, 50
  stem of, **386**, 387
    protoxylem in, **135**
  stomata of, **386**
  vascular bundles in, 307, 326
Ergastic substances, 13, 19
Ericaceae, anatomy of ovary in, **354**
  chloroplasts in, 17
  leaf traces in, 148
  placental bundles in, 355
  traces to lost carpels in, 356
Erythrina, epidermis of corolla of, **358**
Eucalyptus, wood of, **45**, 225
Euonymus, corky-winged stem of, 250
Eupatorium, persistent root hairs of, 280
Euphorbia, latex cells in cortex of, **115**
  secretory cells in nectary of, **115**
Euphorbiaceae, medullary sheath in, 156
  multinucleate stages of vessel elements in, 101
  nectaries in, 117
  nonarticulate latex ducts in, 119
Evergreen trees, time of cambial activity in, 195
  (*See also* Leaf, evergreen)
Exfoliation, of bark, 259, 260
Exocarp, 360, **364**, 367
Exodermis, 291
Expansion theory, 142

F

Fagaceae, sieve tubes in, 107
  vestigial traces in flower of, 356
Fagus, sclereids in secondary phloem of, 241, 246
  periderm in, 255

Fats, in fruits, 365
Ferns, anomalous structure in, 315
  apical cells in, 65, 67, 76
  bundles in, amphicribral, 140
    arrangement of, 152
  leaf in, 317, 325, 339
  leaf gaps in, 145
  mesarch xylem in, 129
  mucilage in cells of, 116
  pits in tracheids of, 92
  stelar types in, 142, 304
Festuca, sclerenchyma in leaf of, **336**
Fiber-tracheid, 93, **94,** 95
  in angiosperm wood, 216
  in gymnosperm wood, 216
  pits in, 93
  septate, **94,** 95
  wall, minute structure of, **31**
    fibrils in, 32
Fibers, **86, 87, 88,** 93, **94,** 95
  in angiosperm wood, 216
  bast, 112, 246
  cortical, 112, 163
  cotton, 171
  development of, 88
  in fruits, 367
  gelatinous wall in, 49, 87, 93, **94,** 215
  in gymnosperm wood, 216
  in leaf, 331
  libriform wood, 93, **94,** 215
  multinucleate, 87
  pericyclic, 131, 157, 331
  phloem, 112
  pit-pairs in, **43, 44,** 87, 93
  sheaths of, around vascular bundles, 138, 308, 309
  size of, 9
  (*See also* Phloem, fibers of,)
Fibrils, 32
  orientation, in fiber-tracheids and wood fibers, 32
  in latewood tracheids, **33**
Fibrovascular bundle, 138, 308
Ficus, cork layers on petiole of, 261
  fruit of, 369
  latex, source of rubber in, 118
Flavones, 18
Flax (*see Linum*)
Fleshy fruits, 363–367
  epidermal cells of, 362

Fleshy fruits, pericarp of, 362, **363, 364**
  ripening of, 365–366
  stony layer of, 367
  texture of, 366
Floerkia, fleshy pericarp of, **363**
Floral apex, **71,** 74
Floral appendages, abscission of, 270–271
  in fruit formation, 367
  fusion of, 355
  traces of, 342, 356
    (*See also* Carpels; Petal; Sepal; Stamen)
Flower, 341–360
  apetalous, unisexual, 356
  chromoplasts in, 17
  cuticle on, 53
  epidermis of, 166
  hairs on, **170,** 171
  nectaries of, 117
  ontogeny of, 341, 342
  organs of, 356, **357, 358,** 359–360
  stele in, 342
  traces of floral appendages, 342
  vascular skeleton of, 342–360
  vascular structure of, **343,** 345–360
Foliar traces (*see* Leaf traces)
Follicle, reduction of, 356, **357**
Food storage, by phelloderm, 252
  in roots, 277, 278
  by secondary phloem, 244
  by secondary xylem, 204, 217
Forsythia, chromoplasts in petal of, **16**
Fossil plants, polysteles in, 142
  protosteles in, 142
Fragaria, achene of, **357**
  fruit of, **364,** 369
  tannin in, **21**
Fraxinus, cambial initials in, 191
  leaf traces in, 148
  sieve tubes in, 107
  vessels in, 98
  wood of, 214, 223, 225
    parenchyma in, 216
    pit-pairs in, **44**
Freesia, bundle end in petal of, **326**
Frost as a factor, in autumn color, 18
  in fall of leaves, 268
Fruits, 360, 361, **362–364,** 365–369
  abscission of, 267, 275
  accessory structures, 368

Fruits, cuticle on, **51,** 54
  dehiscence of, 369
  dry, 362, 367–368
  epidermis of, 34, 166, 362
  fleshy, 363–367
  hairs on, **170,** 171
  morphology of, 360
  multiple, 369
  oil canal in, **115**
  ontogeny of, 360–361
  parenchyma in pulp of, 83
  periderm in, 249, 261, 363
  placenta in, 368
  ripening, 361, **364,** 365, 366
  sclerenchyma in, **86,** 88, 89
  shape of, 361
  stony layer in, 367
  texture of, 366
  vascular skeleton of, 361, 362
  wax on, 54
  winged, 368
Fumaria, root-stem transition in, 295
Fundamental tissue, 67
Fungus, erosion in cell wall, **44**
  in formation of mycorrhizae, 394, 395
  Fusion, under adnation, **351,** 353, **354,** 355
  under cohesion, **347–351**
  in floral vascular skeleton, 347

G

Galls, 201
Gaultheria, fruit of, 366
Gaylussacia, glandular hair of, **116**
  ovary of, **354**
Gentiana, seed coat of, **374**
Gentianaceae, internal phloem in, 305
  roots in, 291
Geum, achene of, **357**
  stem of, 299
  vestigial interfascicular cambium in, 178
Ginkgo, lenticels in, 265
  seed coats of, 375
Glands, 116
  digestive, 117
  hydathodes, 117–118
  laticiferous ducts, 118
  nectaries, 117

Glands, resin, oil, and gum ducts, 118
Gleditsia, persistent root hairs in, 280
  root lenticels in, 290
Gliding growth (*see* Growth)
Gnetales, vessels in, 98, 216
Gossypium, seed coats of, **373,** 375
Graft hybrids, plasmodesmata in, **37**
Grafting, cambium in, 201
Grain in wood, 227
  bird's-eye, 227
  curly, 190, 227
  silver, 227
  spiral, 190
Gramineae, secondary suberization in, 260
Grasses, anastomoses at nodes of, 301
  leaf of, 335–338
    bundle sheaths in, 336
    epidermis of, 337
    hydathodes on, 118, 327
    intercalary meristem in, 323
    mesophyll in, 337
    mestome sheath in, 161
    sclerenchyma in, 338
    stomata in, 338
  meristem in, 66, 69, **73,** 75
  mineralization of cell wall in, 50
Ground cherry (*see* Solanaceae)
Growth, 60
  apical, of axis, 123
  cellular adjustment during, 10
  centripetal and centrifugal, 128, 129
  gliding (or sliding), 10, 191, 194
  intrusive, 11, 194
    plasmodesmata changes during, 37
  primary, 3, 123, 175
  in root, 75
  secondary, 3, 175
    burial of leaf-trace bases by, **181,** 182
    in monocots, 301, 302
    relation to leaf and branch gaps of, 186
    in roots, **285,** 286
  symplastic, 12, 194
    plasmodesmata distribution during, 37
  time of cambial activity, 195–197
Growth layer (*see* Growth rings)
Growth rings, **198,** 208
  double (or multiple), 214
  false, 214

Guaiacum, libriform fiber in, **94**
Guard cells, **52**, 167, **168**, 169
Gummosis, 49, 228–229
Gums, 19, 50, 118
    formation of, in gummosis, 228
    in heartwood, 223
    in phloem parenchyma, 246
    in tyloses, 220
    in xerophytic leaves, 384
    in xylem, 103
Gymnocladus, seed coat of, **370**
Gymnosperms, albuminous cells of, 242
    anomalous, 323
    cambium in, 190, 191
    collateral bundles in, 139
    ectophloic siphonostele in, 142
    endodermis in, 161
    leaf of, 387
    leaf gaps in, 145
    leaf traces in, 147
    ontogeny of vascular tissue in, 194
    origin of lateral roots in, **287**
    perforated pit-membrane in, 39, 92
    persistence of leaves in, 338
    phloem in, 104, 105, 232, 239, 241
    plasmodesmata in, 35
    protoxylem in, 133
    root apex in, 76, 78
    size of fibers in, 9
    transfusion tissue in, 113
    xylem rays in, 205, **206**, 218
        (*See also* listed genera of gymnosperms)

H

Hairs, 1, **2**, **170**, 171
    on fruit, **362**, 368
    glandular, **116**, 171, **358**
    on leaves, 338
    multicellular, 171
    root, 166, **167**, 277, 279, 280
    on seed coat, **373**, 375
    stinging, **116**, 171
    on xerophytes, **383**, 384
Halophytes, 381
Haustoria, of parasite, **394**
Hedera, inferior ovary of, **352**
Helianthus, fusion of corolla of, **348**
Heliotropium, hairs on calyx of, **170**

Helleborus, follicle of, **357**
Hemicellulose, in cell wall, 32
Hemlock (*see Tsuga*)
Hemp (*see Cannabis*)
Herbaceous plants, abscission in, leaf, 270
    stem, 271
    cambium initials in, 190, 191
    leaf development in, 323
    medullary rays in, 205, 208
    primary phloem of, 232, 240
    root germs in, 290
    secondary phloem of, 231, 237, **238**, 239, 240
    stomata of, 169
    tyloses in vessels of, 222
    vascular bundles in, 307, 309
Hevea, articulate latex ducts in, 120
    latex, source of rubber, 118, 246
Hibiscus, secondary phloem in, **238**
Hilum, 22
Histogen, **68**, 75
Histogen theory, 68, 75
Horsetails (*see Equisetum*)
Humulus, stem of, 302
Hydathodes, 393
Hydrastis, follicle of, **357**
Hydrocharis, root apex of, 78
Hydrophytes, 380, 388, **389**, 390, **391**, 392
    sclereids in, 89
    vessels absent in some, 216
Hypericaceae, seed coats of, 371
Hypericum, lateral root in, **287**
    primary vascular cylinder in, 154
Hypodermis, 163, **281**, 364, **382**, **383**, 384, **386**, 387

I

I-beam structure, in leaf, 331, **336**
Ice crystals, in separation layer, 268
Impatiens, collenchyma in, 85
    discrete vascular bundles in, 177
    primary skeleton of, 149
Impregnation of wood (*see* Wood, impregnation of)
Inclusions, nitrogenous, 22
    of protoplast, 13, 19
Inflorescences, abscission of, 275
    in fruit formation, 360, 369
    supra-axillary, 355

Initials, of cambium, 61, **99**, 176, **192**
   of endodermis, 163
   of floral apex, 71, 74
   fusiform, 187, **192**
   of leaf, **324**
   of ray, 187
   of root, **75**
   of stem, **69–71, 72**–74
      (*See also* Apical cells; Cambium; Meristem)
Insectivorous plants, 117, 393
Integuments, 369, **370,** 371, **372,** 373
Intercellular layer, 24, 28
Intercellular spaces, in collenchyma, 85
   in epidermis of petals, **358,** 359
   formation of, 29
   in fruit, 365, 368
   in hydrophytes, 390
   kinds of, 8
   in palisade mesophyll, **320**
   in parenchyma, **83, 100**
   in root cortex, **281**
   in seed, 376
      (*See also* Air chambers)
Intercellular substance, 24, 28
Internode, meristem in, 66
   telescoping in floral apex, 74, 341
   vascular structure of, **151**
Interxylary phloem (*see Phloem*)
Intracellular fluid, 365
Intraxylary phloem (*see Phloem*)
Intrusive growth (*see* Growth)
Intussusception, 35, 220
Invasion theory, 142
Ipomoea, leaf of, **330**
Iresine, primary vascular system of, **150**
Iridaceae, seed coats of, 372
Iris, intercalary meristem in, 66, 323
   inverted bundles in leaf of, **335**
   ovary of, **354**
Isoëtes, stem apex of, 72

J

Jeffersonia, leaf of, **392**
Juglandaceae, placental bundles in, 356
Juglans, adnation in, 355
   cambium in, **189,** 190, 191
   crystals in, **20**
   diaphragmed pith in, 155

Juglans, fruit of, 363, 367
   inferior ovary in, **352**
   leaf traces in, 148
   phellogen in, 253
   phloem in, 105, **106, 233,** 237, 239, 246
   tyloses in, 221
   wood of, 214
      durability of, 222
Jujube (*see Zizyphus*)
Juncaceae, secondary suberization of cells in, 260
Juniperus, phloem fibers in, 241
   wood parenchyma in, 216

K

Kapok, 368
Kauri gum (*see Agathis*)
Kerria, cuticle on twigs of, 54
Kinoplasm, 27
Kinoplasmasomes, **25,** 27
Knots, in lumber, 197
Krugiodendron, wood of, 225

L

Labiatae, intercalary meristem in, 66
   leaf traces in, 148
   stem of, 298
   traces to lost stamens in, 356
Lactuca, collenchyma in, 85
   seed coat of, **373**
   stomata in, **168**
Lacunae, 8, **389,** 392
Larix, cambial cells in, 90, 190
   orientation of cellulose in cell wall of, **33**
   pit-membrane perforations in, **41**
   wood of, 215, 216
      terminal parenchyma in, 216
Latex, articulate ducts of, **119,** 120
   cells and vessels of, 9, **115**
   in fruit, 364
   inclusion of protoplast, 19
   intrusive growth in cells of, 12
   nonarticulate ducts of, 119
   in pericycle, 158
   in phloem parenchyma, 110, 246
Lathyrus, root-stem transition in, 296
Lattices, 105, **106,** 108, **236,** 237

Lauraceae, leaf traces in, 148
   traces to stamen in, 343
Leaf, 1, **2**, 317–339
   abscission of, 267–270, **269**
   collenchyma in, **330**
   cuticle on, **51, 52,** 53
   cystoliths in, 50
   endodermis in, **159,** 161
   epidermis of, 332–333
   evergreen, extension of traces in, **185**
   of grasses, 335–338
   hairs on, **116, 170**
   hydathodes in, 117
   of hydrophytes, **389,** 390, 391
   meristem in, 66, 67, 319
   mesophyll of, 331, 332
   of monocots, 333–338
     bundles in, **335**
     ontogeny of, **334**
   ontogeny of, 319, **320,** 321, **323, 324**
   of parasites, 393
   persistence of, 338, 339
   phyllotaxy of, 318
   primordia of, 63
   of saprophytes, 394
   sclerenchyma in, 89, 330, 331
   shade, **392**
   stomata in, **168,** 169, 333
   structure of, **325, 328, 329**
   tip, 131
   transfusion tissue in, 113
   vascular structure of, 128, 139, **318, 322,** 324–330
   wax on, 54
   of xerophytes, **382, 383,** 384, **386, 387**
   (*See also* Grasses, leaf of; Mesophyll)
Leaf gap, **143,** 144, **145,** 146, **149, 183**
   relation of secondary growth to, 186
   in woody stem, 297
Leaf-scar periderm, 268, **269**
Leaf traces, 142, **143–145, 147,** 148, **149, 152, 153, 183,** 314
   burial of bases of, **181,** 182, **184**
   extension of, **185**
   in ferns, 325
   girdling, **147,** 148
   in monocots, 301
   in petiole, 325
   rupture of, 183, **184**
   tyloses in, after fall of leaf, 222

Leaf traces, in woody stems, 297
Leersia, hairs on leaf of, 338
Legumes (*see* Leguminosae)
Leguminosae, cell division in cambium of, 192
   seed coat of, 374
   sieve tubes in, 107
   stems of, 298
   vestured pits in, 48
Lenticels, 262–266
   closing cells and layers of, **263–265**
   complementary cells and tissue of, **263–265**
   in cork, **257**
   distribution of, 262
   duration of, 266
   formation of, **263**
   in fruits, 363
   origin of, 263–264
   of roots, 290
   structure of, 264–266
Lenticular scale, 255
Lepidium, seed coats of, **370**
Leptocarpus, stem of, **385,** 386, 387
Leucoplasts, 14, 15, **16**
   amyloplast, 16
   elaioplast, 16
   in epidermal cells, 165
   in sieve elements, 104
Lianas, **310,** 312
"Light line," **370,** 374
Lignification, in abscission, **269**
   in cell wall, 34, 49
   in fruit, 367
   of phloem parenchyma and ray cells, 246
   in root of pteridophytes, 291
   in xerophytes, 382, 384
Lignin, in cell wall, **31,** 32
   in wood, 224
Lignocellulose, in wood, 224
Lignosuberin, 158
   in abscission, 268, **269**
Lignum vitae (*see Guaiacum*)
Ligule, 335
Lilac (*see Syringa*)
Liliaceae, fruit of, 362
   roots in, 291
   seed coats of, 372
Lilies (*see Liliaceae*)

Liliiflorae, secondary growth in, 302
Linum, compression in outer tissues of, 182
  epidermis of corolla of, **358**
  fibers in, 88, 111, 157
  procambium in, **126**
  seed coat of, 373, 375
Liquidambar, stem of, 250, **298**
Liriodendron, buried branch base of, **200**
  cambial cells in, 190
  diaphragmed pith in, 155
  fiber-tracheid in, **94**
  oils in bud scales of, **115**
  phloem cells of, 104, **106, 110**
  pit-pairs in vessel of, **43**
  secondary phloem in, **234,** 237, 240, 241
  tyloses in, 221
  vessel element of, **96**
  winter buds in, 321
  wood of, 215
Lobelia, endodermis in, **159**
  primary xylem in, **130, 136**
  vessel element of, **96**
Lonicera, adnation in, **351,** 353
  root germs in, 290
  woody stem of, **298**
Lumen of the cell, 6
Lychnis, traces to lost stamens in, 356
Lycopersicon, fruit of, 361, 368
  chromoplasts in, **16**
  seed coats of, **373,** 375
Lycopodium, bundles in leaves of, 326
  stem apex of, **70,** 72
Lycopsida, absence of leaf gaps in, 145
Lysigenous cavities, 8, **115,** 118
Lysigenous ducts, in fruit, 364
Lysimachia, petal of, **358**
  traces to lost stamens in, 356
Lythraceae, seed coats, 372
Lythrum, aerenchyma in, 390

M

Maclura, adnation in, 353
  durability of wood of, 222, 226
  "multicellular" tyloses in, 221
  sieve plate in, 237
Magnolia, bark of, 258
  bundles in stamen of, 359
  cortex of, **164**

Magnolia, phellogen in, 253
  pit-pairs in, **43, 44**
  seed coats of, **370,** 375
Magnoliaceae, seed coats of, 372
Mahogany (see Swietenia)
"Malpighian cell," 374
Malus pumila, crystals in phloem of, **20**
  development of xylem and phloem from cambium in, **193**
  fruit of, 362–363, 365–367
    abscission of, 275
    cuticle on, **51**
    periderm in, **249**
  leaf of, bundle ends in, **326,** 327
    bundle sheaths in, 328
    ontogeny and structure of, **320, 325, 328**
    stomata in, 168, 333
    vascular skeleton of, **318, 322**
  leaf and branch traces in, **183**
  ovary of, **354**
  phellogen in, 253
  phloem of, parenchyma in, **110**
    fibers in, **111**
    rays in, 241
    sieve element in, **106**
  rootings of, 290
  secretory tissue in nectary of, **115**
  seed coat of, **370**
  vascular floral skeleton of, 361, **366**
  wood of, diffuse-porous, **212**
    heavy, **212,** 225
    pit-pairs in fiber of, **43**
    starch and tannin in, 21
    wet heartwood, 223
  xylem of, **212**
    fiber-tracheid in, **94**
    parenchyma in, **102, 212**
    vessel element in, **96**
Malvaceae, seed coat of, 371
  source of "bast," 246
Mangifera, fruit of, 367
Mango (see Mangifera)
Manila hemp, development of fibers in, 88
Marattia, plasmodesmata in, **36**
Maturation, acropetal, of meristem, 66
  basipetal, 66
  in leaf, 323
  order of, in xylem and phloem 128

Maturation, physical and chemical changes in cell wall during, 34
  of procambial cells, 176
  of xylem, 128, 323
Medeola, stomata in, **168**
Medicago, root-stem transition in, 296
"Medullary" bundles, **313,** 314
Medullary ray, 205, 208
  of vines and herbs, 303, 304
Medullary sheath, 156
Medullary spots, 228
Melastomaceae, cortical bundles in, 314
Melilotus, collenchyma in, 85
Menispermum, stem of, 302
Meristem, 60–78
  classification of, 61
    function, 67
    history of initiating cells, 63
    position in plant body, 64
    stage or method of development, 61
  development from promeristem, 124, **125**
  "embryonic," 62
  file, 62
  in fruit, 369
  ground-tissue, 67
  intercalary, **64,** 66
    in leaves, 323, **334**
  lateral, **64,** 65, 67, 248
  marginal, 67, 319, **320**
  mass, 62
  plate, 62
  primary, 63
  "primordial," 61, 124
  rib, 62
  theories of structural development and differentiation of, 67–74
    apical cell, 67
    histogen, 68
    tunica-corpus, 69–74
  "Urmeristem," 61
  (*See also* Apical meristems; Maturation)
Meristematic tissues, 60, 61, 82
  mature protophloem in, 131
Mesocarp, 360, 368
Mesophyll, 331–332
  cell of, **7, 14**
  in cotyledon, 376
  in grass leaf, 337

Mesophyll, ontogeny of, 319, **323**
  of petal, **358,** 359
  in shade leaves, **392**
  spongy and palisade parenchyma in mesophytic leaves, **320, 325, 328,** 331–332
  in xerophytic leaves, **382, 383,** 384, **386,** 387, 388
Mesophytes, 380
Mestome sheath, 161, 337
Methods of study, 4
Micellae, structure of fibril, 32
Middle lamella, chemical changes in, 34
  formation of, **24,** 28, **29, 34**
  loose use of the term, 33
Mineralization of cell wall, 50
Mints (*see* Labiatae)
Mirabilis, root-stem transition in, 295
Mitchella, adnation in, 353
Mitochondria, 17
Monarda, fusion in calyx of, **347,** 348
Monocotyledons, endodermis in, 159, 161
  fibers in, size of, 9
  intercalary meristems in, 66
  leaf in, 317, 333–338
    bundles in, **335**
    grass, 335–338
    ontogeny of, **334**
    persistence of, 339
  primary phloem in, 132, 232
    companion cells in, 109
  protective layers in, 260, **261**
  root in, 291
    apex of, **77,** 78
    phloem in, 285
    secondary root hairs on, 280
  root cap in, 74
  root-stem transition in, 296
  secondary growth in, 175, 201, **202,** 217, 301, 302
  seed coats of, 371
  septal nectaries in, 117
  stem of, **299,** 301, 302
    xerophytic, **385**
  storied cork in, 260, **261**
  vascular bundles in, 140, **306,** 307, 309
    arrangement of, 152
  vessels in, 98
Monostele, 141
Monotropa, 394

Moraceae, cystoliths in, 50
  sieve tubes in, 107
  source of "bast," 246
"Morphological roots," 289
Morus, abscission of stems in, 271
  durability of wood of, 222
  fruit of, 369
  lenticels in root of, 262, 265, 266, 290
Motor cell, 338, 384
Mucilages, in cell wall, 34
  in endodermis, 158
  in epidermis, 165
  in fruit, 364
  in phloem parenchyma, 110
  in protoplast, 19
  in secretory cells, 115
  in seed coat, **373**, 375
  in xerophytic leaf, **383**, 388
Musa, cuticle on leaf of, **52**, 53
  endodermis in root of, **160**
  fruit of, 364
  leaf of, 335
  starch in pericarp of, 21
  vascular bundles in, 308
Musaceae, articulate latex ducts in, 120
  traces to stamen in, 343
Mycorrhizae, 394, 395
Myrica, wax on fruit of, 54
Myriophyllum, leaf of, 390
Myrtaceae, internal phloem in, 305
  vestured pits in, 48

N

Narcissus, leaf of, 335
Nectary, secretory tissue of, **115**, 117
  septal, 117
Nepenthes, digestive glands of, 117
Nepeta, collenchyma in, 85
  fusion in calyx of, **347**, 348
Nettles (see *Urtica*)
New Zealand flax, 88
Nicotiana, ontogeny of leaf of, **323**
  root apex of, **77**
Nitrogen, in soil, 197
Node, in monocot stems, 301
  vascular structure of, **151**
Nucellus, 369, **370**, 372
Nucleus, 13
  in cell development, 10

Nucleus, division of, 24, **25, 26**
  of mesophyll cell, **7**
Nyssa, diaphragmed pith in, 155
  marginal ray cells in, 220, 242
  phloem ray cells in, 243

O

Oak (see *Quercus*)
Obliteration, of phloem, 232
  of sieve tubes, 245
Ochroma, wood of, **213**, 225
Oil ducts, 118
  globules, 13, 19
Oils, 50
  in bud scale, **115**
  in endosperm, 376
  in fruit, **115**
  in heartwood, 223
  in marginal ray cells, 220
  in phloem parenchyma, 110
  in xylem, 103
Onagraceae, internal phloem in, 305
  seed coats of, 372
Onopordum, hairs on stem of, **170**
Ontogeny, of axis, 124, **125**
  of collenchyma, 85
  of epidermis, 166
  of flower, 341
  of fruit, 360–361
  of leaf, 319, **320**, 321, 323, 324
  of parenchyma, 83
  of phellogen, 250
  of phloem, **189**, 240, 304
  of pith, 155
  of procambium cylinder, 127
  of protoxylem elements, 136
  of root, 279
  of root tip, **77**
  of secondary vascular tissue, 194, **195**
  of stem tip, **69**
  of stoma, 169
  of vascular cylinder, 152
  of vessel, 98, **99, 100**, 101
Ophioglossum, plasmodesmata in cortex of, **36**
Orchids, 388
  seeds of, 373, 376
Organs of plant, 4

Orientation, of bundles in leaf, 324–326, **329**
  of bundles in monocot leaves, **335**
  of carpel bundles, 344
  of chloroplasts, 332
"Ornamentation," of fruits, 368
  of seeds, **374**
Orobanche, 393
Osmunda, pit-pairs in tracheid of, **45**
Ostrya, strength and pliability of wood of, 226
Ovary, development of fruit from, 360
  glandular hair of, **116**
  inferior, **352**, 353, **354**, 355
  superior, **354**
  syncarpous, **349**, 355
  vascular structure of, **349**
Ovules, reduction of, **357**
  trace of, 344
  traces to lost, 356
  vascular supply of, 355, 356

P

Palms, "horny" endosperm in, 376
  leaf of, 335, 339
  secondary growth in, 302
Papaveraceae, articulate latex ducts in, 120
  seed coats of, 372
Papaya (see *Carica*)
Parasites, 393, **394**, 395
  vessels lost in, 98
Parenchyma, **17**, 82, **83**, 86
  in cortex, 163, 164
  crystals in, **20**
  in fruits, **363**, 364, 367, 368
  palisade, 331
  pit-pairs in, 42, **44**
  in primary xylem, **130**, 132, **136**
  in seed coats, 375
  size of cells in, 9
  spongy, 331
  tannin in, 22
  wood, **102**, 103, 216
    (*See also* Mesophyll; Phloem, parenchyma)
Peach (see *Prunus*)
Pectin, in cell wall, 32, 84

Pectin, in fruits, 365
  in middle lamella, 34
Pedicel, abscission zone in, 275
Pelargonium, collenchyma in, 85
  epidermis of corolla in, **358**
Perforation, bars, 97
  multiple, 97
  plate, 97, **100**
  reticulate, 98
  rim, 97, **100**, 101
  scalariform 97, 102
  simple, 97, 102
  of vessel element, **96**, **97**, **99**, **100**
    (*See also* Pit membrane)
Periblem, **68, 76**
Pericambium, 157
Pericarp, 360, **362**, **363**, 368
  bundles in, 362
  of dry fruit, **362**, 368
  fleshy, **363, 364**
  pit-pairs in, **44**
  sclerenchyma in, **86**
  stony, 367
Pericycle, **3**, 123, 131, 157
  effect of cambial growth on, 181–182
  in monocot stem, 301
  origin of accessory cambia in, 312
  origin of lateral roots in, 277, **287**
  origin of phellogen in, 254
  parenchyma in, **83**
  of root, **77**, 283
Periderm, **164**, 248–266, **249, 252**
  in abscission zone, 267
  in branch scar, **274**
  commercial cork, **257**
  duration of, 255–256
  extent of, 255
  formation, **253**, 254, 255
  in fruits, 363
  layering in, **265**
  in leaf scar, 268, **269**
  lenticels in, 262–266
  in monocots, 260, **261**
  morphology of, 258–260
  origin of, 253–255
  of roots, 290–291
  scaling of, **259**
  structure of, 248–253, **253**
  wound, 262
Perimedullary zone, 156

Persea, fruits of, 364, 366
  seed of, 376
Persimmon (*see Diospyros*)
Petal, 356, **358**, 359
  bundle end in, **326**
  reduction of bundles in, 356
  traces of, 342
  traces to lost, 356, 360
Petiole, 317
  bundles in, 325, **329**
  cork layer on, 261
  cuticle on, 53
  ontogeny of, 321
  structure of, **329**
Phaseolus, root-stem transition in, 296
  seed coat of, **370**
  starch in cotyledon of, **21**
Phellem, 248, **249**, 250, **251**, **252**, **257**, **274**, 390
  (*See also* Cork)
Phelloderm, 248, **249**, 252
Phellogen, 248, **249**, 250, **252**, 253
  (*See also* Cork cambium)
Philadelphus, lenticels absent in, 262
Phloem, 2, 82, 103–113
  crushing of, **125**, 131, 132
  definition of, 113
  fibers of, 104, 110, **111**, 157, **164**, 241, 331
    ontogeny of, 131
  gelatinous cell walls in, 49
  internal (or intraxylary), 139, 304, **305**
  interxylary, 311–312
  mature, **119**
  metaphloem, 130, 131
  mother cell, 186, 196
  orientation in leaf of, 324
  parenchyma, 83, 104, **110**, 239–240
    crystals in, **20**, 246
    formation of, **189**
    gums in, 246
    lignification of, 246
    ontogeny of, 131
    resins in, 246
    sieve-plate-like pits in, **236**
    storage in, **119**
    tannin in, **21**, 246
  primary, **3**, 123, **125**
    in dicots, 132, **164**
    effect of cambial activity on, 181–182

Phloem, primary, in leaves, 326
  metaphloem, 131
  in monocots, 132
  ontogeny of, **127**
  order of maturation in, **128**, 130–132
  in roots, **282**, 283, **285**
  pit-pairs in, **45**
  secondary, **3**, **125**, 132, **164**, **193**, 231–246
    extent and amount of, 231
    formation of, **186**, 187, **196**, 197
    ontogeny of, **189**, **195**
    periderm in, **249**, 254
    "pockets" of, in crotch angle, **198**, 199
    rays in, 205, 241–242
    in roots, **285**
    seasonal rings in, 243
    structure of, **232–236**, **238**
    (*See also* Companion cells; Protophloem; Sieve element)
Phloëoterma, 159
Phlox, floral apex of, **71**
  leaf and branch traces in, **144**
Phoenix, fruit of, 361
  plasmodesmata in endosperm of, 35, **36**
  root-stem transition in, 296
  seed of, 376
Phoradendron, 393
Photosynthesis, 169, 332
  in leaf, 317
  in phelloderm 252
  in stem, **385**, **386**, 387
Phragmoplast, 24, **25**, 27, **119**
  rings, **25**, **26**
Phragmosphere, 27
Phryma, hairs on corolla of, **170**
Phyllocladus, leaf-like branchlets in, 317
Phyllode, 317, 333
Phyllotaxy, 318–319
  leaf traces showing, **153**
  shape of pith showing, 154
Phylogeny, of cambial cell types, **192**
  of fiber, 93
  of herbaceous stem in angiosperms, 300
  of parenchyma, 82
  of phloem, 103
  of sieve tubes, 107
  of siphonostele, 142
  of venation, 334
  of vessel, 95, 98

Phylogeny, of vessel perforation, 102
  of xylem, 92, 93, 129
Physocarpus, ovary of, **354**
Physostegia, fusion in calyx of, **347, 348**
Phytelephas, "horny" endosperm in, 376
Phytolacca, lignified vessels in, 225
Picea, leaf of, 387
  spruce gum from, 246
  wood of, **206, 213,** 215, 217, 223
Pilea, collenchyma in, 85
  primary vascular system in, **149**
Pineapple (*see Ananas*)
Pinguicula, glandular hairs on leaf of, **116,** 117
  petal of, **358**
Pinus, absence of wood parenchyma in, 102
  buried branch bases in, **199**
  cambial cells of, 190
    cell division in, **26**
  checking of cell wall in, **44**
  compression wood in, 228
  dormant phloem in, 245
  endodermis in, **159**
  intercalary meristem in, 66, 323
  leaf of, **386,** 387
  pit-pairs in, **41, 45**
  resin canal in, 8, **115,** 118
  rhytidome of, 256
  secondary phloem of, **235,** 240, 241
  secondary xylem of, 218, 219, 225
  stem apex of, 72
  tannin in, 21
  tracheids in, **91**
Piper, anomalous stem in, **310**
  endodermis in, 161
Piperaceae, medullary bundles in, 314
Pistia, root apex of, 78
Pisum, phloem in, **238**
  starch in cotyledon of, **21**
  stem of, 302, 303
  stipules in, 317
  vascular bundles in, **308**
Pit aperture, 38, **40**
  inner and outer, 39, **47**
Pit canal, 39, **47**
Pit cavity, 38, **40**
Pit chamber, 39, **47**
Pit-fields, in sieve elements, 107
  primary, 28, **188**

Pit membrane, 38, 39, 41
  formation of tyloses by, 220, 221
  penetrability of, 226, 227
  perforations in, 39, **41,** 90, 92
Pit-pairs, 38–48
  parts of, 38, **47**
  types of, 38
    bordered, **38,** 39, **40, 41**
      variations in size and structure of, 42, **43–47**
    half-bordered, **38,** 39, **45**
    simple, **38, 39,** 42, **45**
      clustered, **44**
      vestured, **44**
    (*See also* Pits; Pitting)
Pith, 2, **3,** 68, 123, **125,** 155–157
  crystals in, **20**
  diaphragmed, 155–156
  duration of, 156
  effect of cambial activity on, 180
  extrastelar *versus* stelar, 142
  form of, **153,** 154
  in fruit, 369
  hollow, 156
  in monocot stem, **299,** 301
  ontogeny of, 155
  parenchyma in, **83**
  of roots, 156, **282**
  structure of, 155
  tannin in, **21**
Pith ray (*see* Xylem, rays)
Pith-ray flecks, 228
Pits, 37–48
  blind, 12, 42
  cribriform, 48
  in fibers, 87
  formation of, 28
  fused, **86,** 89
  in phloem, 111, 112
  in primary and secondary walls, 42
  primordial, 28
  in sclereids, 89
  in vessels, 95
  simple and bordered, 38, 39
  vestigial, in libriform wood fibers, 47
  (*See also* Pit-pairs)
Pitted cell, in abscission zone, **273**
  of metaxylem, 133
  of phelloderm, 252
  (*See also* Pit-pairs; Pits)

Pitting, reticulate, 273
  scalariform, 92, **273**
  unilateral compound, 42
    (*See also* Pit-pairs; Pits)
Placenta, in fruits, 368
  vascular supply of, 355
Placentation, 356
Plant body, cellular complexity of, **7**
  constitution of, 4
  fundamental parts of, 1, **2**, 123
  meristems in, 64
  primary, 3, 123, 175
  primary *versus* secondary, 137n, 175
  secondary, 204
Plantaginaceae, seed coat in, 372
Plantago, endodermis in leaf of, 161
  intercalary meristem in, 66
  seed coats of, **370**
Plasma membrane, 6, 13, **24**
Plasmodesmata, 6, 28, 35, **36**, 37
Plastids, 10, 13, 14, 16
Platanus, bark of, 260
  hairs on leaf of, **170**
  pit-pairs in, **45**
  sclereids in, 241, 246
  stem of, **298**
  wood of, 226
Plerome, **68, 76**
Podocarpus, fungus erosion in cell wall of, **44**
  wood parenchyma in, 216
Podophyllum, primary phloem in, 132
Polemoniaceae, seed coat of, 372
Pollen grain, wall of, 49
Polygonaceae, placental bundles in, 356
Polygonatum, root of, **282**
  stomata in, **168**
Polygonella, leaf of, 387
  stem of, **385**, 387
Polypodium, endodermis in, **159**
  vascular bundles in, **306**
Pomegranate (*see Punica*)
Pontederia, air chambers in 390, **391**
Poppy (*see* Papaveraceae)
Populus, abscission in, leaf, **269**, 270
    stem, 271, **272**, **273**, 274
  cambial initials in, 191
  crystals in, **20**
  periderm of, **249**, 255
  petiole of, **329**

Populus, phellogen in, 253
  phloem of, dormant, 245
    fibers in, 241
    lattices in, 105, **236**, 237
    secondary, **232**
    sieve plates in, 105, **236**
  pith of, 154
  primary vascular cylinder of, **149, 153**
  root of, **281, 284**
  root germs in, 290
  wood of, **210**, 225, 226
    durability of, 226
    heartwood in, 223
    pit-pairs in, **43**
    rays in, **210**, 217
    secondary, **210**, 225, 226
    specific gravity of, 225
    tyloses in, **210**, 220
Portulaca, tyloses in, 222
Potamogeton, hydrophyte, 388
  leaf of, **389**, 390, 392
  stem of, **389**
Potato (*see Solanum tuberosum*)
Potentilla, achenes of, **357**
  phloem-ray cells of, 242
Primary body, 123, 175
  effect of cambial activity on, 180
  versus secondary body, 137n
Primary cell wall, chemical changes in, 34
  formation of, **24**
  pits in, 42
  structure of, 28, **29, 30, 34**
    (*See also* Cell wall)
Primary protective layer, 267, **269**
Primary tissues, 123, 127
  cell arrangement in, 136, 137
  effect of cambial activity on, 180–181
  vascular, 68, 126–127, **128**, 129–140, 296
Primulaceae, roots in, 291
Procambium, 62, 67, **125, 126**, 127, 131, 175–176
Promeristem, 61, **68**, 124, **125, 127**
Proplastids, **14**, 15
Prosenchyma, 83
Protective layers, of abscission zone, 267, 268, **269**
  in monocots, 260
Protein granules, 13
Protoderm, 67

Protophloem, 129, 131
　elongation of, 134
　fibers in, 111, 131
　in meristematic regions, 66, **77**, **125**
Protoplast, 12, 13, 19
Protoxylem, 129, **130**, 132, **135**, **136**, **143**
　annular element in, **130**, 133, 134, **135**, **136**
　　rings of, 48, 132
　elongation of, 134
　lacunae in, **135**, **136**
　in meristematic regions, 66, **77**, **125**
　ontogeny of, 136
　reticulations in, 48
　scalariform elements in, **119**, **130**, 133
　spirals in, 30, 48, 132, 136
　vessels in, 134
Provascular meristem, 126n
Provascular tissue, 126n
Pruning, 199, 200, 208
Prunus, abscission of leaves in, 268
　"annual rings" in, 256
　fruit of, **363**, 364–367
　gummosis in, 228, 229
　periderm of, cork cells in, 250
　　formation of, 255
　　lenticels in, **264**, **265**, 266
　sieve tubes in, 107
　wood of, 227, 228
Pseudotsuga, compression wood in, 228
Psilotaceae, apical cells in, 67
Pteridium, bundle arrangement in, 152
　medullary bundles in, 314
　sieve element in, **106**
　vessels in, 98
Pteridophytes, apical cell in, 65, 67
　leaf traces in, 147
　origin of lateral roots in, 287
　phloem in, 104
　　sieve areas in, 105
　primary body only in, 175
　primary type of endodermis in, 159, 161
　protoxylem in, 133
　roots in, 291
　vascular supply of leaves in, 326
Pteris, root apex of, **76**
Pteropsida, leaf gaps in, 145
Pterospora, seed of, **372**
Pulvinar band (or swelling), 53

Punica, seed of, **372**, 375
Pyrola, flower of, **343**, 346
　ovary of, **354**
Pyrus, cambium in, **189**, 190, 191
　fruit of, 362–368
　obliteration of sieve tubes in, 245
　rootings of, 290
　scale bark of, **259**
　sclerenchyma in, **86**, 364

Q

Quercus, abscission in, inflorescence, 275
　stem, 271, 274
　bark of, 258
　cambial cells in, 194
　leaf in, **328**
　leaf traces in, **153**, 154
　phloem of, fibers in, 246
　　obliteration of sieve tubes in, 245
　　parenchyma in, 110
　　sclereids in, 110, 246
　　tannin in, 22
　traces to lost petals in, 356
　wood of, **211**, 214, 215
　　durability of, 222
　　gelatinous fibers in, **94**
　　libriform fibers in, **94**
　　parenchyma in, **102**, **211**
　　rays in, **211**, 217, 218
　　silver grain in, 227
　　specific gravity of, 225
　　tracheids in, **91**
　　tyloses in, **211**, 220
　　vessels in, **96**, 98
Quercus suber, cork of, "annual rings" in, 256
　commercial use of, **257**, 258
　elasticity of, 252
　formation of, 250, **251**
　lenticels in, 266
　phellogen in, 253
Quince (*see Chaenomeles*)
Quinine, 246

R

Radial bundles, **139**
Radial dots, 158, **159**
Ramular traces (*see* Branch traces)

Ranales, secondary phloem in, 237
  traces to stamens, 343
  vessels lacking in, 216
Ranunculaceae, floral organs of, 342
  fruit of, 362
  medullary bundles in, 314
  seed coat of, 372
  traces of lost ovules in, 356
  vascular bundles in, 307
Ranunculus, achenes of, **357**
  companion cells in, 239
  medullary ray in, 205, 208
  primary phloem in, 132
  primary skeleton in, 149, 177
  root of, **281, 282**
Raphides, 19, **20**, 364
Raspberry (*see Rubus*)
Ray cells, "dentations" in, **45**
  initials of vascular, 187, 191
  marginal, 220
    conterminous and interspersed, 220
  pit-pairs in, **44, 45**
Rays, interfascicular, 149
  vascular, 191, 195
    absence of, 208
    aggregate, 217
    compound, 217
    formation of, 205
    (*See also* Xylem, rays)
Receptacles, in flower, 345, 346, **354**
  in fruit, 369
Reduction of vascular supply, 356
  in flowers, **348**, 356, **357**
  in hydrophytes, 391, 392
  in parasites, 393
  in saprophytes, 394
  in xerophytes, 387
Redwood, 228
Reseda, cohesion of carpels, **350**
Resin, 19, 50
  in buried branch bases, **199**
  canal, **115, 386**
  ducts, 118
  in gymnosperms, 216
  in heartwood, 223
  in phloem parenchyma, 246
  in tyloses, 220
  in xylem, 103
Rhamnus, traces to lost petals in, 356
Rheum, petiole of, **330**

Rhizomes, endodermis in, **159**, 161
  metaxylem in, 130
  in saprophytes, 394
Rhododendron, xerophytic leaves of, 385
Rhus, tyloses in, 221
Rhytidome, **256**
  formation of, **253**, 254
  in monocots, 261
  in roots, 291
Ribes, phellogen in pericycle of, 254
  root germs in, and rooting of, 290
Rims of Sanio, 49
Rings, in protoxylem, **135**
Rise of sap, 103
Robinia, cambium in, **188**, 190, 194
  ontogeny, of secondary vascular tissue in, **195**
  of vessel element in, **99**
  phloem of, fibers in, **111**
    parenchyma in, **110**
    secondary, **235, 236**, 239–241, 245
    sieve element in, **106**
    simple sieve plate in, 237
  wood of, clustered vessels in, 215
    durability of, 222, 226
    "multicellular" tyloses in, 221
Root, 1, **2**, 123, 277–283, **284**, 285–291
  adventitious, meristem of, 123
    origin of, 157, 289
  annular element in, 284
  anomalous structure in, 313, **314**, 315
  apex, 63, 65, 74, **75–77**, 78, 131
  cambium in, **119**, 176, 179
    monocot, 201
    time of activity of, 195
  cortex of, 280, 283
  endodermis of, **160**, 161
  epidermis of, 166, **167**
  lateral, formation of, 286, **287, 288**, 289
    initials of, 163
    pericyclic origin of, 277, **287**
  lenticels in, 263, 265
  morphology of, 278
  ontogeny of, 279
  pericycle in, 157
  periderm in, 255, **281**, 290, 291
  phloem in, first-formed, 128
  pith in, 156
  primary, 278
  primary vascular tissues of, 140, 283

Root, in saprophytes, 394
  secondary, 278
  secondary growth in, **285**, 286
  stele of, 142, **282**
  versus stem, 277, 293
  structure of, **281**
  xylem in, exarch, 129, 277
    first-formed, 128
    metaxylem in, 130
Root cap, 74, **75, 77**, 78, 277, 279
Root germs, 289, 290
Root hairs (*see* Hairs)
Rootings, 290
Rosa, ovary of, 353, **354**
Rosaceae, flower of, fusion in, 351
  organs of, 342
  traces to lost ovules in, 356
  fruit of, epidermis in, 362
  leaf traces in, 148
  sieve tube in, 107
  wood of, chloroplasts in, 16
    pith-ray flecks in, 228
Royal palm (*see Roystonia*)
Roystonia, persistent periderm in, 260
Rubber, 19
  in secondary phloem, 246
  latex, a source of, 118
Rubus, fruit of, 362, **364**, 365, 369
  hairs on fruit of, **170**
  ovary of, **354**
  rootings of, 290
Ruellia, seed coat of, **374**
Rumex, collenchyma in, 85
  cuticle on stem of, 53
  tyloses in, 222
Rupture, of endodermis in lateral root formation, 287
  of epidermis after periderm formation, 253
  of leaf traces, **184**
  of outer tissues after periderm formation, 254
Russian dandelion (*see Taraxacum kok-saghyz*)
Rutaceae, traces to lost carpels in, 356

S

Sabal Palmetto, vascular bundles in, **306**
Sagittaria, petiole bundles in, **329**
Saintpaulia, rootings of, 290

Salicornia, fleshy xerophyte, 388
Salix, abscission in, inflorescence, 275
  leaf, 270
  twig, 275
  cambial initials in, 191
  leaf traces in, 148
  periderm of, **249**
  lenticels in, 265
  phloem of, crystals in, **20**
    fibers in, **111**
    parenchyma in, **110**
    secondary, **235**, 241, 242
    sieve plates and lattices in, 107
  root of, **282**
    lateral, **288**
  root germs and rooting, 290
  traces to lost petals in, 356
  wood of, marginal ray cells in, 220
    pith-ray flecks in, 228
    pit-pairs in, **45**
Salvia, fusion of calyx of, **347**, 348
Sambucus, collenchyma in, 85
  phloem parenchyma in, 240
Sansevieria, development of fibers in, 88
Santalaceae, ovary in, 353
Sapodilla (*see Achras*)
Sapote (*see Calocarpum*)
Saprophytes, 393, 394
Sarracenia, digestive glands of, 117
Sassafras, oils in secretory cells of, 115
  wood of, fusiform parenchyma cell in, **94**
  marginal ray cells in, 220
  tyloses in, 221
Scale bark, 254, **259**
Schizaea, persistent root hairs in, 280
Schizogenous ducts, 116
Schizogenous spaces 8
Schizolysigenous spaces, 8
Scirpus, primary vascular system in, **150**
  vascular bundles in, **306**
Sclereids, **86**, 88, 89, 110, 111
  conversion from parenchyma, 51, 110, 112, 258
  in cortex, 163
  in fruits, 364, 366, 367
  in leaves, 331
  in phelloderm, 252
  pit-pairs in, 42, **45**, 89
  in secondary phloem, 241

Sclereids, types of, astrosclereids, 89
  brachysclereids, 89
  macrosclereids, 89
  osteosclereids, 89
  trichosclereids, 89
Sclerenchyma, 82, **86,** 87–89
  crystals in, 246
  in fruits, 363, 364, 366–368
  in gymnosperms, 241
  in hydrophytes, 391
  in leaves, 330, 331, **336**
  in parasites, 393
  in roots, **282**
  in saprophytes, 394
  seasonal bands of, 243
  in seed coats, 374, 375
  sheath, **306**
  in xerophytic leaves, **382, 383,** 384, **386,** 387
  (*See also* Fibers; Sclereids)
Sclerification, of secondary phloem, 246
  of secondary phloem and cortex, 258
Sclerotic cells, 88
Scrophulariaceae, fruit of, 362
  traces to lost stamens in, 356
Seasonal rings in secondary phloem, 243
Secondary body, 123
  origin and development of, 175–202
  versus primary body, 137*n*
Secondary cell wall, chemical changes in, 34
  erosion by fungus hyphae in, **44**
  formation of, **24,** 29
  microscopic checking of, **44**
  minute structure of, **31, 33, 34**
  pits in, 42
  protoxylem-like thickenings in, 48
  with tertiary spirals, **30**
Secondary cortex, 163, 252
Secondary endodermis, 158, 159
Secondary growth (*see* Growth)
Secondary meristems, 63, 248, 312
Secondary phloem (*see* Phloem)
Secondary tissues, 123, 175
  cell arrangement in, 136, 137
  cork, 123, 175
  in monocot stem, **202**
  ontogeny of, 194, **195**
  in roots, **281, 285,** 286
  in stems, 297

Secondary tissues, vascular, **175**
Secondary xylem (*see* Xylem)
Secretory cells, 114, 115, **116**
  mucilage in, 116
  oils in, 115
  in pericycle, 158
Secretory chamber, 358
Secretory tissue, 114, **115,** 116–118, **119,** 120
Section, planes of, **4,** 5
Securidaca, anomalous stem in, **310**
Sedges, mineralization of cell wall in, 50
Sedum, 388
Seed coats, 361–369, **370,** 371, **372–374,** 375
Seedlings, endodermis in, 161
  secondary xylem of, 239
Seeds, 369–376
  aleurone grains in, 22
  embryo and endosperm, **370,** 376
  gelatinous cell wall in, 49
  macrosclereids in, 89
  nitrogenous inclusions in, 22
  osteosclereids in, 89
  plasmodesmata in endosperm of, 35
  vascular bundles in, **372,** 376
  wall of epidermal cells of, 34
Segmentation of apex, 67
Selaginella, apical cells in, 67, 76
  stem apex of, 72
  vessels in, 98
Senecio, fusion in corolla of, **348**
Sepal, 356, 359
  trace, 342
Separation layer, 267, **269**
  gelatinous cell wall in, 268
Sequoia, inclusions in heartwood cells of, 19
  microscopic checking of tracheid wall in, **44**
  perforated pit membranes in, 90
  stem apex of, **70, 71,** 72
  wood of, **207,** 216
    durability of, 226
    specific gravity of, 225
Serjania, anomalous stem in, **310**
Shagbark hickory (*see Carya*)
Shell bark, 254
Sieve areas, 105
Sieve cell, 104, 109, 232, 237
  (*See also* Sieve element)

Sieve element, 103, 104, **106**
  functioning life of, 245
  ontogeny of, 107–108, 127, 131
    (*See also* Sieve cell; Sieve tube)
Sieve field, 107
Sieve plate, 105, **106**, 232, **236**, 237
Sieve tube, 103, 104, 107. 232, **236**, 237, **238**
  duration of, 108
  formation of, **189, 195**
  obliteration of, 108, 109
  ontogeny of, in root tip, **77**
  phylogeny of, 107
  in root, **119**
    (*See also* Sieve element)
Silica, in cell wall, 50
Sinocalamus, stem apex of, **70, 73**
Siparuna, wall of fiber-tracheid in, **31**
Sisal, 88
Skunk cabbage (*see Symplocarpus*)
Slime plugs, 104, **236**
Smilacina, raphides in fruit cell, **20**
Smilax, endodermis in, **160**
  root of, **282**
Solanaceae, fruit of, 364
  internal phloem in, 305
Solanum Dulcamara, periderm in, **249**
  phellogen in, 253
  stem of, 302
Solanum tuberosum, cambial initials in, 191
  collenchyma in, **84**, 85
  nitrogenous inclusions in tuber of, 22
  periderm of tuber in, **252**, 261
  phellogen of tuber in, 253
  phloem of, companion cells in, 239
    internal, **305**
    primary, 232
    sieve elements in, **106**
  primary vascular system in, **151**
  starch in tuber of, 22
  stomata in, **168**
Solidago, collenchyma in, 85
Sorbus, ovary of, **354**
Sparganium, adnation in, 355
Spartina, leaf of, **383**, 384
Spiraea, ovary of, **354**
  petiole bundles in, **329**
  stomata in, 333

Spiral elements, in primary xylem, **130**, 133
  of roots, **284**
Spiral thickenings, 30
Spondias, fruit of, 367
Spores, wall of, 49
Stalace, 279
Stamen, 359
  reduction of bundles in, 356
  trace of, 343
  trace to lost, 356, 360
Starch, 13, **21**, 22, **83**
  in endodermis, 158, 162
  in fruits, 365
  in guard cells, 169
  in phelloderm, 252
  in phloem parenchyma, 110, 113, 244
  in roots, 278
  in seeds, 376
  in tyloses, 220
  in xylem, 103
Stele, 1, 140–142
  in flower, 342
  in herbaceous stems, 154, **299**
  in hydrophytes, **389**
  monocot, **299**, 301
  in roots, **282, 284**
  types of, monostele, 141
    polystele, 141
    protostele, 140, **141**, 142, **281, 284**
    siphonostele (or solenostele), 140, **141**, 142
      amphiphloic, 141, 142, 304
      ectophloic, 141, 142
  in vine stems, 302, 303
  in woody stems, **298**
Stellaria, stem of, 299
Stem, 1, **2**, 123, 293–315
  abscission of, 271–275
  adventitious, 123, 157
  anomalous structure in, 309, **310**, 311, 312, **313, 314**, 315
  apex, 60, 63, 65, **69–71**, 72, **126**, 131
  cambium in, 178, 195
  cortex of woody, **164**
  cuticle on, **51**, 53, 54
  endodermis in, 157, **159**, 161
  epidermis of, 166
  hairs on, **170**
  of hydrophytes, **389**, 390

Stem, intercalary meristem in, 66
  of microphyllous xerophytes, **385,** 387
  origin of, 293
  pericycle in, 157
  periderm of, **249**
    formation in woody, 253
    lenticels in, 262, **264, 265**
  phloem of, fibers in, 111
    first-formed, 128
    internal, **305**
    secondary, **238**
  primary vascular system in, 139, 142, **151, 153**
  root-stem transition, 293–296
  of saprophytes, 394
  sclereids in, 89
  stomata on, 169
  types of, 296–302
    herbaceous, 297, 298, **299,** 300, 303, 304
    monocot, **299,** 301, 302
    vine, 302, 303
    woody, 297, **298**
  versus roots, 277, 293
  xylem of, 129
    first-formed, 128
    protoxylem in, **135**
  (*See also* Cortex; Periderm; Phloem; Primary body; Xylem)
Stem bundles, 152
Stigma, bundles in, 359
Stipules, 317
  ontogeny of, 321
  vascular supply of, 144, **147,** 326
Stomata, 166–167, **168,** 169, 333, 338
  accessory cells of, 167, 170, 386
  on fruits, 362
  on petals, 359
  on stamens and carpels, 360
  on xerophytic leaf, **383,** 384, **386,** 387
  on xerophytic stem, **385, 386,** 387
Stomatal opening (or aperture), 167
Stone cells (*see* Sclereids)
Storied cambium, 190, 202
Storied cork, 260, **261**
Strawberry (*see Fragaria*)
Streptopus, adnation in, 355
Stretching, of cork cells, 252
  of phloem, 244

Stretching, of protoxylem cells, 129, **135**
  (*See also* Elongation)
Striations, of cell wall, 3
Strychnos, interxylary phloem in, 311
Suberin, in cell wall, 32, 49
Suberin lamella, in monocots, 260
  in phellem, 252
Suberization, of cell wall, 34, 49, 252, 260
Substitute fiber, 94
Sumach (*see Rhus*)
Sun scald, 196
Swietenia, heartwood in, 223
  inclusions in cells of, 19
  septate fiber-tracheid in, **94**
Symbiosis, 394, 395
Symplocarpus, venation of, 334
Syncarpy, stages in, **394**
Syringa, inflorescences lost by weathering, 275

T

Tannins, 19, **21,** 32, 50
  in autumn color, 18
  in endodermis, 158
  in epidermis, 165
  in fruit, 362, 363, 365
  in heartwood, 223
  in phloem parenchyma, 110, 246
  in xerophytic leaf, **383,** 384
Taraxacum, chromoplasts in corolla of, 16
  collenchyma in, 85
  intercalary meristem in, 66
Taraxacum kok-saghyz, cambium region in, **119**
Taxodium, stem apex of, 72
Taxus, absence of wood parenchyma in, 102
  wall of tracheid in, **30**
Tecoma, absence of lenticels in, 262
  internal phloem in, 304
Tephrosia, stele of root in, 284
Tertiary spirals, 30
Tetracentraceae, absence of vessels in, 98
Thinouia, anomalous stem in, **310**
Thuja, diffuse wood parenchyma in, 216
  phellogen in, 254
  phloem fibers in, 241
  phloem rays in, 242
  primary vascular system of, **150**

Thuja, woody stem of, 297
Tilia, adnation in, 355
  crystals in, **20**
  fibers of, 112
  secondary phloem in, 237, 240, 242, 245
  vessel of, **30**
  wood of, 216, 225, 226
Tiliaceae, "bast" in, 246
  seed coats of, 371
Tillandsia, hairs of, 388
Timber, durability of heartwood, 223
  penetrability of green and seasoned, 226, 227
  structure of, 204
  value of sapwood as, 224
  (*See also* Wood)
Tissue initiation, 60
Tissue systems, absorbing, 113
  epidermal or tegumentary, 114
  fundamental or ground, 114
  mechanical, 113
  storage, 113
  vascular, 114
Tissues, 3, 4
  classification of, 81
    kind of constituent cells, 82
    stage of development, 81
  complex, 82, 89–113
  interfascicular, 312
  intermediate, 312
  permanent, 61, 82, 124
  secretory, 114
  simple, 82–89
    (*See also* Primary tissues; Secondary tissues)
Tobacco (*see Nicotiana*)
Todea superba, chloroplasts in, **15**
Tomato (*see Lycopersicon*)
Torus, **38**, 39, **40, 41**
Trabeculae, **48**
Traces (*see* Branch traces; Leaf traces)
Trachea, 97, 134
Tracheid, 90–93, **91**
  erosion by fungus hyphae in wall of, **44**
  first-formed, 127
  gelatinous wall in, 95
  microscopic checking in wall of, **44**
  pit-pairs in, **40, 43, 45,** 90, 92
  in protoxylem, 132, 133
  scalariform-pitted, 92

Tracheid, in secondary xylem, 194, 216, **219**
  tertiary spirals in, **30**
  trabeculae in, **48**
Tradescantia, primary vascular system of, **150**
  stamen hair in, **17**
Tragopogon, latex vessel of, **115**
Trailing arbutus (*see Epigaea*)
Transfusion cells, of endodermis, 158
Transfusion tissue, 113, 387
Transition cell, 327
Transition region, 278, 293–296, **294**
Traumatic tissue (*see* Wound tissue)
Trichomes, 170
Trichostema, fusion in calyx of, **347,** 348
Trifolium, stem of, 299
  vascular bundles in, **308, 318**
Trillium, stem of, **299**
Triticum, stem of, **299**
  stem apex of, 73
Trochodendraceae, absence of vessels in, 98, 216
Trollius, follicle of, **357**
Tropaeolum, root-stem transition in, 296
Tsuga, sclereids in, 241
  sclerenchyma in, **86**
  sieve element of, **106**
  tannin in phloem of, 22
  terminal wood parenchyma in, 216
Tunica, 69, 70, 73, 74
Tunica-corpus theory, 69–74
Tyloses, 220, **221,** 222
Tylosoids, 221
Typhaceae, secondary suberization of cortical cells in, 260

U

Ulmaceae, sieve tubes in, 107
Ulmus, bark of, 258
  cambium of, **188, 189,** 190, 191
  phellogen in, 253
  phloem of, lattice in, 237
    simple sieve plate in, **236,** 237
    slime plugs in, **236,** 237
  phyllotaxy of, 154
  pith of, 154
  stem of, 250
    abscission in, 271, **274**
  wood of, clustered vessels in, 215

INDEX   425

Ulmus, wood of, ring-porous, frontispiece
  strength and pliability of, 226
  wet heartwood in, 223
Umbelliferae, leaf traces in, 148
  medullary sheath in, 156
  oil ducts in, 118
  seed of, 371
  traces to lost ovules in, 356
Unisexual flowers, vestiges of lost organs in, 356
Urtica, stinging hair of, **116**
Urticaceae, cystoliths in, 50
  fibers, size of, 9
  nonarticulate latex ducts in, 120
  vestigial floral traces in, 356
Utricularia, leaf of, 390

V

Vaccinium, cuticle on stem of, **51**
  fruit of, 361, 362, 365, 368
  ovary of, **354**
  seed coats of, **370**
Vacuoles, **7**, 10, **17**, **25**, 61
Vacuome, 17
Valerianaceae, traces to lost carpels in, 356
Vascular bundles, 305–309, **306, 308**
  definition of, 138, 309
  of flowers, fusion in, **347–351**
    in organs, **357**, 359–361, **366**
    in placenta, 355
    in stigma, 359
    vestigial, 360
  in hydrophytes, **389**, 392
  in pericarp, 362
  of leaf, **322, 327**
    inversion of, **335**
    xerophytic, **386, 387**, 388
  protoxylem in, 130
  secondary growth in monocots, **301**
  of seeds, 376
  types of, amphicribral, **139**, 140, **306**, 325, 359
    amphivasal, **139**, 140, **306**
    bicollateral, **139**
    cauline, **151**, 152, **153**
    collateral, **139, 306,** 307, 308, 325
    common, 151, 152

Vascular bundles, types of, concentric, **139**
  cortical, 314
  occurrence of, 139
  (*See also* Branch traces; Leaf traces; "Medullary" bundles)
Vascular cylinder, primary, **143, 145,** 146, 148, **149, 150,** 152, **153, 154, 177,** 296
  secondary, **177,** 297
Vascular floral skeleton, 342–360, **366**
  fusion in, **347–351, 354,** 355
  inferior ovary, **352,** 353, **354**
  in organs, **357,** 359, 360
  placental supply, 355
  reduction of, 356, **357**
  vestigial vascular tissue in, 356
Vascular skeleton, primary, 140–142, **143,** 144–148, **149–151,** 152–154
  (*See also* Vascular floral skeleton)
Vascular system, primary, **150, 151**
  of monocots, 301
  of root, 284
Vascular tissues, 1
  in leaf, 323, 324, 326–327. **329**
  in parasite and host, **394**
  primary, cell arrangement in, 136
    in roots, **282, 283**
    in root, 277
  secondary, ontogeny of, 194, **195**
    in roots, 286
Vein islets, **318, 322**
Veins, **320, 322,** 326, 327
Velamen, 48, 291
Venation, 318, 334
Veratrum, cambium in, 201
  secondary growth in, 302
Verbascum, color in corolla of, 18
Veronica, stem of, 299
Vessel, 95–102, **96, 99, 100**
  clustered, 215
  element, member, or unit, 97
  in Gnetales, 216
  ontogeny of, **99, 100,** 101–102, 194, **195**
  perforations in, 97, **99, 100**
  pit-pairs in, **43, 44**
  in primary xylem, 127, 132, 133
  in secondary xylem, 216, 217
  segment, 97
  tertiary spirals in, **30**

Viburnum, druse in, **20**
  phellogen in, 253
Vicia, plasmodesmata in root of, **36**
Vinca, stem apex of, **71**, 73
Vines, "medullary rays" of, 303, 304
  stem type, 302, 303
  woody stem of, **298**
Viola, seed coat of, **370**
Violaceae, seed coats of, 371
Viscum, half parasite, 393
Vitis, absence of lenticels in, 262
  fruit of, 362, 364
  periderm in, 255, 256
  "ring bark" of, 260
  stem of, 302
  veins in leaf of, **322**

W

Waldsteinia, follicle of, **357**
Water pores, 117
Water stomata, 117
Watermelon (*see Citrullus*)
Wax, 54, 374, 382, 383
Wax palms, 54, 383
Weathering, inflorescences lost by, 275
  leaves lost by, 339
  of outer tissues, **253**
  phloem lost by, 231, 244, **256**
Welwitschia, intercalary meristem in, 323
Willow (*see Salix*)
Winteraceae, absence of vessels in, 98
Wood, of angiosperms, 216
  of commerce, 204
  compression, 228
  diffuse-porous, **210**, **212**, 214
  durability of, 222, 225–226
  early, 209
  fall, 209
  fibers, **94**
    fibrils in wall of, 32
    libriform, pits in, 47
    pit-pairs in, **43, 45**
    wall structure of, **34**
  of gymnosperms, 216
  heartwood, **200**, 223–224
    durability of, 223
    penetrability of, 227
    tyloses in, 222

Wood, impregnation of, 41, 224
  late, 209
  light and heavy, **213**, 224, 225
  parenchyma, **102**, 103
    in crotch angle, **198**
    crystals in, **20**
    distribution of, **215**
    fusiform, **94**
    in gymnosperms and angiosperms, 216
    metatracheal, 216
    ontogeny of secondary, 194
    paratracheal, 216
    simple pit-pairs in, **44**
    terminal, **215**, 216
    vasicentric, **215**, 216
    wood-ray, 102
  penetrability of, 226–227
  properties and uses in relation to microscopic structure, 224
  ring-porous, Frontispiece, **211**, 214
  sapwood, **200**, 223, 224
    penetrability of, 227
    tyloses in, 222
  specific gravity of, 225
  spring, 209
  strength of, 225
  summer, 209
    (*See also* Grain in wood; Timber; Xylem)
Woodwardia, tracheid of, **91**
Woody plants, abscission of stems in, 271–275
  gummosis in, 229
  periderm in, 255
  roots of, **281, 284**
  secondary phloem in, 231
  stomata in, 169
Woody stems, 297, **298**
Wound cork, 262
Wound roots, 289
Wound tissue, adventitious roots and stems from meristem in, 123
  cambium growth about wounds, 200–201
  meristems in, 63
  pith-ray flecks in, 228
  tyloses in, 222
Wrightia, seed coat of, **374**

X

Xanthium, fusion in corolla of, **348**
Xanthophyll, 18
Xerophytes, 380–381, **382, 383, 385, 386,** 387–388
  accessory cells in, 167
  endodermis in leaves of, 161
  fleshy, **383,** 388
  leaves of, 337, **382, 383,** 384, **386,** 387
  lignification and cutinization in, 382
  malacophyllous, **383,** 388
  microphyllous, **385, 386,** 387
  sclerophyllous, **383,** 384
  stem of, **385**
  stomata in, 333
  trichophyllous, 384
  vessels lacking in, 98, 216
Xylem, 2, 82, 90–103
  in abscission zone, **273**
  centrifugal development in, **128,** 129
  centripetal development in, **128,** 129
  elements, 133
  endarch, **128,** 129, 131, 137
  exarch, **128,** 129, 131, 137, 277
  in leaf, 324, 326–327
  mesarch, **128,** 129, 131, 137
  metaxylem, **130,** 131, 133
  mother cell, 186, 194
  ontogeny of, **127**
  in root, **77**
  parenchyma in, 83
  primary, **3,** 123, **125, 127**
    effect of cambial activity on, 180
    order of maturation in, **128, 130,** 131–140
    types of, 137
  rays, 205
    heterogeneous, 217

Xylem, rays, homogeneous, 217
    multiseriate, 217
    pit-pairs in, **45,** 204, 205, 217–220, **219**
    tracheids in, 218–220
    uniseriate, 217
  secondary, 3, 125, **164,** 204–205, **206, 207, 210–213, 219,** 220–229
    developing and mature, **186,** 187, **193, 196**
    gross structure of, 204–214
    histological structure of, 214–224
    in roots, **284, 285**
  tyloses, in, 220–222
    (*See also* Protoxylem; Tracheid; Vessel; Wood)

Y

Yucca, cambium in, 201
  secondary growth in, 302

Z

Zamia, stem apex of, **70**
Zea, aleurone grains in endosperm of, **23**
  leaf of, **328**
  leucoplasts in, **16**
  parenchyma in, **83**
  proplastids in, **14**
  protoxylem in, **136**
  seed coat in, 371
  stem of, **299**
  stomata in, 168
  vascular bundles in, 307
  vessel elements in, 98, **100**
Zingiber, oils in secretory cells of, 115
Zizyphus, fruit of, **363,** 366
Zonation in meristems, 71